SYMBIOTIC NITROGEN FIXATION

Developments in Plant and Soil Sciences

VOLUME 57

The titles published in this series are listed at the end of this volume.

Symbiotic Nitrogen Fixation

Proceedings of the 14th North American Conference on Symbiotic
Nitrogen Fixation, July 25–29, 1993, University of Minnesota,
St. Paul, Minnesota, USA

Edited by
P. H. GRAHAM, M. J. SADOWSKY and C. P. VANCE

Partly reprinted from *Plant and Soil*, Volume 161, No. 1 (1994)

Springer Science+Business Media, B.V.

A C.I.P. Catalogue record for this book is available from
the Library of Congress.

ISBN 978-0-7923-2781-3   ISBN 978-94-011-1088-4 (eBook)
DOI 10.1007/978-94-011-1088-4

Contents

Preface VII

Acknowledgements VIII

*1. Evolution and diversity in the legume-rhizobium symbiosis: Chaos theory?
   by J.I. Sprent 1

*2. Recent developments in *Rhizobium* taxonomy
   by E. Martínez-Romero 11

*3. Host range, RFLP, and antigenic relationships between *Rhizobium fredii* strains and *Rhizobium* sp. NGR234
   by H.B. Krishnan and S.G. Pueppke 21

*4. Rapid identification of *Rhizobium* species based on cellular fatty acid analysis
   by B.D.W. Jarvis and S.W. Tighe 31

*5. Ammonium sensing in nitrogen fixing bacteria: Functions of the *glnB* and *glnD* gene products
   by C. Kennedy, N. Doetsch, D. Meletzus, E. Patriarca, M. Amar and M. Iaccarino 43

*6. Regulation of nodulin gene expression
   by F.J. de Bruijn, R. Chen, S.Y. Fujimoto, A. Pinaev, D. Silver and K. Szczyglowski 59

*7. Synthesis, release, and transmission of alfalfa signals to rhizobial symbionts
   by D.A. Phillips, F.D. Dakora, E. Sande, C.M. Joseph and J. Zoń 69

*8. Role of rhizobial lipo-oligosacharides in root nodule formation on leguminous plants
   by O. Geiger, T. Ritsema, A.A.N. van Brussel, T. Tak, A.H.M. Wijfjes, G.V. Bloemberg, H.P. Spaink and B.J.J. Lugtenberg 81

*9. Sucrose transport and hydrolysis in *Rhizobium tropici*
   by V.I. Romanov and E. Martínez-Romero 91

*10. Shoot/root assimilate allocation and nodulation of *Vigna unguiculata* seedlings as influenced by shoot light environment
   by M.J. Kasperbauer and P.G. Hunt 97

*11. Mechanism of osmotically regulated N-acetylglutaminylglutamine amide production in *Rhizobium meliloti*
   by L. Tombras Smith, A. Ameer Allaith and G.M. Smith 103

* Chapters indicated with an asterisk are reprinted from *Plant and Soil*, Volume 161, No. 1. (1994).

*12. What triggers the regulation of nitrogenase activity in forage legume nodules after defoliation?
by U.A. Hartwig and J. Nösberger   109

*13. Nodulation and nitrogen fixation in extreme environments
by L.M. Bordeleau and D. Prévost   115

*14. Analysis and regulation of legume inoculants in Canada: The need for an increase in standards
by P.E. Olsen, W.A. Rice, L.M. Bordeleau and V.O. Biederbeck   127

*15. Recent developments in the actinorhizal symbioses
by A.M. Berry   135

16. Characterization of *Bradyrhizobium japonicum* chromosomal genes involved in the regulation of
hup gene expression in free-living conditions and in controlling hydrogenase activity
by C. van Soom, J. Vanderleyden and A.P. van Gool   147

17. Characterization of *Rhizobium galegae* by REP-PCR, PFGE and 16S rRNA sequencing
by I. Huber and S. Selenska-Pobell   153

18. An hypothesis for the role of malic enzyme in symbiotic nitrogen fixation in soybean nodules
by D.A. Day, R.G. Quinnell and F.J. Bergersen   159

19. Eastern Canadian soybean field trials of rhizobial strain NS 1 in two commercial carriers
by R.E. Sanders   165

20. Nitrogen fixation efficiency of cold-adapted rhizobia on sainfoin (*Onobrychis viciifolia*):
Laboratory and field evaluation
by D. Prévost, L.M. Bordeleau, R. Michaud, C. Lafrenière, J. Waddington and V.O. Biederbeck   171

21. Survival of *Bradyrhizobium japonicum* in pig slurry used as carrier for soil inoculation
by G. Ciafardini and G.C. Turtura   177

22. Plasmid DNA content of several agronomically important *Rhizobium* species that nodulate alfalfa,
berseem clover, or *Leucaena*
by F.M. Hashem and D. Kuykendall   181

23. International FAO/IAEA programmes on biological nitrogen fixation
by G. Hardarson   189

Preface

Since its inception, more than 26 years ago, the North American Symbiotic Nitrogen Fixation Conference (formerly the *Rhizobium* Conference) has been a major vehicle by which the different disciplines have exchanged current information regarding symbiotic N_2 fixation. Subject matter has ranged from the most basic to the very applied. The informal atmosphere of the meeting has provided an ideal forum for scientists and graduate students to present research reports as oral presentations and/or posters. This year's conference, held in Minneapolis, Minnesota, brought together over 200 scientists from 22 different countries to discuss their latest research progress in 53 lectures and on over 80 posters.

Over the last three decades there has been a large amount of research on biological nitrogen fixation, in part stimulated by increasing world prices of nitrogen-containing fertilizers and environmental concerns. In the last several years, research on plant-microbe interactions, and symbiotic and asymbiotic nitrogen fixation has become truly interdisciplinary in nature, stimulated to some degree by the use of modern genetic techniques. These methodologies have allowed us to make detailed analyses of plant and bacterial genes involved in symbiotic processes and to follow the growth and persistence of the root-nodule bacteria and free-living nitrogen-fixing bacteria in soils. Through the efforts of a large number of researchers we now have a better understanding of the ecology of rhizobia, environmental parameters affecting the infection and nodulation process, the nature of specificity, the biochemistry of host plants and microsymbionts, and chemical signalling between symbiotic partners.

This volume gives a summary of current research efforts and knowledge in the field of biological nitrogen fixation. Since the research field is truly diverse in nature, this book presents a collection of papers in the major research area of physiology and metabolism, genetics, evolution, taxonomy, ecology, and international programs. The editors hope that this book will serve as useful reference source and serve to point out areas in need of further research.

St. Paul, Minnesota
December 1993

M. Sadowsky
P. Graham
C. Vance

Acknowledgements

Organization of a meeting such as the 14th North American Symbiotic Nitrogen Fixation Conference, and now the publication of the conference proceedings, requires help and support from many people. We are particularly fortunate to have had so many colleagues, students and friends with whom to share the load.

As organizers of the meeting we would like to acknowledge this support, and to thank all who gave so willingly.

Especially, we would like to thank the following organizations for their generous support:

Canadian Seed Coaters, Ltd., Brampton, ON, Canada
CelPril Industries, Manteca, CA, USA
Esso Ag Biologicals, Saskatoon, SK, Canada
LiphaTech. Inc., Milwaukee, WI, USA
Microbio Rhizogen, Saskatoon, Saskatchewan, Canada
Minnesota Crop Improvement Assoc., St. Paul, MN, USA
Minnesota, Soybean Growers Assoc., Mankato, MN, USA
Morgan Scientific Co., Andover, MA, USA
Pioneer Hi-Bred International, Inc., Johnston, IA, USA
Precision Seed Coaters, Inc., Phoenix, AZ, USA
Samuel Roberts Noble Foundation, Ardmore, OK, USA
Trace Chemical, Inc., Pekin, IL, USA
United States Department of Agriculture,
    Agricultural Research Service
United States Department of Agriculture
    Cooperative States Research Service
    National Research Initiative
University of Minnesota
    College of Agriculture
    Department of Agronomy and Plant Genetics
    Department of Plant Pathology
    Department of Soil Science
    Graduate School
Urbana Laboratories, St. Joseph, MO, USA
Vigro Industries, Inc., West Lafayette, IN, USA

We are also indebted to the reviewers of the papers which appear in the conference proceedings. They kept well to the narrow time frame we could allow, were careful in their evaluations, and generally very helpful in suggesting improvements.

We owe a special debt to our respective department heads for their support and tolerance; to Nancy Harvey and Eugene Anderson of the Educational Development System for the organization of the meeting; to Kristine Kirby for graphic design; to Marjorie Smith for secretarial help; and to our graduate and post-doctoral students for helping to resolve the myriad problems that occur in the course of a conference.

P. Graham
M. Sadowsky
C. Vance

Plant and Soil **161**: 1–10, 1994.

Evolution and diversity in the legume-rhizobium symbiosis: chaos theory?

Janet I. Sprent

Department of Biological Sciences, University of Dundee, Dundee DD1 4HN, Scotland, UK

Key words: *Bradyrhizobium*, diversity, evolution, legumes, N_2 fixation, *Rhizobium*

Abstract

Diversity in both legumes and rhizobia is discussed, in the light of evolution of nodulation. An hypothesis is developed that two separate nodulation events occurred in the humid tropics during the evolution of legumes in the late Cretaceous. One of these involved an ancestor of *Rhizobium* and a root infection. This was initially parasitic and provided little benefit until bacteria were released from infection threads as in modern crop species. The other involved a photosynthetic ancestor of *Bradyrhizobium* with a wound infection on stems, and has never involved infection threads. As continents moved and climates changed to a seasonal type, involving either rainfall or temperature extremes, further constraints were imposed. The argument is pursued for the case of acacias and their rhizobia in arid regions. Here selection pressures on rhizobia led to the evolution of stress tolerant forms, not all of which are capable of symbiosis, and where symbiotic genes may be an expensive encumbrance. Lateral transfer of material on megaplasmids led to a wide range of symbiotic and non-symbiotic forms in response to local pressures. When environmental constraints are superimposed on initial evolutionary developments, the result is an apparently chaotic situation where there is no obvious pattern of co-evolution between hosts and rhizobia. Evidence of such coevolution may still be buried in this chaos and may be amenable to molecular analysis.

Introduction

If we take the legumes as a single family, the Leguminosae, with three sub-families, the likelihood is that its first ancestors appeared in the late Cretaceous-early Tertiary period, 60–70 Ma before present. Certainly by about 50 Ma ago there is fossil evidence, mainly from wood, that all 3 extant sub-families were present (Herendeen and Dilcher, 1992). However, the earlier idea that the Caesalpinioideae was the base sub-family from which the Mimosoideae and Papilionoideae evolved (Polhill and Raven, 1981) is not supported by recent evidence, including that on nodule structure. This brings us to one of the major questions addressed here, was the ability of legumes to nodulate the result of a single event or has it occurred more than once? The other major question to be considered is what is the evidence for co-evolution of legumes and rhizobia? In attempting to answer these questions it is necessary to examine diversity in both host and bacteria. Because legumes evolved in the tropics, most examples will be taken from tropical and near-tropical regions where diversity in both symbionts is proving to be very large.

The genera used as examples here may be unfamiliar to most workers outside the tropics. Apart from interest in how legumes evolved with its possible relevance to development of new symbioses (Sprent, 1989), many have proven or potential economic importance. If they are to be managed properly, it is essential to know whether they **can** nodulate and if so, how much nitrogen they fix. The following comments refer to some of the nodulated and non-nodulated genera discussed in this paper. For further details readers are referred to the fascinating book of Allen and Allen (1981) which should be consulted in conjunction with Polhill and Raven (1981) and Faria et al. (1989) for more recent plant taxonomic affinities.

Cassia (sensu lato, see below) contains many trees grown as ornamentals, for shade and other purposes. Many species produce drugs, the laxative from senna leaves (*C. fistula*, now transferred to *Senna*) probably being the best known. *C. rotundifolia* (now *Chamaecrista*) is a small shrub and the only nodulated caesalpinioid legume known to the author to be used as a browse species, being recommended for cattle farmers is large areas of central Queensland, Australia (Oram, 1990). *Vatairea* produces high quality wood, resistant to decay and insect attack. *Andira* tree species product high quality wood and are also used as ornamentals, for nectar (bees, butterflies, hummingbirds) and as a shade plant for coffee. The shrubby/herbaceous *Neptunia* may be free-floating in tropical rivers such as the Amazon; it is commonly used in such areas as cattle fodder. In spite of being named after the sea god Neptunia, its species extend to arid regions in Australia.

Selection pressures on the host

We have recently argued (Raven and Sprent, 1989; Sprent et al., 1993) that when plants colonised land their ability to photosynthise resulted in nitrogen limitation. Good root systems to scavenge for combined N in soil were thus essential, and early legumes might have had the added benefit of housing nitrogen-fixing bacteria. All woody (and most other) nodulated legumes for which there is sufficient evidence appear to be very efficient at exploiting soil N. We further supported the idea that some primitive nodules may have been more parasitic than mutualistic, and raised the possibility that two types of initial

nodulation event may have occurred (Sprent et al., 1993). Here major types of nodule structure are discussed from the standpoint of two separate lines of evolution.

The first case is based on observations of extant nodulated genera of the Caesalpinioideae and a few woody members of the Papilionoideae. In these, bacteria are retained within threads bounded by host wall material into their nitrogen-fixing stage. This appears to be the forerunner of most (but not all) nodules found in familiar crop plants. The possible sequences of events can be seen in species of the genus *Chamaecrista*.

Chamaecrista: evolution in action? The genus *Cassia* was divided into *Cassia*, *Senna* and *Chamaecrista* by Irwin and Barneby (1982). All confirmed nodulating species occur in *Chamaecrista* a genus of some 250 species, including shrubs, herbs and rarely trees. It is mainly neotropical, with a major radiation in eastern Brazil (Lewis, 1987). There are no confirmed reports of lack of nodulation in this genus. Many of the 30 *Cassia* species are large trees, frequently occurring in impoverished soils. None is confirmed to be nodulated, but they are effective in scavenging soil N and their leaves often have N contents similar to that of N_2-fixing species. For this reason they may be used in agroforestry, sometimes under the mistaken impression that they can fix nitrogen! *Senna* is a larger genus of about 240 spp. whose habit ranges from large trees through shrubs to herbs; again no species is confirmed to be nodulating. In terms of flower structure there is a gradation from zygomorphy (*Cassia*) through variable (*Senna*) to a highly asymmetrical perianth (Irwin and Barneby, 1982). Further, *Cassia* has only one form of stigma whereas the other genera have two (Owens and Lewis, 1989). For these and other reasons *Cassia* sens.lat. is not thought to represent a single evolutionary line. Consistent with taxonomic observations we have found much greater levels of variation in terms of number of products generated by RAPD (Randomly Amplified Polymorphic DNA) analysis in *Chamaecrista* and *Senna* than in *Cassia* (Whitty, Powell, and Sprent, unpublished obser-

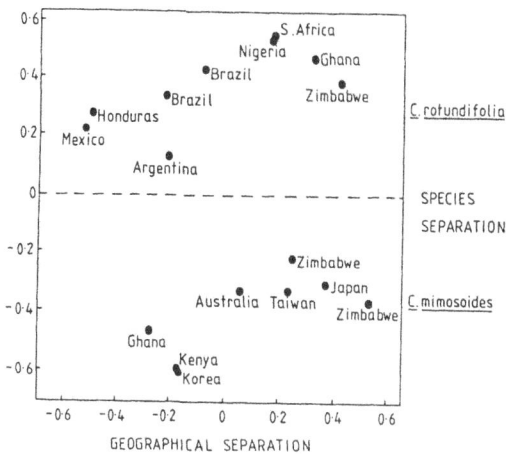

Fig. 1. Separation of species and geographical provenances of species of *Chamaecrista* using Randomly Amplified Polymorphic DNA (RAPD) profiles. Products from ten, ten-base oligonucleotide primers were subject to analysis using a Genstat programme, based on Principal Component Analysis. The scales are arbitrary and cannot be ascribed to particular features. Unpublished data of Whitty, Powell and Sprent.

vations) and have been able to separate the three genera, and species within a genus. As an example of data generated by RAPD analysis, Figure 1 shows the separation of *C. rotundifolia* from *C. mimosoides* along one axis and separation of samples from different geographical regions along the other. Geographical separation was more successful for *C. rotundifolia* and *C. mimosoides*, possibly because the seed of *C. mimosides* was not always obtained from a country in which it is endemic.

Infected cells from nodules of some species of *Chamaecrista* species show bacteroids retained within infection threads throughout the period of nitrogen fixation, while in other species, the rhizobia are released from threads during differentiation, as found in familiar crop species (Naisbitt et al., 1992). Broadly speaking this change parallels differences in plant habit from woody to herbaceous, but with the added feature that "persistent infection threads" have only been found in species from Brazil (all genera so far described with this feature, whether caesalpinioid or papilionoid are from tropical South America (Polhill and Raven, 1981). We have preliminary evidence that RAPD analysis will enable us to locate markers for these nodule features.

We hypothesised previously (Sprent et al., 1994) that high rates of nitrogen fixation may be more essential for annual than for perennial species. Evidence that the number of bacteroids per unit of nodule tissue was greater in those *Chamaecrista* species where bacterioids were released from the infection threads was also given. We now have evidence that the higher rates of acetylene reduction per unit of nodule mass are obtained in species where the bacteroids are released (Minchin, Naisbitt and Sprent, unpublished observations). Evidence from *Chamaecrista* is consistent with (a) an evolution of nodulation relatively early in the evolution of legumes (b) evolution from a more parasitic to a more symbiotic state, coupled with evolution from a woody perennial to an herbaceous annual habit (Table 1). Parker (1957) and others have also hypothesised that rhizobia evolved from a plant parasite. Trinick (1982) in an excellent critical review of evidence to that date also suggested that persistent infection threads are a primitive state: at that time they were only known in the non-legume genus *Parasponia*.

Papilionoid legumes The sequence of events seen in extant species of *Chamaecrista* may be mirrored in some groups of papilionoid legumes. Nodules where rhizobia remain bounded by infection threads even during the period of nitrogen fixation have been found in five woody genera from the tribes, Millettieae and Dalbergieae. In the Dalbergieae we appear to have a set of closely related genera (Polhill, 1981) with wide variation in nodulation characters (Sprent and Raven, 1992). Of the 19 genera currently recognised in the Dalbergieae: *Andira* and *Hymenolobium* retain bacteria in infection threads throughout the life of the nodule (Faria et al., 1987), and must be considered primitive in nodulation traits, while eight genera (listed in Sprent et al., 1989) have aeschynomenoid nodules. If non-nodulation in some of the remaining genera is confirmed (and the position is now quite strong for *Vatairea* and *Vataireopsis*, have these genera lost the character or did they never acquire it? This is a question we cannot answer at present. The best known

Table 1. Some features of nodules on plants of two species of *Chamaecrista*. From Naisbitt et al., 1992, and Minchin, Naisbitt and Sprent (unpublished)

Character	Species	
	ensiformis	*fasciculata*
Plant habit	woody shrub	annual herb
Walls of threads in infected tissue	thick	absent
Infected: uninfected cells	0.31	1.43
Bacteria/100 μm^{-2} of infected tissue	10	130
Nitrogenase activity	very low	significant
Nature of symbiosis	more parasitic	more mutualistic

example of the aeschynomenoid type of nodule is found in *Arachis hypogaea* (groundnut, peanut) and it appears typical for the tribe Aeschynomeneae, in which *Arachis* is placed. In this genus, infection occurs through wounds where lateral roots emerge. Although this appears to be more primitive than the highly complex hair infection pathway, and in many ways is similar to the pathogenic situation, the final nodule is arguably more advanced (Faria and Sprent, 1994). It is certainly very efficient in the absence of interstitial cells and (usually) the lack of vacuoles in infected cells, contributing to a high number of bacteria housed per unit volume of infected tissue. It is difficult to see how the processes of development in aeschynomenoid nodules could be derived from a system depending upon infection thread formation. We have argued elsewhere (Sprent and Raven, 1992) that it may have been a separate evolutionary event which occurred in flooded areas, leading initially to the formation of stem nodules, with photosynthetic rhizobia. When plants colonized drier areas, stem nodules would have been surrounded by atmospheres at less than 100% relative humidity, which in turn would have reduced their growth and activity (Parsons et al., 1993; Table 2). Gradual migration of nodulation to below ground, coupled with loss of photosynthetic activity in the bradyrhizobia could have followed (Sprent and Raven, 1992).

Mimosoid legumes We have examined many nodulated genera within this group and so far have found neither persistent infection threads nor aeschynomenoid nodules. However, the subfamily shows three types of infection pathway, via root hairs, via lateral root junctions (so far only seen in *Neptunia*, (James et al., 1992)) and between intact epidermal cells (Faria et al., 1988). Although the last two processed do not initially involve infection threads, infection threads are formed later in both processes and they carry rhizobia to newly formed cells behind the nodule meristem, where they are subsequently released into peribacteroid units. This sequence of events has not so far been observed in papilionoid or caesalpinioid legumes.

A further complication observed in *Neptunia plena* is that when grown hydroponically it does not produce root hairs, infection taking place through adventitious or lateral root junctions: when grown on solid substrates hairs may form, and hair infection occur (James et al., 1993).

Evolution of rhizobia

Swimming against the tide, Postgate has consistently argued for nitrogen fixation in prokaryotes being a relatively modern phenomenon. He has recently (Postgate, 1992) suggested that if nitrogen-fixing prokaryotes had existed in an ear-

Table 2. Effect of relative humidity on growth and nitrogenase activity of stem nodules of *Sesbania rostrata*. Data from Parsons et al. (1993)

	Relative humidity (%)				
	15	30	58	85	100
Stem nodule dry wt. (mg plant^{-1})	9.5 ± 6.7	12.3 ± 6.3	20.5 ± 4.8	19.4 ± 6.5	18.1 ± 4.6
μmol C_2H_4 g^{-1} d wt nod^{-1} h^{-1}	116 ± 67	197 ± 42	237 ± 49	283 ± 16	290 ± 14

lier age, eukaryotes would have 'acquired' them in an endosymbiotic way giving rise to a nitrogen-fixing organelle or diazoplast - analogous to a chloroplast. He suggests that, so far, selection pressure has not been sufficiently strong for plants to acquire *nif* genes in this way. On the other hand bacteria, by their very nature adapt quickly and could, by lateral transfer, acquire the ability to fix nitrogen when faced with N-limited environments such as may arise following major disturbances (fire, glaciation, drought, vulcanization).

In an attempt to marry some of the arguments of Postgate (1992) and Sprent et al. (1993) I should like to consider the genus *Acacia* and its nodulating partners. *Acacia* is one of the largest genera of the Leguminosae and is likely soon to be subdivided. The diversity in the host plant is arguably equalled or exceeded by that of the rhizobia with which it nodulates.

Acacia satisfies our criterion that it should be a good scavenger. Many species produce vast root systems, exploiting both deep water tables and large areas of surface soil. They may associate with VA- or ectotrophic mycorrhizal fungi or both, and some are known to have cluster roots. These features are consistent with evolution at a time and place when, climates were seasonally arid and soil nutrients in short supply. Although there are several central American species and a few in Asia, most acacias are found in Africa and Australia. Most species examined can nodulate, but there are some, for example *A. brevispica* from Africa, which apparently cannot (Odee and Sprent, 1992).

Most of the early work on acacia rhizobia was carried out in Australia. It is clear that a wide range of strains ranging from fast- to slow- and even very slow-growing may form effective nodules (e.g. Barnet and Catt, 1991). The fast-growing strains are very diverse and some may even be classified in a new genus, closer to *Bradyrhizobium* than *Rhizobium* (Barnet et al., 1993). These distinct fast-growing strains are heat-, salt- and desiccation tolerant, a feature coupled with use of trehalose as a stress protective solute (see also Smith, 1994) (many salt-tolerant rhizobia accumulate glutamate, Graham, 1992).

In a more general study on Rhizobia isolated from trees, Zhang et al. (1991) concluded that 'fast-growing tree rhizobia stem from a reservoir of taxonomically diverse but ecologically well-adapted strains'. In particular, Sudanese strains have very high T_{max} values and high salt tolerance, in agreement with the work of Barnet's group in Australia. In a wide ranging study of Kenyan rhizobia, Odee (1993) found overall that fast-growing (*?Rhizobium*) strains were generally more salt tolerant (3% NaCl) than slow-growing (*?Bradyrhizobium*) strains.

Thus current evidence from the tropics and sub-tropics suggests that many of the most desiccation and salt stress-tolerant rhizobia nodulating woody legumes are fast-growing. If these strains generally have host-range (and presumably other symbiotic genes) on plasmids (Zhang et al., 1991) then we have potentially a very dynamic situation. Stouthamer and Kooijman (1993) have recently formulated an interesting model, well summarised in the title of their paper 'Why it pays for bacteria to delete disused DNA and to maintain megaplasmids'. They argued for natural conditions where short periods of high growth rate may alternate with long periods of low growth rate that a large genome size is a selective disadvantage. If it is necessary to maintain a large genome then a moderate-sized chromosome plus a megaplasmid would allow a faster rate of DNA replication than a single very large chromosome, thus enabling cells to maintain a favourable surface: volume ratio for nutrient uptake (assuming cells normally grow in volume until DNA replication is complete). Implicit in this general argument is that the material on the megaplasmid is not absolutely essential for growth. This of course is true for symbiotic genes. Unlike Postgate (1992), Stouthamer (1992) believes that the ability to fix nitrogen, as well as a whole suite of other complex metabolic pathways, evolved very early on, and has subsequently been lost in many extant bacteria. Presumably the losing process can still go on! The difficult question is whether or not there is an acquiring process. Two possibilities have been suggested. The first may be regarded as (in evolutionary terms) a short term option. It suggests that the codes for disused proteins are maintained as cryptic genes which may be reactivated by mutation (Stouthamer and Kooijman, 1993 and references therein). The second which may be long or short term is that lateral transfer of genetic material occurs within populations of soil microorganisms. This is an option favoured by Postgate (1992).

Either (or both) could explain the situation currently observed with fast-growing tree rhizobia. If the genetic material for symbiosis is incomplete (first option) then a wide range of effectivity would be expected. This is generally well-supported by the literature. Barnet and Catt (1991) in their extensive Australian study, found that strains isolated from acacias varied (on the same hosts) from fully effective to totally ineffective. Coupled with evidence that the occurrence of tree rhizobia in soils relates more to soil conditions than host species, one regretfully concluded that in this case the ability to fix nitrogen is not a high priority for rhizobia - survival in the soil is much more important. Although for crop species such a soybean (Kuykendall, 1989; Kuykendall et al., 1982) there is evidence that multiplication within a nodule is important for maintaining soil populations of infective rhizobia, information on populations of non-symbiotic rhizobia (i.e. in this case those which have lost at least the relevant parts of their megaplasmids) in soils is not available. New molecular techniques should fill this gap shortly and, further enable the testing of hypotheses on change of genetic signature in for example, drought following unstressed years. We shall then be in a better position to assess the extent to which loss of and acquisition of plasmids is important. Arguments against lateral transfer (cited by Stouthamer, 1992) are largely for rather dissimilar organisms. Between related organisms it may be common and certainly has been implied in fast-growing rhizobia from temperate soils (Young, 1985).

Soil considerations

Combined nitrogen Widely grown and selected crop (grain and forage) legumes are generally herbaceous papilionoid species such as beans and clovers. Nitrogen fixation in these is usually inhibited by combined N, nitrate being the species most commonly studied (Sprent and Sprent, 1990). However, if we accept the arguments that legumes should be good scavengers of soil N and that early symbioses were inefficient, then the ability simultaneously to use soil N and fixed N would be expected. Recent evidence with acacias and other mimosoid trees suggests that plants can nodulate and fix nitrogen in relatively high levels of soil ammonium (Goi et al., 1992; Goi, 1993). *Acacia auriculiformis* appeared unable to assimilate significant amounts of nitrate - this species tolerates a variety of soil pH conditions from acid to alkaline (Anon, 1980) and may be exposed to ammonium and/or nitrate. Provenance variation may thus be expected. Sun et al. (1992) found that four of the ten provenances of *A. mangium* tested showed a strong preference for ammonium.

Other factors

In such a complex environment it is not surprising that plants may not nodulate in soils known to contain compatible rhizobia (e.g. Barnet and Catt, 1991 for *Acacia* spp.). Soil factors such as low pH and high Al are known to affect *nod* gene induction (Richardson et al., 1988). Induction of *nod* genes can be effected by a variety of different flavonoids and other similar compounds, produced by legumes and non-legumes (Peters et al., 1986). We are thus faced with the possibility that soils may contain many induced rhizobia in the absence of a host. For example, nodulated and non-nodulated acacias may occur in the same location. Table 3 gives data for 5 strains of rhizobia, isolated from hosts other than acacias, whose *nod* genes could be induced by substances from both nodulated and non-nodulated acacias (or *Faidherbia*). For nodulation to proceed in compatible hosts, correct nod factors must be produced by induced rhizobia.

General nutrient status of soils may also affect nodulation. Flooded soils in the Brazilian Amazon, which are rich in nutrients other than N, supported more nodulated legumes than non-flooded terrafirme soils, which are nutrient poor (Moreira et al., 1992).

Co-evolution of hosts and rhizobia

Young (1993) argued that *Bradyrhizobium* and *Rhizobium* diverged long before nodules evolved, but, since their *nod* genes are clearly related, lateral transfer must have occurred. As Young pointed out this poses many questions. In that *Rhizobium* is closely related to the tumour-forming *Agrobacterium*, there is some logic in *nod* genes originating in *Rhizobium* and being transferred to *Bradyrhizobium* (it is even possible that tumour-inducing DNA was transferred from *Agrobacterium* to legume). This suggestion is the opposite of

that of Norris (1965) who proposed slow-growing strains to be ancestral. In an extensive study of woody plants from the Amazon, Moreira and Franco (1994) found one slow-growing cluster of alkali-forming strains to nodulate members of all three subfamilies. Similarly they found all subfamilies to be nodulated by fast-growing strains. The inevitable conclusion was that there is no evidence for co-evolution, in contrast to Young and Johnston (1989). Can both sides be correct?

Let us suppose that there were two (or more) evolutionary events leading to nodulation by *Rhizobium* and *Bradyrhizobium*-like organisms. As legumes evolved, as continents moved and environmental conditions changed, additional constraints were imposed upon the underlying need for N_2-fixation. Populations of soil rhizobia in the wide sense (i.e. not constrained by the requirements for symbiosis) would adapt to various soil conditions such as heat and salinity. Lateral transfer of various genetic elements has continued so that in the extant situation host-rhizobial combinations are selected for natural conditions, rather than those under which the original evolutionary event occurred. This would accommodate observations such as heat-tolerant tree rhizobia nodulating beans (Hungria et al., 1993) and the huge diversity of rhizobia nodulating acacias. It would also account for cases where the same rhizobial strain may nodulate species from different sub-families. As more information is collected from the vast reservoir of diversity in the tropics, these ideas can be modelled and tested. Until such understanding is achieved, successful inoculation programmes in many parts of the world which most need them must pose acute problems.

Conclusions and general discussion

There is increasing evidence to support two types of evolutionary event leading to the formation of legume nodules. Legumes were thought to have evolved in the humid tropics and since nodulated caesalpinioid genera such as *Campsiandra* are found in flooded areas such as the Orinoco basin, it is reasonable to hypothesis that this process, like

Table 3. Induction of *nod* genes in various rhizobial isolates by aqueous fractions of roots of two nodulating and two non-nodulating mimosoid legume trees. Unpublished data of J. Shaw. A low level of induction is indicated by +, moderate by ++ and high by +++

Isolate	Host of origin	Host species[a]			
		nodulating	non-nodulating		
		s	a	b	g
1.	*Leucaena leucocephala*	+	+	+	++
2.	*Mimosa scabrella*	++	nt[b]	++	+++
3.	*Pithecellobium tortum*	++	+++	++	nt
4.	*Aeschynomene indica*	+++	+++	+++	+++
5.	*Leucaena salvadorensis*	+++	+++	+++	+

[a]s = *Acacia seyal*; a = *Faidherbia (Acacia) albida*; b = *Acacia brevispica*; g = *Acacia greggii*

[b]not tested.

stem nodulation may have evolved in the humid tropics. Whether the event occurred more than once is a moot point.

If the caesalpinioid nodule type is very inefficient, it must impose a drain on the plant's resources: similarly a highly efficient stem nodule may be expensive in terms of plant photosynthate (Parsons et al., 1992). Would this have been a major disadvantage? It could be argued that in high light tropical environments, with a plentiful supply of water, photosynthate might be in excess of normal requirements. It could even be argued that in stem nodulated plants, nitrogen fixation may be in excess of requirements, since *Sesbania* and *Aeschynomene* residues have a high N content which is rapidly mineralised.

Accepting the early separation of *Rhizobium* and *Bradyrhizobium* and their involvement in early nodulation events, how do we explain the apparent lack of evidence for co-evolution of host and rhizobia as found by Moreiro and Franco (1994). I suggest that this arose from the movement of legumes from the humid tropics into new, seasonal environments. These would have imposed new constraints on both host and rhizobia, such that the benefits of nodulation may have changed. Such selection pressures would have been, and still are, very site-specific. Thus the arguments put forward for the case of *Acacia* and its rhizobia would be superimposed on the original

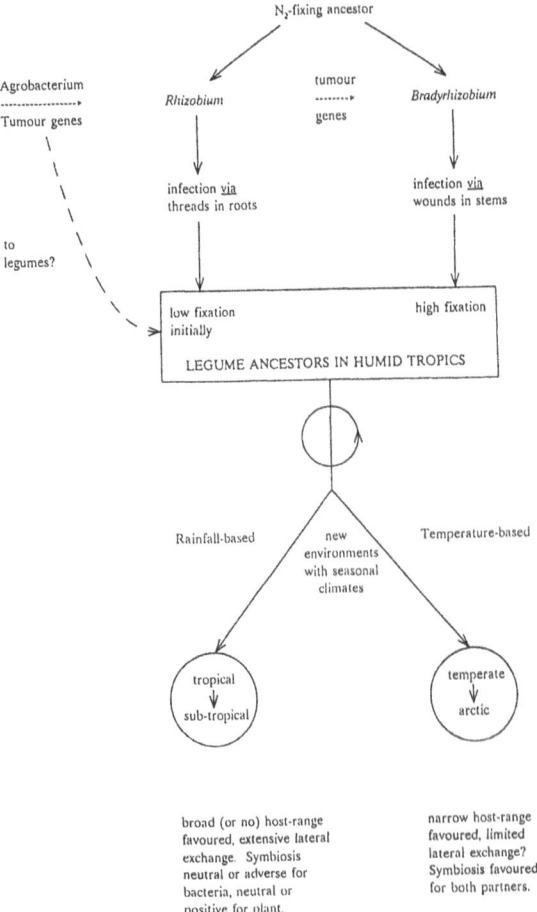

Fig. 2. Diagram illustrating possible events during the evolution of nodulated legumes. Note that the lower part is free to rotate with respect to the upper part.

events of nodulation. The situation, summarised in Figure 2 has many elements of chaos theory which has been defined as 'stochastic behaviour occurring in a deterministic system' or more simply 'lawless behaviour governed entirely by law' (Stewart, 1989).

Acknowledgements

I am grateful to all my colleagues, especially those in Dundee who allow me back from administration into the laboratory from time to time. Our work has been generously supported by the European Commission, British Council, British Overseas Development Administration, Agricultural and Food Research Council, Natural Environment Research Council, Science and Engineering Research Council, Conselho Nacional de Desenvolvimento Cientifico e Technologics (Brazil), Royal Society and others. Referees of this paper are thanked for their constructive suggestions.

References

Allen O N and Allen E K 1981 The Leguminosae. University of Wisconsin Press, Madison. Macmillan Publishing Company, London. 812 p.

Anon 1980 Firewood crops. National Academy of Sciences, Washington D.C. 236 p.

Barnet Y and Catt P C 1991 Distribution and characteristics of root nodule bacteria isolated from Australian *Acacia* spp. Plant and Soil 135, 109–120.

Barnet Y, Catt P C, Jenzaniontham R and Mann K 1993 Fast growing root-nodule bacteria from Australian *Acacia* spp. *In* New Horizons in Nitrogen Fixation. Eds. R Palacios, J Mora and W E Newton. p 594. Kluwer Academic Publishers, Dordrecht.

Faria S M de, Hay G T and Sprent J I 1988 Entry of rhizobia into roots of *Mimosa scabrella*. Bentham occurs between epidermal cells. J. Gen. Microbiol. 134, 2291–2296.

Faria S M de, Lewis G P, Sprent J I and Sutherland J M 1989 Occurrence of nodulation in the Leguminosae. New Phytol. 111, 607–619.

Faria S M de, McInroy S G and Sprent J I 1987 The occurrence of infected cells, with persistent infection threads, in legume root nodules. Can. J. Bot. 65, 553–558.

Faria S M de and Sprent J I 1994 Legume nodule development, an evolutionary hypothesis. *In* Advances in Legume Systematics 5. The Nitrogen Factor. Eds. J I Sprent and D McKey. Royal Botanic Gardens, Kew. (*In press*).

Goi S R, Sprent J I, James E K and Jacob-Neto J 1992 Influence of nitrogen form and concentration on the nitrogen fixation of *Acacia auriculiformis*. Symbiosis 14, 115–122.

Goi S R, 1993 Nitrogen nutrition of woody legumes. PhD thesis, University of Dundee. 199 p.

Graham P H 1992 Stress tolerance in *Rhizobium* and *Bradyrhizobium* and nodulation under adverse conditions. Can. J. Microbiol. 38, 475–484.

Herendeen P S and Dilcher D L (Eds.) 1992 Advances in Legume Systematics, part 4. The Fossil Record. Royal Botanic Gardens, Kew. 326 p.

Hungria M, Franco A A and Sprent J I 1993 New sources of high-temperature tolerant rhizobia for *Phaseolus vulgaris*. Plant and Soil 149, 103–109.

Irwin H S and Barneby R C 1982 The American Cassiinae: a synoptical revision of leguminosae tribe Cassiiae subtribe Cassiinae in the New World. Memoirs of the New York Botanic Garden 35, 1–454.

James E K, Shaw J E, Catellan A J, Faria S M de and Sprent J I 1993 The infection of aquatic and terrestrial *Neptunia* species by *Rhizobium In* New Horizons in Nitrogen Fixation. Eds R Palacios, J Mora and W E Newton. p 351. Kluwer Academic Publishers, Dordrecht.

James E K, Sprent J I, Sutherland J M, McInroy S G and Minchin F R 1992 The stucture of nitrogen fixing roote nodules on the aquatic mimisoid legume *Neptunia plena*. Ann. Bot. 69, 173–180.

Kuykendall L D 1989 Influence of *Glycine max* nodulation on the persistence in soil of a genetically marked *Bradyrhizobium japonicum* strain. Plant and Soil 116, 275–277.

Kuykendall L D, Devine T E and Cregan P B 1982 Positive role of nodulation in the establishment of *Rhizobium japonicum* in subsequent crops of soybean. Current Microbiol. 7, 79–91.

Lewis G P 1987 Legumes of Bahia. Royal Botanic Gardens, Kew. 369 p.

Moreira F M de Souza, de Silva M F and Faria S M de 1992 Occurrence of nodulation in legume species in the Amazon region of Brazil. New Phytol. 121, 563–570.

Moreira F and Franco A A 1993 Rhizobia-host interactions in tropical ecosystems in Brazil. *In* Advances in Legume Systematics 5, the Nitrogen Factor. Eds. J I Sprent and D McKey. Royal Botanic Gardens, Kew. (*In press*).

Naisbitt T, James E K and Sprent J I 1992 The evolutionary significance of the legume genus *Chamaecrista*, as determined by nodule structure. New Phytol, 122, 487–492.

Norris D P 1965 *Rhizobium* relationships in legumes. Proc. 9th Int. Grassl. Congr. Sao Paulo 2, 1087–1092.

Odee D W 1993 The ecology of nitrogen-fixing symbioses under arid conditions of Kenya. PhD thesis, University of Dundee. 145 p.

Odee D W and Sprent J I 1992 *Acacia brevispica*: a non-nodulated mimosoid legume? Soil Biol. Biochem. 24, 717–719.

Oram R N 1990 Register of Australian Herbage Plant Cultivars, CSIRO Publications, Parchment Press, Melbourne.

Owens S J and Lewis G P 1989 Taxonomic and functional implications of stigma morphology in species of *Cassia*, *Chamaecrista* and *Senna* (Leguminosae: Caesalpinioideae). Pl. Syst. Evol. 163, 93–105.

Parker C A 1957 Evolution of nitrogen-fixing symbiosis in higher plants. Nature 179, 593–4.

Parsons R, Raven J A and Sprent J I 1992 A simple open flow system used to measure acetylene reduction activity of *Sesbania rostrata* stem and root nodules. J. Exp. Bot. 43, 595–604.

Parsons R, Sprent J I and Raven J A 1993 Humidity and light affect the growth, development and nitrogenase activity of stem nodule of *Sesbania rostrata* Brem. New Phytol. 125, 749–755.

Peters N K, Frost J W and Long S R 1986 A plant flavone, luteolin, induces expression of *Rhizobium meliloti* nodulation genes. Science 233, 977–980.

Polhill R M 1981 Dalbergieae. *In* Advances in Legume Systematics. Eds. R M Polhill and P H Raven. pp 233–242. Royal Botanic Gardens, Kew.

Polhill R M and Raven P H (Eds.) 1981 Advances in Legume Systematics, Part 1. Royal Botanic Gardens, Kew. 425 p.

Postgate J 1992 Bacterial evolution and the nitrogen-fixing plant. Phil. Trans. Roy. Soc. London **B**, 338, 409–416.

Raven J A and Sprent J I 1989 Phototrophy, diazotrophy and palaeoatmospheres: biological catalysts and the H, C, N and O cycles. J. Geol. Soc. 146, 161–170.

Richardson A E, Simpson R J, Djordjevic M A and Rolfe B G 1988 Expression of nodulation genes in *Rhizobium leguminosarum* bv *trifolii* is affected by low pH, Ca and Al. Appl. Environ. Microbiol. 54, 2541–2548.

Smith L T, Allaith A A and Smith G M 1994 Mechanism of osmotically regulated N-acetylglutaminylglutamine amide production in *Rhizobium meliloti*. Plant and Soil 161, 103–108.

Sprent J I 1989 Which steps are essential for the formation of functional legume nodules? New Phytol. 111, 129–153.

Sprent J I, Minchin F R and Parsons R 1993 Evolution since Knoxville: were rhizobia wise to inhabit land plants? *In* New Horizons in Nitrogen Fixation. Eds. R. Palacios, J Mora and W E Newton. pp 65–76. Kluwer Academic Publishers, Dordrecht.

Sprent J I and Raven J A 1992 Evolution of nitrogen fixing symbioses. *In* Biological Nitrogen Fixation. Eds. G Stacey, R H Burris and H J Evans. pp 461–496. Chapman and Hall, NY.

Sprent J I and Sprent P 1990 Nitrogen Fixing Organisms: pure and applied aspects, 256 p. Chapman and Hall, London.

Sprent J I, Sutherland J M and Faria S M de 1989 Structure and function of root nodules from woody legumes. *In* Advances in Legume Biology. Ed. C H Stirton and J L Zarucchi. Monogr. Syst. Bot. Missouri Bot. Gard. 29, 559–578.

Stewart I 1989 Does God Play Dice? Penguin Books, London. 317 p.

Stouthamer A H 1992 Metabolic pathways in *Paracoccus denitrificans* and closely related bacteria in relation to the phylogeny of prokaryotes. Antonie van Leeuwenhoek 61, 1–33.

Stouthamer A H and Kooijman S A L M 1993 Why it pays for bacteria to delete disused DNA and to maintain megaplasmids. Antonie van Leeuwenhoek 63, 39–43.

Sun J S, Sands R and Simpson R J 1992 Genotypic variaton in growth and nodulation by seedlings of *Acacia* species. For. Ecol. Manage. 55, 209–223.

Trinick M J 1982 Biology. *In* Nitrogen Fixation. Eds. W J Broughton. pp 76–146. Oxford, Clarendon Press.

Young J P W 1985 *Rhizobium* population genetics: enzyme polymorphism in isolates from peas, clover, beans and lucerne grown at the same site. J. Gen. Microbiol. 1331, 2399–2408.

Young J P W 1993 Molecular phylogeny of rhizobia and their relatives. *In* New Horizons in Nitrogen Fixation. Eds. R Palacios, J Mora and W E Newton. pp 587–592. Kluwer Academic Publishers, Dordrecht.

Young J P W and Johnston A W B 1989 The evolution of specificity in the legume-rhizobium symbiosis. Trends Ecol. Evol. 4, 341–349.

Zhang X, Harper R, Karsish M and Lindström K 1991 Diversity of *Rhizobium* bacteria isolated from root nodules of Leguminous trees. Inter. J. Syst. Bacteriol. 41, 104–113.

Plant and Soil **161**: 11–20, 1994.

Recent developments in *Rhizobium* taxonomy

Esperanza Martínez-Romero
Centro de Investigación sobre Fijación de Nitrógeno, UNAM, Apdo. postal 565-A, Cuernavaca, Mor., México

Key words: *Bradyrhizobium*, genetic diversity, *Rhizobium*, systematics, taxonomy

Abstract

Recent developments in *Rhizobium* taxonomy are presented from a molecular and evolutionary point of view. Analyses of ribosomal RNA gene sequences provide a solid basis to infer phylogenies in the Rhizobiaceae family. These studies confirmed that *Rhizobium* and *Bradyrhizobium* are only distantly related and showed that *Rhizobium* and *Bradyrhizobium* are related to other groups of bacteria that are not plant symbionts. *Rhizobium* and *Agrobacterium* species are intermixed. Differences in plasmid content may explain to a good extent the different behavior of *Rhizobium* and *Agrobacterium* as symbionts or pathogens. Other approaches to identify and classify bacteria such as DNA-DNA hybridization, fatty acid analysis, RFLP and RPD-PCR techniques and phylogenies derived from other genes are in general agreement to the groupings derived by ribosomal sequences. Only a small proportion of nodulated legumes have been sampled for their symbionts and more knowledge is required on the systematics and taxonomy of *Rhizobium* and *Bradyrhizobium* species.

Introduction

New molecular approaches that analyze the bacterial genome are renewing our interest in bacterial systematics and taxonomy, and broadening the perception that man has of microbes. These approaches have not only revealed unsuspected relationships among apparently unrelated bacteria, but also demonstrated the existence of marked genetic diversity within groups of microorganisms. These analyses could help to understand mechanisms that operate in bacterial evolution, they provide tools to confidently identify bacteria, and also provide a solid reference framework for other type of studies.

DNA-DNA or DNA-RNA hybridization, restriction fragment length polymorphism analysis of DNA (RFLP), RPD's (de Bruijn, 1992),

DNA sequencing as well as other approaches such as the electrophoretic analysis of metabolic enzymes and numerical taxonomy have proven valuable in bacteria taxonomy and systematics (Selander et al., 1986; Schleifer and Stackebrandt, 1983). Both the conservation of ribosomal RNA due to its structural constraints in ribosomes, and the existence of variability in some domains, render ribosomal RNA genes sequences (5S, 16S, and 23S) as very good choices to compare organisms and to infer phylogenies (Woese, 1987).

Phylogenies of the Rhizobiaceae based on 16S rRNA sequences

Bradyrhizobium, *Azorhizobium* and *Rhizobium* species form nodules in the roots or stems of legumes where they fix atmospheric nitrogen. These species and their host plants are listed in Table 1. Analysis of 16S ribosomal sequences revealed that *Rhizobium* and *Bradyrhizobium* are only distantly related, but that each has close relationships to other groups of bacteria that are not plant-symbionts (Sawada et al., 1993; Willems and Collins, 1993; Yanagi and Yamasato, 1993; Young et al., 1991). *Bradyrhizobium* spp., including the phototrophic strain BTail are more related to *Rhodopseudomonas*, to *Afipia* and to *Blastobacter denitrificans*. (Willems and Collins, 1992; Young et al., 1991). *Rhizobium* is related to *Agrobacterium*, to *Brucella*, to *Rochalimea* and to *Bartonella*. *Phyllobacterium*, one of the other genera of the Rhizobiaceae that forms hypertrophies on leaves, also appears related to *Rhizobium huakuii* and to *R. loti* (Yanagi and Yamasato, 1993).

A summary phylogenetic tree is shown in Fig. 1 which gathers data from published (Sawada et al., 1993; Willems and Collins, 1992; Yanagi and Yamasato, 1993; Young et al., 1991) and unpublished (Hernández-Lucas et al.) genetic distances derived from partial sequences and from full sequences of 16S rRNA genes.

Some differences in phylogenetic trees may be obtained depending on the theoretical analysis performed. Contrast, for example, the Fitch-derived tree and the parsimony analysis- tree derived by Willems and Collins (1993). Some branches of phylogenetic trees can be predicted with high probabilities. For others, alternate node positions having equal probabilities may be possible, making their positions uncertain.

Agrobacterium and *Rhizobium* species are consistently intermingled (Sawada et al., 1993; Willems and Collins, 1993, Yanagi and Yamasato, 1993). *R. galegae* is a branch among other agrobacteria lineages, while the degree of relationship between *R. tropici*, *Agrobacterium rhizogenes* and *Agrobacterium* spp. is remarkable. *R. tropici* is native to South America but has also been encountered in nodules of *P. vulgaris* from acid soils in Kenya (Giller, unpublished). It is a broad host rhizobia that nodulates *P. vulgaris* bean, *Leucaena* spp. and some other legumes

(Martínez et al., 1993); *R. tropici* is able to grow in acid pH, and is tolerant to aluminum (Graham et al., 1982), and of high temperatures. *R. tropici* has been subdivided in two types based on phenotypic differences and differences in ribosomal RNA genes (Martínez et al., 1991). *Agrobacterium* sp. K-Ag-3 was isolated from a tumor from a Kiwi plant in Hiroshima, Japan and Ch-Ag-4 was isolated from cherry in Okayama, Japan (Sawada and Ieki, 1992) and according to these authors they represent two subtypes of unclassified agrobacteria. By the analysis of complete ribosomal 16S RNA genes, K-Ag-3 is indistinguishable from *R. tropici* type B and very close to Ch-Ag-4 and to *R. tropici* type A. Interestingly, the similarities in ribosomal sequences between *Rhizobium* and *Agrobacterium* are in agreement with similarities in colony morphology and growth in different media. *R. tropici* strains do not form tumors in sunflower (Martínez, unpublished). As acid resistance is determined chromosomally in *R. tropici*, (P. Graham, personal communication), we tested Ch-Ag-4 and K-Ag-4 for growth in acid medium. These agrobacteria were able to form colonies in MM in pH 4.5. Differences in plasmid content may explain to a good extent the different behavior of *Rhizobium* and *Agrobacterium* as symbionts or pathogens. Thus, it is important to distinguish between the evolutionary histories of plasmids and the evolutionary histories of chromosomes in phylogenetic studies of bacteria.

Phylogenies of plasmids and evidence of their transfer

In *Rhizobium*, most symbiotic information lies on extrachromosomal elements termed symbiotic (sym) plasmids (Martínez et al., 1990), but other plasmids have also been shown to be involved in the symbiotic process (Brom et al., 1992; Hynes and McGregor, 1990; Martínez and Rosenblueth, 1990). In *Agrobacterium*, plasmids are responsible for tumorigenesis. Non-symbiotic rhizobia (Segovia et al., 1991) and non-pathogenic agrobacteria (Kersters and de Ley, 1984) have been described. Non-symbiotic rhi-

Table 1. Species of root and stem-nodulating bacteria and their hosts

Rhizobium species:	Host legumes:
Rhizobium meliloti	*Medicago, Melilotus, Trigonella*
Rhizobium fredii, R. xinjiangensis	*Glycine max* and *G. soja* and other legumes
Rhizobium leguminosarum	
bv. viciae	*Pisum, Vicia*
bv. *trifolii*	*Trifolium*
bv. *phaseoli*	*Phaseolus*
Rhizobium tropici	*Phaseolus vulgaris, Leucaena* spp.
Rhizobium etli	*Phaseolus vulgaris*
Rhizobium galegae	*Galega officinalis, G. orientalis*
Rhizobium loti	*Lotus* spp.
Rhizobium huakuii	*Astragalus sinicus*
Bradyrhizobium species:	
Bradyrhizobium japonicum	*Glycine max*
Bradyrhizobium elkanii	*Glycine max*
Azorhizobium species:	
Azorhizobium caulinodans	*Sesbania rostrata*

zobia lack symbiotic plasmids, but have a genetic structure and diversity similar to the population of symbiotic rhizobia (Segovia et al., 1991). This would indicate that sym plasmid loss and gain is a continuous and dynamic process in rhizobia. Furthermore, the acquisition of genetic information to become a pathogen or a symbiont seems to be a very recent event for some lineages of the *Agrobacterium - Rhizobium* cluster as discussed before.

R. leguminosarum bv. *phaseoli* seems to be the result of plasmid transfer in historic times. *P. vulgaris* is native to the Americas (Gepts, 1990), and was only introduced into Europe in the XVI century. There are no indigenous species of *Phaseolus* in Europe. Segovia et al. (1993) suggest that *R. etli* strains were introduced with beans to Europe at the same time as their host. Some strains remained as such, but in others sym plasmid transfers occurred into other rhizobia having different chromosomal DNA.

RFLP analysis of *R. galegae* has also shown evidence of plasmid transfer within two geograph-

ically distant populations of *R. galegae. R. galegae* nodulates two species of goat's rue, *Galega officinalis* and *G. orientalis*, with patterns of *nod-* and *nif* (nitrogen-fixing) genes linked to host-plant specificity. A different grouping was obtained when chromosomal probes were analyzed, some strains of *R. galegae* from *G. officinalis* being more closely related to strains isolated from *G. orientalis* (Kaijalainen and Lindström, 1989). Interstrain transfer of symbiotic sequences in the course of evolution is the most plausible explanation for this.

Rhizobium and *Agrobacterium* strains readily interchange plasmids under laboratory conditions. Different *Rhizobium* species containing Ti (tumor-inducing) plasmids from *Agrobacterium tumefaciens* are tumor-inducing, though the tumors formed are smaller in size (Hooykaas et al., 1977). *Agrobacterium tumefaciens* with symbiotic plasmids from *Rhizobium* form nodules on the corresponding host legume (Hooykaas et al., 1982, 1985; Kondorosi et al., 1982; Truchet et al., 1984; Van Brussel et al., 1982). When *R.*

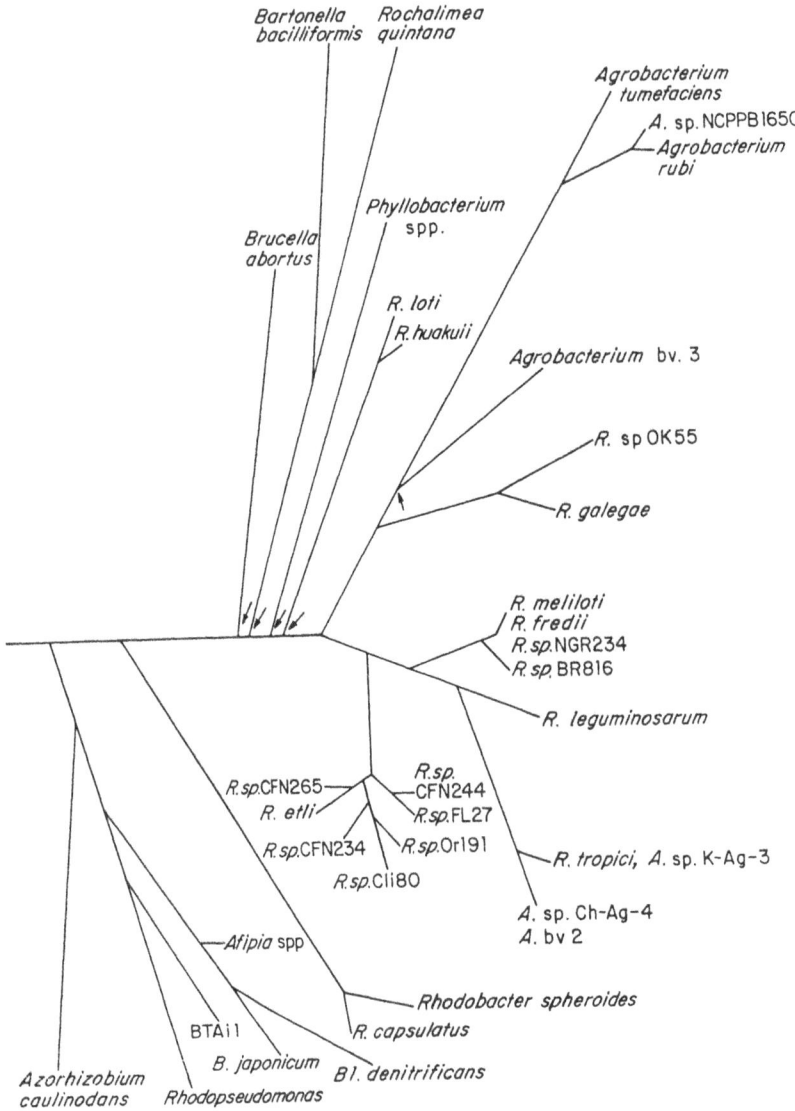

Fig. 1. Phylogenetic tree derived from results obtained by Willems and Collins, 1993; Yanagi and Yamasato, 1993; Sawada et al., 1993; Young et al., 1991 and Hernandez Lucas et al., unpublished. Genetic distances were used to construct the tree by Neighbor-Joining method (Saitou and Nei, 1987). Position of nodes indicated with arrows is not definitive.

tropici sym plasmid was transferred to *A. tumefaciens* plasmid-less strain GMI9023, *A. tumefaciens* transconjugants nodulated and fixed nitrogen in bean, albeit at a reduced level (Martínez et al., 1987). The transconjugants also nodulated *Leucaena* (Fig. 2). As mentioned above, *R. tropici*'s closest relatives are *Agrobacterium* spp.

It would be interesting to have more information about *Phyllobacterium* in regard to the existence of plasmids and sequences of putative *nod* genes.

An evolutionary hypothesis

It has been suggested that *Rhizobium* and *Bradyrhizobium* lineages diverged before the origin of legumes (Ochman and Wilson, 1987), and that subsequently, the information required for nodule formation was passed from one genus to the other (Young, 1993). I will present some facts and ideas suggesting that the information flow was from *Bradyrhizobium* to *Rhizobium*. A different hypothesis has been proposed by Janet Sprent (this volume). *Bradyrhizobium* species in general

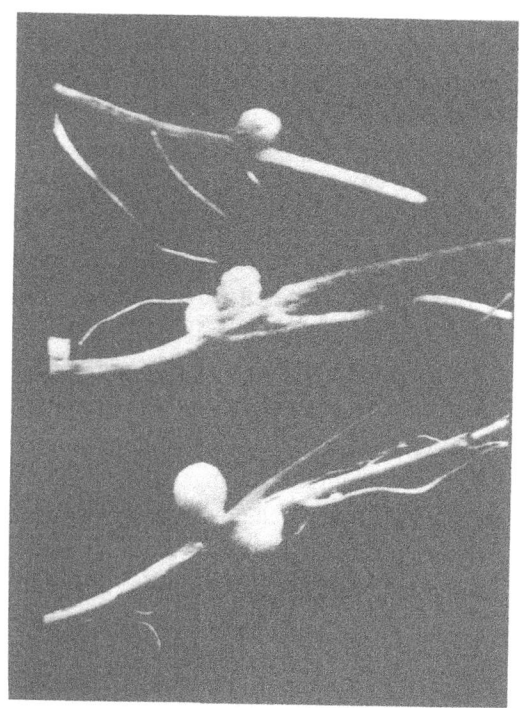

Fig. 2. Leucaena esculenta nodules induced by *R. tropici* CFN299 (bottom), *Agrobacterium tumefaciens* GMI9023 transconjugants harboring: pSym and plasmid b from CFN299 (middle) and psym fromCFN299 (up)

have a broader host-range than *Rhizobium*, leading Norris (1956) to propose that *Bradyrhizobium* was the more primitive symbiont. Symbiotic information for nodule formation in legumes could have been transferred from *Bradyrhizobium* to a proto-*Agrobacterium radiobacter*, then after this "catastrophic" event, further distributed with the *Agrobacterium-Rhizobium* chromosomal lineages. Transfer and recombination of symbiotic information could have been the basis for an accelerated evolution that led to *Rhizobium* speciation in relation to legume specificity.

Azorhizobium caulinodans is perhaps more related to *Bradyrhizobium* than to *Rhizobium* by the analysis of 16S ribosomal RNA-genes (Sawada et al., 1993; Willems and Collins, 1993) but *nod*-gene structural similarity is higher between *Bradyrhizobium* and *Rhizobium* than between *Azorhizobium* and *Bradyrhizobium* (Goethals et al., 1989), *nod*-gene information could have originated in an ancestor of the *Azorhizobium* -

Bradyrhizobium branch, and then diverged in both lineages. Although plasmids are quite scarce in *Bradyrhizobium* , some sequences that are on the chromosome in one strain may be plasmid borne in another (Haugland and Verma, 1981). Interestingly pJP4 and r68.45 can be transferred between populations of *Bradyrhizobium* in nonsterile soil with transfer frequencies higher than previously reported for in vitro transfer (Kinkle et al., 1993). It is worthy of mention that repetitive sequences have been found close to gene regions containing the symbiotic information in *B. japonicum*, and this may promote some instability (Hahn and Hennecke, 1988).

Other markers and their linkage in the genome

It is agreed that there is not extensive recombination between *Rhizobium* chromosomes; thus bacteria behave as clones, with linkage between different genetic markers (Piñero et al., 1988; Souza et al., 1992). Thus, it is only necessary to screen specific gene regions to obtain a good image of the whole genome. GSII seems to be a good marker of groups or species in the Rhizobiaceae. *R. meliloti* strains, *R. etli*, *R. tropici* types A and B, *R.leguminosarum* strains have been analyzed in western blots and the isoelectric point of GSII has been determined (Taboada et al., 1993). Bacteria are correctly classified by these means, indicating most probably a common ancestor for each group. Similarly, analysis of fatty acid profiles allows an adequate grouping of rhizobia (Jarvis and Tighe, 1994).

REP- and ERIC-PCR techniques are also useful tools for *Rhizobium* classification. De Bruijn (1992) showed that results from REP-PCR and ERICS-PCR are in agreement with phylogenies derived from multilocus enzyme electrophoresis. Classification of genetically related strains of *Bradyrhizobium japonicum* serocluster 123 by the patterns of their repetitive sequences was correlated with RFLP's (Judd et al., 1993). However repetitive DNA may change faster than the genome as a whole, as it seems to be involved in recombination and amplification events (Flores et al., 1988). Otherwise, REP-PCR and ERIC-PCR are advantageous because they allow recognition of closely related strains and they are easy and fast to perform.

RFLP analysis of ribosomal genes or PCR-fragments of ribosomal genes are useful to distinguish groups. *Bradyrhizobium* specific probes, *Rhizobium* and *Bradyrhizobium* species-specific and even strain specific probes are starting to be developed (Bjourson et al., 1992; Ludwig, pers. commun.; Wheatcroft and Watson, 1988).

Other DNA sequences

A better sample of the genome would always be more convenient, and this undoubtedly will come in the future, as DNA sequencing becomes more routinely used. In *Salmonella*, it has been found that trees derived from a single gene are not always enough to describe phylogenies (Nelson et al., 1991).

23S rRNA are larger molecules than 16S rRNA, they contain more genetic information that may be useful in phylogenetic analysis (Ludwig et al., 1992). 23S rRNA gene sequences of *Bradyrhizobium* and *Rhodopseudomonas* have been analyzed (Ludwig, personal communication), it would be interesting to have more 23S rRNA sequences from other *Rhizobium* spp.

Phylogenetic trees derived from citrate synthase gene sequence are in general agreement to phylogenies derived from ribosomal genes (Pardo et al., 1994). More sequences of citrate synthase gene from different *Rhizobium* species would be required to draw a complete scheme. *nif* gene phylogeny in *Rhizobium* is linked to the chromosome (Hennecke et al., 1985; Young, 1992).

DNA-DNA homology

Nucleic acid hybridization is considered a reliable means of establishing the relationship between bacterial species, though not of sufficient accuracy. Classically, genomic species encompass strains with approximately 70% or greater DNA

relatedness, although the exact level below which organisms are considered to belong to different species varies.

Total DNA-homology as revealed by DNA-DNA hybridization seems not to be in close agreement with 16S ribosomal sequence phylogeny in some cases. This is evident in Table 2 which shows DNA-DNA hybridization results for some of the rhizobia depicted in Fig. 1. DNA-DNA hybridization experiments take into account DNA borne on plasmids. In some *Rhizobium* species, (e.g. *R. etli*) this may represent up to 45% of the genome. Since this extrachromosomal DNA most probably undergoes change faster than core chromosomal DNA, it can contribute to values in DNA homology which are not in clear agreement with other criteria for estimating strain relatedness. We suppose this is, in part, the explanation for the discrepancies between Table 2 and Fig. 1, and for the low DNA:DNA hybridization values reported here.

Bacterial taxonomy on trial

Claims to revise the genus *Agrobacterium* in view of its close relationships to *Rhizobium* have been raised (Sawada et al., 1993; Willems and Collins, 1993). While new species are being proposed, clouds of related rhizobia are starting to emerge, raising questions on realistic limits between species. For example, according to 16S ribosomal RNA partial sequences (Eardly et al., 1992, Hernández-Lucas et al., unpublished; Laguerre et al., 1993) *R.etli* is a branch among other rhizobia with different specificities (Fig. 1). *R. etli* differs from these *R.* spp. in many plasmid-borne traits. We have discussed previously that *R. tropici* is overlapped with *Agrobacterium* spp. In *R. meliloti*, the existence of two highly differentiated evolutionary lineages has been shown. One is adapted to annual medic species of the Mediterranean basin. The genetic distance is so large that it could warrant different species. The extensive genotypic diversity among strains of *R. meliloti* is associated with the unusually high level of species

Table 2. Relative levels of homology at 65°C between the DNA from selected *Rhizobium* species

	Hybridization[a] %
Between *R. tropici* type A strains (average)	91.7%
Between *R. tropici* type B strains (average)	81.4%
R. tropici type A with	
R. tropici type B	39%/36%[b]
R. tropici type A with	
Agrobacterium sp Ch-Ag-4	24%
Between *R. etli* strains (average)	70%
R. etli and *R*.spp. related to *R. etli*[c]	28.3%
R. etli[T] CFN42 with	
R. leguminosarum bv. *viciae*	48%/45%[b]
R. etli[T] CFN42 with	
R. leguminosarum bv. *trifolii*	49%

[a]Average estimated from Martínez-Romero et al., 1991; Segovia et al., 1991; Martínez et al., unpublished.

[b]Independent result obtained by a different hybridization method by Laguerre and Amarger, INRA, 17 Rue Sully BP 1540, 21034 Dijon-Cédex- France.

[c]R. spp. related to *R. etli*: CFN234, CFN244, CFN265, Cli80 and FL27.

diversity in the genus *Medicago* (Eardly ct al., 1990).

While there seems to be an agreement that a biological meaningful classification of *Rhizobium* should be based on chromosomal genes rather than on plasmid-encoded symbiotic characteristics (Young et al., 1993), it seems that bacterial taxonomy has to be deeply changed. We are perhaps waiting for a comprehensive view of the genomes and a more complete scope of existing microorganisms to set the rules, but changing seems difficult. The true impact of taxonomy would be not only to give names but to provide a true conceptual framework for research. The concept of genetic isolation is certainly not true in bacterial species, and different species sharing plasmids would be perhaps not uncommon.

The known microorganisms are only a very small proportion of existing organisms (Torsvik et al., 1990). This is specially true for *Rhizobium* and *Bradyrhizobium* where only a small number of nodulating legumes have been sampled for their symbionts. It has been estimated that at least 2800 species of legumes form nodules (Allen and Allen, 1981), yet the 8 *Rhizobium* species and two *Bradyrhizobium* species listed in Table 1 represent less than 1% of the nodulated species of legumes. A number of tree and tropical legumes may be nodulated by both *Bradyrhizobium* or *Rhizobium* spp. (Martínez et al., 1985; Zhang et al., 1991). Very convenient schemes to characterize and classify such rhizobia have been proposed (Graham et al., 1991) and need to be followed up.

Bacterial diversity is perhaps the most valuable resource for biotechnology. Biologists have only begun to assess the complexity and potentiality of each bacterial species (Bull et al., 1992).

Acknowledgements

I am grateful to Marco A Rogel for help. Partial support was from VLIR-ABOS grant (Belgium) and contract from FAO/IAEA 302-D1-MEX-6319.

References

Allen O N and Allen E K 1981 The Leguminosae. A Source Book of Characteristics Uses and Nodulation. The University of Wisconsin Press, USA.

Bjourson A J, Stone C E and Cooper J E 1992 Combined subtraction hybridization and polymerase chain reaction amplification procedure for isolation of strain-specific *Rhizobium* DNA sequences. Appl. Environ. Microbiol. 58, 2296–2301.

Brom S, Garcia de los Santos A, Stepkowsky T, Flores M, Dávila G, Romero D and Palacios R 1992 Different plasmids of *Rhizobium leguminosarum* bv. *phaseoli* are required for optimal symbiotic performance. J. Bacteriol. 174, 5183–5189.

Bull A T, Goodfellow M and Slater J H 1992 Biodiversity as a source of innovation in biotechnology. Annu. Rev. Microbiol. 46, 219–252.

De Bruijn F J 1992 Use of repetitive (repetitive extragenic palindromic and enterobacterial repetitive intergeneric consensus) sequences and the polymerase chain reaction to fingerprint the genomes of *Rhizobium meliloti* isolates and other soil bacteria. Appl. Environ. Microbiol. 58, 2180–2187.

Eardly B D, Materon L A, Smith N H, Johnson D A, Rumbaugh M D and Selander R K 1990 Genetic structure of natural populations of the nitrogen-fixing bacterium *Rhizobium meliloti*. Appl. Environ. Microbiol. 56, 187–194.

Eardly B D, Young J P W and Selander R K 1992 Phylogenetic position of *Rhizobium* sp. strain Or191, a symbiont of both *Medicago sativa* and *Phaseolus vulgaris*, based on partial sequences of the 16S rRNA and *nif*H genes. Appl. Environ. Microbiol. 58, 1809–1815.

Flores M, González V, Pardo M A, Leija A, Martínez E, Romero D, Piñero D, Dávila G and Palacios R 1988 Genomic instability in *Rhizobium phaseoli*. J. Bacteriol. 170, 1191–1196.

Gepts P 1990 Biochemical evidence bearing on the domestication of *Phaseolus* (Fabaceae) beans. Econ. Bot. 44, 28–38.

Goethals K, Gao M, Tomekpe K, Van Montagu M and Holsters M 1989 Common *nod*ABC genes in *nod* locus 1 of *Azorhizobium caulinodans*: nucleotide sequence and plant inducible expression. Mol. Gen. Genet. 219, 289–298.

Graham P H, Sadowsky M J, Keyser H H, Barnet Y M, Bradley R S, Cooper J E, de Ley D J, Jarvis B D W, Roslycky E B, Strijdom B W and Young J P W 1991 Proposed minimal standards for the description of new genera and species of root- and stem- nodulating bacteria. Int. J. Syst. Bacteriol. 41, 582–587.

Graham P H, Viteri S E, Mackie F, Vargas A A T and Palacios A 1982 Variation in acid soil tolerance among strains of *Rhizobium phaseoli*. Field Crops Res. 5, 121–128.

Hahn M and Hennecke H 1988 Cloning and mapping of a novel nodulation region from *Bradyrhizobium japonicum* by genetic complementation of a deletion mutant. Appl. Environ. Microbiol. 54, 55–61.

Haugland R and Verma D P S 1981 Interspecific plasmid and genomic DNA sequence homologies and localization of *nif* genes in effective and ineffective strains of *Rhizobium japonicum*. J. Mol. Appl. Genet. 1, 205–217.

Hennecke H, Kaluza K, Thöny B, Fuhrmann M, Ludwig W and Stackebrandt E 1985 Concurrent evolution of nitrogenase genes and 16S rRNA in *Rhizobium* species and other nitrogen fixing bacteria. Arch. Microbiol. 142, 342–348.

Hooykaas P J J, den Dulk-Ras H, Regensburg-Twink A J G, van Brussel A A and Schilperoort R A 1985 Expression of a *Rhizobium phaseoli* sym plasmid in *R. trifolii* and *Agrobacterium tumefaciens*: incompatibility with a *R. trifolii* sym plasmids. Plasmid 14, 47–52.

Hooykaas P J J, Klapwijk P M, Nuti M P, Schilperoort R A and Rorsch A 1977 Transfer of the *Agrobacterium tumefaciens* Ti plasmid to avirulent agrobacteria and to *Rhizobium ex planta*. J. Gen. Microbiol. 98, 477–484.

Hooykaas P J J, Snijdewint F G M and Schilperoort R A 1982 Identification of the Sym plasmid of *Rhizobium leguminosarum* strain 1001 and its transfer to and expression in other rhizobia and *Agrobacterium tumefaciens*. Plasmid 8, 73–82.

Hynes M F and McGregor N F 1990 Two plasmids other than the nodulation plasmid are necessary for formation of nitrogen-fixing nodules by *Rhizobium leguminosarum*. Mol. Microbiol. 4, 567–574.

Judd A K, Schneider M, Sadowsky M J and de Bruijn F J 1993 Use of repetitive sequences and the polymerase chain reaction technique to classify genetically related *Bradyrhizobium japonicum* serocluster 123 strains. Appl. Environ. Microbiol. 59, 1702–1708.

Kaijalainen S and Lindström K 1989 Restriction fragment length polymorphism analysis of *Rhizobium galegae* strains. J. Bacteriol. 171, 5561–5566.

Kersters K and de Ley J 1984 *Agrobacterium. In* Bergey's Manual of Systematic Bacteriology. Eds. N R Kreig and J Holt. pp 244–254. Williams and Wilkins, Baltimore.

Kinkle B K, Sadowsky M J, Schmidt E L and Koskinen W C 1993 Plasmids pJP4 and r68.45 can be transferred between populations of Bradyrhizobia in nonsterile soil. Appl. Environ. Microbiol. 59, 1762–1766.

Kondorosi A, Kondorosi E, Pankhurst C E, Broughton W J and Banfalvi Z 1982 Mobilization of a *Rhizobium meliloti* megaplasmid carrying nodulation and nitrogen fixation genes into other rhizobia and *Agrobacterium*. Mol. Gen. Genet. 188, 433–439.

Laguerre G, Fernández M P, Edel V, Normand P and Amarger N 1993 Genomic heterogeneity among French *Rhizobium* strains isolated from *Phaseolus vulgaris* L. Int. J. Syst. Bacteriol. 43, 761–767.

Ludwig W, Kirchhof G, Klugbauer N, Weizenegger M, Betzl D, Ehrmann M, Hertel C, Jilg S, Tatzel R, Zitzelsberger H, Liebl S, Hockberger M, Shah J, Lane D, Wallnöfer P R and Scheifer K H 1992 Complete 23S ribosomal RNA sequences of gram-positive bacteria with a low DNA G+C content. System. Appl. Microbiol. 15, 487–501.

Martínez E, Palacios R and Sánchez F 1987 Nitrogen-fixing nodules induced by *Agrobacterium tumefaciens* harboring *Rhizobium phaseoli* plasmids. J. Bacteriol. 169, 2828–2834.

Martínez E, Pardo M A, Palacios R and Cevallos M A 1985 Reiteration of nitrogen fixation gene sequences and specificity of *Rhizobium* in nodulation and nitrogen fixation in *Phaseolus vulgaris*. J. Gen. Microbiol. 131, 1779–1786.

Martínez E, Poupot R, Promé J C, Pardo M A, Segovia L, Truchet G and Dénarié J 1993 Chemical signaling of *Rhizobium* nodulating bean. *In* New Horizons in Nitrogen Fixation. Eds. R Palacios, J Mora and W E Newton. pp 171–175, Kluwer Academic Publishers. Dordrecht.

Martínez E, Romero D and Palacios R 1990 The *Rhizobium* genome. Crit. Rev. Plant Sci. 9, 59–93.

Martínez-Romero E and Rosenblueth M 1990 Increased bean (*Phaseolus vulgaris* L.) nodulation competitiveness of genetically modified *Rhizobium* strains. Appl. Environ. Microbiol. 56, 2384–2388.

Martínez-Romero, E, Segovia L, Martins Mercante F, Franco A A, Graham P and Pardo M A 1991 *Rhizobium tropici*, a novel species nodulating *Phaseolus vulgaris* L. beans and *Leucaena* sp. trees. Int. J. Syst. Bacteriol. 41, 417–426.

Nelson K, Whittam T S and Selander R K 1991 Nucleotide polymorphism and evolution in the glyceraldehyde-3-phosphate dehydrogenase gene (*gap*A) in natural populations of *Salmonella* and *Escherichia coli*. Proc. Natl. Acad. Sci. USA. 88, 6667–6671.

Norris D O 1956 Legumes and the *Rhizobium* symbiosis. Empire J. Experimental Agric. 24, 246–270.

Ochman H and Wilson A C 1987 Evolution in bacteria: evidence for a universal substitution rate in cellular genomes. J. Mol. Evol. 26, 74–86.

Pardo M A, Lagúnez J, Miranda J and Martínez E 1994 Nodulating ability of *Rhizobium tropici* is conditioned by a plasmid-encoded citrate synthase. Mol. Microbiol. 11, 315–321.

Piñero D, Martínez E and Selander R K 1988 Genetic diversity and relationships among isolates of *Rhizobium leguminosarum* biovar *phaseoli*. Appl. Environ. Microbiol. 54, 2825–2832.

Saitou N and Nei M 1987 The neighbor-joining method: a new method for reconstructing phylogenetic trees. Mol. Biol. Evol. 4, 406–425.

Sawada H and Ieki H 1992 Phenotypic characteristics of the genus *Agrobacterium*. Ann. Phytopath. Soc. Japan 58, 37–45.

Sawada H, Ieki H, Oyaizu H and Matsumoto S 1993 Proposal for rejection of *Agrobacterium tumefaciens* and revised descriptions for the genus *Agrobacterium* and for *Agrobacterium radiobacter* and *Agrobacterium rhizogenes*. Int. J. Syst. Bacteriol. 43, 694–702.

Schleifer K H and Stackebrandt E 1983 Molecular systematics of prokaryotes. Annu. Rev. Microbiol. 37, 143–187.

Segovia L, Piñero D, Palacios R and Martínez-Romero 1991 Genetic structure of a soil population of nonsymbiotic *Rhizobium leguminosarum*. Appl. Environ. Microbiol. 57, 426–433.

Segovia L, Young J P W and Martínez-Romero E 1993 Reclassification of American *Rhizobium leguminosarum* biovar *phaseoli* type I strains as *Rhizobium etli sp. nov.* Int. J. Syst. Bacteriol. 43, 374–377.

Selander R K, Caugant D A, Ochman H, Musser J M, Gilmour M N and Whittam T S 1986 Methods of multilocus enzyme electrophoresis for bacterial population genetics and systematics. Appl. Environ. Microbiol. 51, 873–884.

Souza V, Nguyen T T, Hudson R R, Piñero D and Lenski R E 1992 Hierarchical analysis of linkage disequilibrium in *Rhizobium* populations: evidence of sex? Proc. Natl. Acad. Sci. USA 89, 8389–8393.

Taboada H, Encarnación S, Vargas M C, Narváez V, Mora Y, Martínez E and Mora J 1993 Glutamine synthetase II as a biological marker of the Rhizobiaceae family. *In* New Horizons in Nitrogen Fixation. Eds. R. Palacios, J Mora and W E Newton. p. 657. Kluwer Academic Publishers. Dordrecht.

Torsvik V, Goksoyr J and Daae F L 1990 High diversity in DNA of soil bacteria. Appl. Environ. Microbiol. 56, 782–787.

Truchet G, Rosenberg C, Vasse J, Julliot J S, Camut S and Dénarié J 1984 Transfer of *Rhizobium meliloti* pSym genes into *Agrobacterium tumefaciens*: host-specific nodulation by atypical infection. J. Bacteriol. 157, 134–142.

Van Brussel A A N, Tak T, Wetselaar A, Pees E and Wijffelman C A 1982 Small leguminosae as test plants for nodulation of *Rhizobium leguminosarum* and other rhizobia and agrobacteria harbouring a leguminosarum sym plasmid. Plant. Sci. Lett. 27, 317–325.

Wheatcroft R and Watson R J 1988 A positive strain identification method for *Rhizobium meliloti*. Appl. Environ. Microbiol. 54, 574–576.

Willems A and Collins M D 1992 Evidence for a close genealogical relationship between *Afipia*, the casual organism of cat scratch disease, *Bradyrhizobium japonicum* and *Blastobacter denitrificans*. FEMS Microbiol. Lett. 96, 241–246.

Willems A and Collins M D 1993 Phylogenetic analysis of rhizobia and agrobacteria based on 16S rRNA gene sequences. Int. J. Syst. Bacteriol. 43, 305–313.

Woese C R 1987 Bacterial evolution. Microbiol. Rev. 51, 221–271.

Yanagi M and Yamasato K 1993 Phylogenetic analysis of the family Rhizobiaceae and related bacteria by sequencing of 16S rRNA gene using PCR and DNA sequencer. FEMS Microbiol. Lett. 107, 115–120.

Young J P W 1992 Phylogenetic classification of nitrogen-fixing organisms *In* Biological Nitrogen Fixation. Eds. G. Stacey, R H Burris and J H Evans. pp 43–86. Chapman and Hall, New York.

Young J P W 1993 Molecular phylogeny of rhizobia and their relatives. *In* New Horizons in Nitrogen Fixation. Eds. R Palacios. J Mora and W E Newton. pp 587–592, Kluwer Academic Publishers. Dordrecht.

Young J P W, Downer H L and Eardly B D 1991 Phylogeny of the phototrophic *Rhizobium* strain BTail by polymerase chain reaction-based sequencing of a 16S rRNA gene segment. J. Bacteriol. 173, 2271–2277.

Young P, Martínez E, Barnet Y, Cooper J and Lindström K 1993 Report from the taxonomy meeting, subcommittee on *Agrobacterium* and *Rhizobium*. *In* New Horizons in Nitrogen Fixation. Eds. R Palacios, J Mora and W E Newton. pp 777–778. Kluwer Academic Publishers. Dordrecht.

Zhang X, Harper R, Karsisto M and Lindström K 1991 Diversity of *Rhizobium* bacteria isolated from the root nodules of leguminous trees. Int. J. Syst. Bacteriol. 41, 104–113.

Plant and Soil **161**: 21–29, 1994.

Host range, RFLP, and antigenic relationships between *Rhizobium fredii* strains and *Rhizobium* sp. NGR234

Hari B. Krishnan and Steven G. Pueppke[1]
Department of Plant Pathology, University of Missouri, Columbia, MO 65211, USA. [1]*Corresponding author*

Key words: host range, *nod* genes, nodulation, *Rhizobium*, symbiosis

Abstract

Rhizobium fredii is a nitrogen-fixing symbiont from China that combines broad host range for nodulation of legume species with cultivar specificity for nodulation of soybean. We have compared 10 *R. fredii* strains with *Rhizobium* sp. NGR234, a well known broad host range strain from Papua New Guinea. NGR234 nodulated 16 of 18 tested lugume species, and nodules on 14 of the 16 fixed nitrogen. The *R. fredii* strains were not distinguishable from one another. They nodulated 13 of the legumes, and in only nine cases were nodules effective. All legumes nodulated by *R. fredii* were included within the host range of NGR234. Restriction fragment length polymorphisms (RFLPs) were detected with four DNA hybridization probes: the regulatory and common *nod* genes, *nodDABC*; the soybean cultivar specificity gene, *nolC*; the nitrogenase structural genes, nifKDH; and RFRS1, a repetitive sequence from *R. fredii* USDA257. A fifth locus, corresponding to a second set of soybean cultivar specificity genes, *nolBTUVWX*, was monomorphic. Using antisera against whole cells of three *R. fredii* strains and NGR234, we separated the 11 strains into four serogroups. The anti-NGR234 sera reacted with a single *R. fredii* strain, USDA191. Only one serogroup, which included USDA192, USDA201, USDA217, and USDA257, lacked cross reactivity with any of the others. Although genetic and phenotypic differences among *R. fredii* strains were as great as those between NGR234 and *R. fredii*, our results confirm that NGR234 has a distinctly wider host range than *R. fredii*.

Abbreviations: RFLP - restriction fragment length polymorphism, TBS - tris-buffered saline, YEM - yeast extract-mannitol

Introduction

The soybean, *Glycine max* (L.) Merr., is a major legume crop in both developed and undeveloped areas of the world. The relationship between this plant and nitrogen-fixing bacteria was first documented in the late nineteenth century, when Kirchner (1895) isolated and cultured a soybean root-nodule symbiont from soil that had been collected in Japan. This slow-growing bacterium is now known as *Bradyrhizobium japonicum*. It has been examined extensively, and for nearly 100 years, bradyrhizobia were considered to be the only organisms capable of nodulating soybean. Two additional groups of soybean symbionts now are recognized. One, for which the name *B. elkanii* has been proposed (Kuykendall et al., 1992), represents a specialized group of slow growing organisms that can be distinguished from *B. japonicum* by several criteria. A second, entirely new group of soybean symbionts was discovered just more than 10 years ago in soil and nodules that had been harvested between 1978 and 1980 in the east central Chinese provinces of Honan, Shandou, Shanghai, and Shansi (Keyser et al., 1982; Keyser and Griffin, 1987). In contrast to bradyrhizobia, which have long generation times and make their growth medium alkaline (Jordan, 1984), the Chinese strains multiply rapidly and acidify their growth medium. These bacteria quickly captured the attention of the scientific community, in part because soybean has its center of origin and diversity in China, and in part because they invalidated the assumption that soybean symbionts all are slow-growing bradyrhizobia.

There has been some controversy about the systematics and taxonomic status of the original fast-growing strains, as well as additional isolates that were recovered later from soybean in China. On the basis of physiological analysis and DNA relatedness, Elkan and associates (Scholla and Elkan, 1984; Scholla et al., 1984) assigned the original group of strains to a new species, *Rhizobium fredii*, and separated it into chemovars *fredii* and *siensis*. Chen et al. (1988) recognized two species on the basis of numerical analysis: *S. fredii*, which included all of the original strains and 16 additional isolates, and *S. xinjiangensis*, which contained only new isolates. This interpretation has been questioned by Jarvis et al. (1992) and Yanagi and Yamasoto (1993), who provided evidence to support the conservation of *R. fredii*. Although the proper status of *R. fredii* remains tentative, it has been noted that the species has affinity to *R. meliloti* (Sadowsky et al., 1987; Yanagi and Yamasoto, 1993) as well as *R. loti*, *R. leguminosarum*, and several strains of fast growing rhizobia from tropical trees (Zhang et al., 1991). Moreover, *Rhizobium* sp. NGR234, a fast-growing isolate from hyacinth bean [*Lablab purpureus* (L.)Sweet] in Papua New Guinea (Trinick, 1980), is serologically related to *R. fredii* (Sadowsky et al., 1987), and its 16S rRNA sequence is identical

to those of *R. fredii* and *R. meliloti* (Jarvis et al., 1992).

Unlike *B. japonicum*, *R. fredii* is markedly cultivar-specific in its interactions with soybean. Many primitive cultivars, as well as the wild progenitor species, *G. soja* Sieb. & Zucc., form nitrogen-fixing nodules with all tested strains (Devine, 1985; Heron and Pueppke, 1984; Keyser et al., 1982; Keyser and Cregan, 1984). Agronomically improved cultivars also are nodulated by some strains, notably USDA191, but many strains are incompatible with these genotypes–nodules are either absent or abnormal root proliferations are induced (Balatti and Pueppke, 1992; DuTeau et al., 1986; Hattori and Johnson, 1984; Israel et al., 1986; Jansen van Rensburg et al., 1983; Keyser et al., 1982). In contrast to such genotype specificity with its host of isolation, *R. fredii* is promiscuous in its ability to nodulate other legumes (Heron and Pueppke, 1984; Morrison et al., 1986; Stowers and Eaglesham, 1984). Thus, this organism is a useful model to simultaneously examine both broad and narrow host range.

Our laboratory is particularly interested in the genetic and biochemical basis of legume specificity of *R. fredii*, especially strains USDA191 and USDA257 (Heron and Pueppke, 1984, 1987; Heron et al., 1989; Krishnan and Pueppke, 1991a, b, c, 1993; Meinhardt et al., 1993). During the course of our studies, it became apparent that the symbiotic relationships among *R. fredii* strains are incompletely documented and often based on comparisons of one or two strains under nonuniform conditions. In addition, the affinities between *R. fredii* and the well known broad host range strain NGR234 (Stanley and Cervantes, 1991) have been inadequately described. We report here the results of a series of systematic experiments with 10 of the original *R. fredii* strains, as well as NGR234. We have examined host range with 18 legume species, and we confirm heterogeneity at genetic loci involved in nodulation and nitrogen fixation. We also present data on antigenic relationships among the strains.

Materials and methods

Bacteria and plants

Rhizobium fredii USDA191, USDA192, USDA193, USDA201, USDA205, USDA206, USDA208, USDA214, USDA217, and USDA257 were from the culture collection of the U.S. Department of Agriculture, Beltsville, MD (Keyser and Griffin, 1987). Broad host range strain NGR234 was provided by W. J. Broughton of the University of Geneva, Switzerland. Bacteria were maintained in storage at -70°C in 15% glycerol. Working stocks were kept on slants of YEM agar (Vincent, 1970) at 4°C, and cultures were grown in liquid YEM medium. Nodulation of vermiculite-grown plants was assessed in duplicate experiments as described previously (Krishnan and Pueppke, 1991c). Seeds of birdsfoot trefoil [*Lotus corniculatus* L.], guar [*Cyamopsis tetragonolobus* (L.)Taub, cv. Essex], and Illinois bundleflower [*Desmanthus illinoensis* (Michx.)MacM.] were from the Department of Agronomy, University of Missouri. Seeds of the remaining test plants were from our laboratory collection or that of W.J. Broughton at the University of Geneva.

RFLP analysis of symbiosis loci

Total genomic DNA was isolated as described previously (Heron et al., 1989). Five DNA fragments were used separately as ^{32}P-labeled hybridization probes under conditions of standard stringency (Heron et al., 1989). The *nod* probe was a 2.2-kb *Bam*HI/*Hind*III fragment containing *nodAB* and portions of *nodC* and *nodD1* of *R. meliloti* strain 1021 (Egelhoff et al., 1985). The *nif* probe was the 6.9-kb *Eco*RI insert of pSa30, which contains *nifKDH* of *Klebsiella pneumoniae* (Cannon et al., 1979). A probe containing the soybean cultivar specificity genes *nolBTUVWX* of USDA257 was prepared from the 8.0-kb insert of plasmid pRfDH410 (Meinhardt et al., 1993). The probe for *nolC*, another soybean cultivar specificity gene of USDA257, was obtained from a gel-purified 1053-bp *Bgl*II fragment that had been excised from plasmid pHBK130 (Krishnan and

Pueppke, 1991a). A hybridization probe for sym plasmid-borne repetitive sequences of USDA257 was prepared from the 16-kb RFRS1 fragment (Krishnan and Pueppke, 1993), which had been excised and gel-purified from cosmid pRFRS1 (Krishnan and Pueppke, 1991b).

Immunological cross-reactivity of strains

Bacterial cells were grown for 3 days in YEM media, centrifuged at $8,000 \times g$ for 10 min, and adjusted with TBS (20 mM Tris HC1, pH 7.5, containing 500 mM NaC1) to a turbidity of 0.8 at 625 nm (ca. 1.5 X 10^9 cells/mL). Aliquots of 5 μL of undiluted suspensions and of suspensions diluted 1:2, 1:5, and 1:10 were spotted onto nitrocellulose sheets, which had been equilibrated previously in TBS. Membranes were air-fried and immersed in a blocking solution consisting of 5% powdered dry milk in TBS for 30 min. Membranes then were incubated with antibodies that had been diluted 1:100 in TBS containing 5% powdered milk. After 1 h, the nitrocellulose was washed three times with TBS containing 0.1% Tween-20. Bound antibodies were detected by incubation with ^{125}I-conjugated Protein A, followed by autoradiography (Burnett, 1981). In some experiments, bound antibodies were identified by an alternative procedure that employed horseradish peroxidase-protein A conjugates from Bio-Rad. The antibodies against strains USDA191 and USDA257 (Heron and Pueppke, 1987) and against USDA208 (Zdor and Pueppke, 1991) have been described. Antibodies against cells of NGR234 were generated by the method of Zdor and Pueppke (1991).

Results

The host range of NGR234 is broader than that of R. fredii

Table 1 gives the results of replicated nodulation experiments in which all 11 strains were screened individually with 18 legume species representative of four tribes of the subfamily Papil-

Table 1. Host specificity of *Rhizobium* strains[a]

Host species	Tribe	Nodulation response	
		R. fredii	NGR234
Acacia aroma Gillies ex H & A	Acaceae	0	0
Albizia lebbeck (L.)Benth.	Ingeae	Fix[+]	Fix[+]
Cajanus cajan (L.)Millsp.	Phaseoleae	Fix[+]	Fix[+]
Calopogonium caeruleum (Benth.) Hemsl.	Phaseoleae	0	Fix[+]
Canavalia ensiformis (L.)DC	Phaseoleae	Nod[+]	Nod[+]
Crotalaria sericea Retz.	Genisteae	Nod[+]	Fix[+]
Cyamopsis tetragonoloba (L.)Taub.	Galegeae	0	0
Desmanthus illinoensis (Michx.)MacM.	Eumimoseae	Nod[+]	Fix[+]
Flemingia congesta Roxb.	Phaseoleae	Fix[+]	Fix[+]
Glycine max (L.)Merr. cv. Peking	Phaseoleae	Fix[+]	Nod[+]
Indigofera tinctoria L.	Galegeae	Fix[+]	Fix[+]
Lotus corniculatus L.	Loteae	Nod[+]	Fix[+]
Macrotyloma axillare (E. Mey.)Verdc.	Phaseoleae	Fix[+]	Fix[+]
Phaseolus angularis (Willd.)Wight	Phaseoleae	Fix[+]	Fix[+]
Psophocarpus tetragonolobus (L.)DC	Phaseoleae	Fix[+]	Fix[+]
Pueraria phaseoloides (Roxb.)Benth.	Phaseoleae	0	Nod[+]
Stylosanthes hamata (L.)Taub.	Hedysareae	0	Nod[+]
Tephrosia vogelii Hook. f.	Galegeae	Fix[+]	Fix[+]

[a] Reactions of the 10 individual *R. fredii* strains were indistinguishable from one another and are grouped together. Fix[+] = Healthy plants containing deeply red-pigmented nodules. Nod[+] = Plants containing white or greenish nodules or rudimentary structures. 0 = no nodules.

ionoideae and three of the Mimosoideae. Nodulation phenotypes were classified as Fix[+] if plants were healthy and contained deeply red-pigmented nodules; all other responses, including irregular swollen structures (Balatti and Pueppke, 1992) and nodules with white or greenish interiors, were categorized as Nod[+]. Nodulation phenotyes of *R. fredii* stains were remarkably similar. Nodules were present on roots of 13 of the species, and nine interactions were Fix[+]. NGR234 nodulated 16 species, and 12 of them were rated Fix[+]. The two mimosoid species, *Albizia lebbeck*, a tropical tree, and *Desmanthus illinoensis*, and herbaceous North American perennial, were nodulated by all of the strains, but *R. fredii* strains failed to fix nitrogen in association with the latter. All legumes that were nodulated by *R. fredii* also were nodulated by NGR234, and with the exception of soybean, no host was rated Fix[+] with *R. fredii* but only Nod[+] with NGR234.

Symbiosis-related loci are polymorphic

RFLP data were generated for five sets of sequences that function in symbiosis or that are known to be linked physically to these genes in *R. fredii*. These include *nodDABC*, a group of genes involved in perception of host signals and synthesis of factors that initiate meristematic activity in the host (Dénarié et al., 1992); *nifKDH*, which encodes the nitrogenase enzyme complex responsible for reduction of N_2 to ammonia (Cannon et al., 1979); *nolC*, a negatively acting chromosomal gene that regulates cultivar-specific nodulation of soybean (Krishnan and Pueppke, 1991a); *nolB-TUVWX* a negatively acting cluster of sym plasmid genes that also defines cultivar-specificity in soybean (Meinhardt et al., 1993); and RFRS1, the largest member of a family of repetitive sequences that is reiterated on the symbiosis plasmid of USDA257 (Krishnan and Pueppke, 1991b, 1993).

Polymorphisms were apparent at each of the tested loci, with the exception of *nolBTUVWX*.

Two copies of *nodD* are known to be present in strain USDA191–*nodD1* is on a 3.0-kb *Eco*RI fragment and *nodD2* on a 6.0-kb *Eco*RI fragment (Appelbaum et al., 1988; Krishnan and Pueppke, 1991c). Neither of these is linked to *nodABC*, which is on a 9.6-kb *Eco*RI fragment in this strain (Appelbaum et al., 1985). Figure 1A shows that nine of 10 *R. fredii* strains yielded a fragment indistinguishable from the *nodD1* fragment of USDA191 (Fig. 1A). The exception is USDA192, where this fragment is only 2.5-kb in size. Polymorphisms in the *nodD2* fragment appear as a uniform change in fragment size from 6.0 kb to 4.5 kb for USDA205, USDA206, and USDA208. The *nodABC* fragment is atypical in just one strain, USDA257, where it is 9.2 instead of 9.6 kb. Although NGR234 contains a fragment indistinguishable from the *nodD1* fragment of most *R. fredii* strains, both of the other NGR234 fragments with homology to *nodDABC* hybridization probe were unique.

In *R. fredii*, the basic RFLP pattern for *nifKDH* consisted of four fragments of 4.8, 4.2, 3.25, and < 1 kb. The two largest and the smallest fragments were present in all strains (Fig. 1B), but the 3.25-kb fragment was absent from USDA193. A fifth weakly hybridizing fragment of 1.95 kb was visible in digests of USDA192 and USDA193 (Fig. 1B). As was the case for *nodDABC*, the NGR234 pattern, which uniquely lacked the largest of the hybridizing fragments, was the most divergent.

The two loci that function in cultivar-specific nodulation of soybean, *nolC* and *nolBTUVWX*, were localized on single *Eco*RI fragments in each of the tested strains (Fig. 1C and 1D). The *nolC* probe hybridized to a 9.4-kb *Eco*RI fragment of USDA257 and to 16.5-kb *Eco*RI fragments of all other strains. *nolBTUVWX*, in contrast, was monomorphic, consisting of an 8.0-kb *Eco*RI fragment in all 11 strains.

We could detect homology between RFRS1 and DNA from NGR234, but it was very weak and is not visible in Figure 1. In contrast, complex hybridization patterns were obtained when this sequence was used to probe DNA from *R.*

fredii strains (Fig. 1E). As many as 12 fragments appeared in dark exposures of gels (data not shown). With the exception of USDA205 and USDA206, each *R. fredii* strain produced a unique pattern of hybridization to RFRS1. The USDA205 and USDA206 patterns were indistinguishable, and indeed, these strains could not be differentiated from one another by any of the hybridization probes (Fig. 1).

Antigenic relationships among the strains

We used sera against cells of three *R. fredii* strains and NGR234 to examine antigenic relationships within *R. fredii* and to relate them to NGR234 (Table 2). Each of the strains gave a strong positive reaction with at least one of the sera, and USDA191 and USDA193 were cross-reactive, testing positive with three and two sera, respectively. Sera against USDA191 and USDA208 both reacted with cells of USDA193, and the anti-USDA257 serum was unique, reacting with a clearly define subset of four strains.

The patterns of serological response to strains used as immunogens was not always reciprocal. Thus cells of USDA191 reacted with sera against USDA208, but cells of USDA208 did not react with sera against USDA191. The anti-NGR234 serum was quite specific, reacting only with NGR234 and USDA191 (Table 2). Anti-USDA191 serum, however, failed to react with NGR234, and thus the USDA191 and NGR234 reactions also were nonreciprocal.

Discussion

Two key, interrelated conclusions can be drawn from the data presented here. The first is that the host ranges of the *R. fredii* strains appear to be uniform and are definable as a subset of that of NGR234. And second, the RFLP heterogeneity among *R. fredii* strains is nearly as great as that between the *R. fredii* strains as a group and broad host range strain NGR234. In conjunction with serological data (Sadowsky et al., 1987), these observations underscore the similarly between *R.*

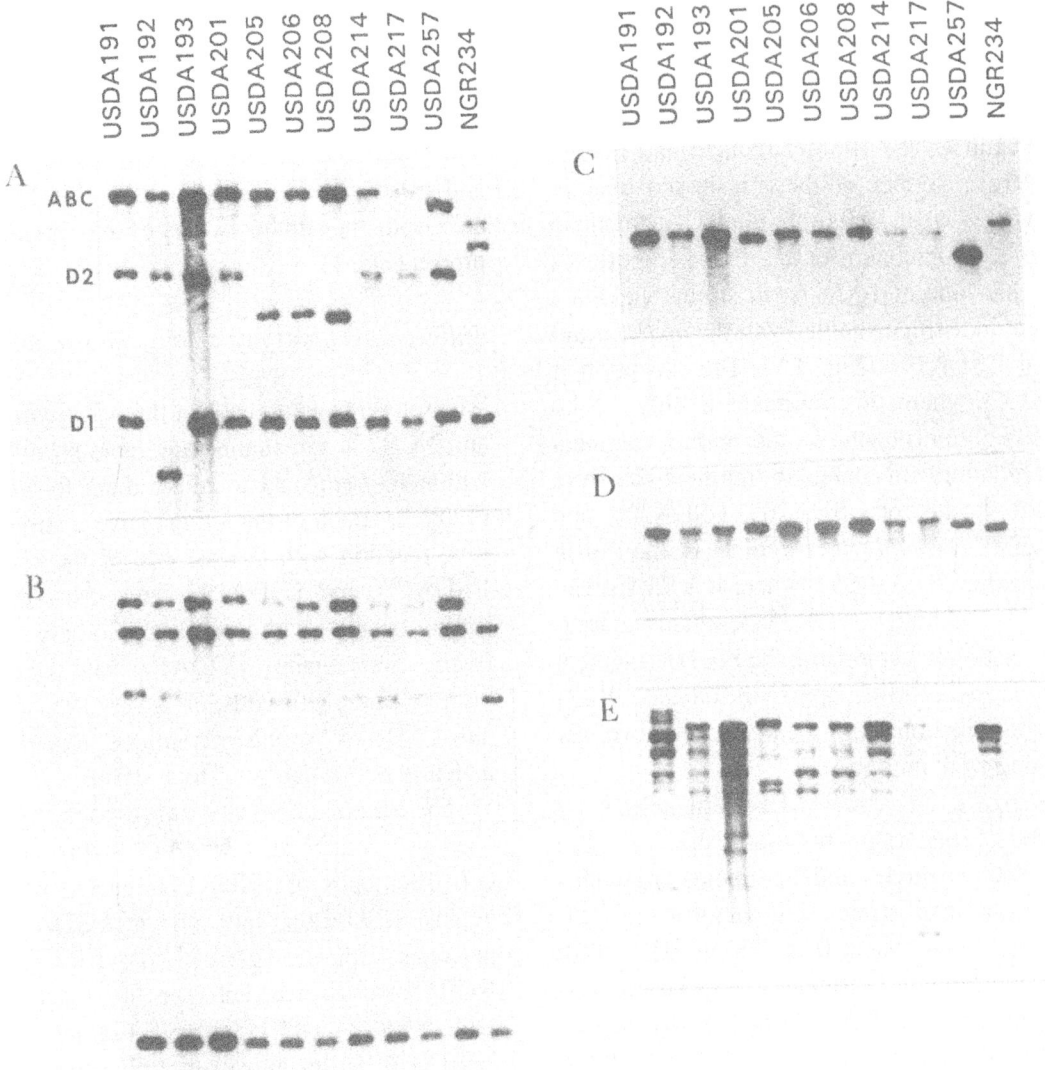

Fig. 1.   Restriction fragment length polymorphisms within *Rhizobium fredii* and broad host range strain NGR234. DNA was purified, digested with *Eco*RI, fractionated on agarose gels, and transferred to membranes as described in Materials and Methods. The hybridization probes are as follows: **(A)** *nodDABC*; **(B)** *nif*KDH; **(C)** *nolC*; **(D)** *nolBTUVWX*; **(E)** the RFRS1 series of repetitive sequences. Bands corresponding to *nodABC*, *nodD1*, and *nodD2* of *R. fredii* USDA191 are identified on the left side of panel (A).

fredii and NGR234 (Jarvis et al., 1992), and they lend little support to proposals to subdivide *R. fredii* (Scholla and Elkan, 1984). In addition, they indicate that more detailed analysis of *R. fredii* versus NGR234 may permit us to refine our present concepts of broad host range.

Because *R. fredii* is a recent discovery, comparatively little is known about its *nod* and *nif* genes in comparison to those of *B. japonicum*,

R. meliloti, and the biovars of *R. leguminosarum* (Dénairé et al., 1992; Martinez et al., 1990; Rodriquez-Quinones et al., 1987). RFLP analysis with *nif*KDH and two soybean cultivar specificity loci failed to detect widespread differences among the strains, even though their abilities to nodulate soybean are distinct (Heron and Pueppke, 1984; Keyser et al., 1982; Morrison et al., 1986). Such uniformity stands in contrast to that of *nodDABC*,

Table 2. Immunological cross-reactivity of strains[a]

Antiserum	Strain										
	USDA 191	USDA 192	USDA 193	USDA 201	USDA 205	USDA 206	USDA 208	USDA 214	USDA 217	USDA 257	NGR234
USDA191	+10		+5			+5		+5			
USDA208	+5		+5		+5		+5				
USDA257		+5		+5					+5	+5	
NGR234	+5										+5

[a] Reactions were assessed by the binding of ^{125}I-labelled antiserum to cells that had been immobilized on nitrocellulose filters as described in Materials and Methods. Numbers refer to the greatest dilution of cells that gave a positive reaction.

which is arranged in five different permutations among the 11 strains. We are now analyzing clones of *nodD* and *nodABC* from USDA191, USDA257, and NGR234 in an attempt to understand relationships among the regulatory *nodD* loci and the inducible *nodABC* genes.

Repetitive sequences are well known in rhizobia, and they are reconsidered to be agents of genetic rearrangement, particularly in sym plasmids (Brom et al., 1991; Rastogi et al., 1992; Romero et al., 1991). The polymorphisms associated with RFRS1, which contains repetitive sequences from USDA257, thus are particularly interesting. Every *R. fredii* strain except USDA205 and USDA206 reacted uniquely with the RFRS1 probe. [USDA205 and USDA206 nevertheless can be resolved from one another by plasmid profiles (Heron and Pueppke, 1984) and serology (Sadowsky et al. 1987)]. Although NGR234 contains only weak homology to RFRS1 sequences, it harbours a set of repetitive sequences of its own (Perret et al., 1991). These observations are in accord with previous analysis with two other members of the RFRS family of repetitive sequences from *R. fredii* (Krishnan and Pueppke, 1991b, 1993), and they point out a distinction that, by analogy with the Ti plasmids of *Agrobacterium tumefaciens* (Otten et al., 1992), may have implications for understanding the origin and evolution of sym plasmids.

The observed relationship between the host range of NGR234 and that of *R. fredii* is most intriguing. Although both organisms are recognized as symbionts of multiple legume species (Keyser et al., 1982; Lewin et al., 1987; Morrison et al., 1986; Stanley and Cervantes, 1991; Trinick, 1980), there is scant information on the comparative host ranges of the two organisms (Broughton et al., 1984; Morrison et al., 1986). Our data with 18 diverse legumes make it clear that the *R. fredii* strains share a common host range and that NGR234 is more promiscuous than *R. fredii*, regardless of whether or not symbiotic effectiveness is taken into account. The two host ranges bear a simple relationship to one another, that of *R. fredii* as a subset and NGR234 as a superset. The only apparent exception to this generality is soybean, one of the original hosts of isolation for *R. fredii* (Keyser et al., 1982). Many cultivars of this species form Fix$^+$ nodules with *R. fredii* yet remain nodule-free or produce irregular swollen structures in response to NGR234 (Balatti and Pueppke, 1992; Balatti and Pueppke, unpublished data; Broughton et al., 1984). With the exception of soybean as a special case, our data support the hypothesis that the NGR234 host range should be viewed as an expanded version of that of *R. fredii*. If this interpretation is correct, then *R. fredii* and NGR234 may share a basic set of *nod* genes that defines their common host range. Other *nod* genes, however, will be present only in NGR234, where they are responsible for nodulation of plants that cannot be nodulated by *R. fredii*. We currently are testing this possibility.

Acknowledgements

This research was supported by funds from the Food for the 21st Century Program, University of Missouri and a competitive research grant from the U.S. Department of Agriculture and is Journal Series No. 11978 of the Missouri Agricultural Experiment Station.

References

Appelbaum E R, Chartrain N, Thompson D, Johansen K, O'Connell M and McLoughlin T 1985 Genes of *Rhizobium japonicum* involved in development of nodules. *In* Nitrogen Fixation Research Progress. Ed H J Evans and P H Bottomley. pp 101–107. Martinus Nijhoff, Dordrecht, Netherlands.

Appelbaum E R, Thompson D V, Idler K and Chartrain N 1988 *Rhizobium japonicum* USDA 191 has two *nodD* genes that differ in primary structure and function. J. Bacteriol. 170, 12–20.

Balatti P A and Pueppke S G 1992 Identification of North American soybean lines that form nitrogen-fixing nodules with *Rhizobium fredii* USDA257. Can. J. Plant Sci. 72, 49–55.

Brom S, de los Santos A G, Girard M L, Dávila G, Palacios R and Romero D 1991 High-frequency rearrangements in *Rhizobium leguminosarum* bv. *phaseoli* plasmids. J. Bacteriol. 173, 1344–1346.

Broughton W J, Heycke N, Meyer H and Pankhurst C E 1984 Plasmid-linked *nif* and "*nod*" genes in fast-growing rhizobia that nodulate *Glycine max*, *Psophocarpus tetragonolobus*, and *Vigna unguiculata*. Proc. Natl. Acad. Sci. USA 81, 3093–3097.

Burnett W N 1981 Western Blotting: electrophoretic transfer of proteins from SDS-polyacrylamide gels to unmodified nitrocellulose and radiographic detection with antibody and radioiodinated protein A. Anal. Biochem. 112, 195–203.

Cannon F C, Riedel G E and Ausubel F M 1979 Overlapping sequences of *Klebsiella pneumoniae nif* DNA cloned and characterised. Mol. Gen. Genet. 174, 59–66.

Chen W X, Yan G H and Li J L 1988 Numerical taxonomic study of fast-growing soybean rhizobia and a proposal that *Rhizobium fredii* be assigned to *Sinorhizobium* gen. nov. Int. J. Syst. Bacteriol. 28, 392–397.

Dénarié J, Debellé F and Rosenberg C 1992 Signalling and host range variation in nodulation. Annu Rev. Microbiol. 46, 497–531.

Devine T E 1985 Nodulation of soybean plant introduction lines with the fast-growing rhizobial strain USDA205. Crop Sci. 25, 354–356.

DuTeau N M, Palmer R G and Atherly A G 1986 Fast-growing *Rhizobium fredii* are poor nitrogen-fixing symbionts of soybean. Crop Sci. 26, 884–889.

Egelhoff T T, Fisher R F, Jacobs T W, Mulligan J T and Long S R 1985 Nucleotide sequence of *Rhizobium meliloti* 1021 nodulation genes: *nodD* is read divergently from *nodABC*. DNA 4, 241–248.

Hattori J and Johnson J A 1984 Fast-growing *Rhizobium japonicum* that effectively nodulates several commercial *Glycine max* L. Merrill cultivars. Appl. Environ. Microbiol. 48, 234–235.

Heron D S and Pueppke S G 1984 Mode of infection, nodulation specificity, and indigenous plasmids of 11 fast-growing *Rhizobium japonicum* strains. J. Bacteriol. 160, 1061–1066.

Heron D S and Pueppke S G 1987 Regulation of nodulation in the soybean-*Rhizobium* symbiosis. Strain and cultivar variability. Plant Physiol. 84, 1391–1396.

Heron D S, Érsek T, Krishnan H B and Pueppke S G 1989 Nodulation mutants of *Rhizobium fredii* USDA257. Mol. Plant-Microbe Interact. 2, 4–10.

Israel D W, Mathis J N, Barbour W M and Elkan G H 1986 Symbiotic effectiveness and host-strain interactions of *Rhizobium fredii* USDA191 on different soybean cultivars. Appl. Environ. Microbiol. 51, 898–903.

Jansen van Rensburg H, Strijdom B W and Otto C J 1983 Effective nodulation of soybeans by fast-growing strains of *Rhizobium japonicum*. S. Afr. J. Sci. 79, 251–252.

Jarvis B D W, Downer H L and Young J P W 1992 Phylogeny of fast-growing soybean-nodulating rhizobia supports synonymy of *Sinorhizobium* and *Rhizobium* and assignment to *Rhizobium fredii*. Int. J. Syst. Bacteriol. 42, 93–96.

Jordan D C 1984 Family III. Rhizobiaceae Conn 1938. *In* Bergey's Manual of Determinative Bacteriology, Vol. 1. Ed. N R Krieg et al. pp 234–256. Williams and Wilkins, Baltimore.

Keyser H H and Cregan P B 1984 Interactions of selected *Glycine soja* Sieb. & Zucc. genotypes with fast- and slow-growing soybean rhizobia. Crop Sci. 24, 1059–1062.

Keyser H H and Griffin R H 1987 Beltsville *Rhizobium* Culture Collection Catalogue. U.S. Dept. of Agriculture, Beltsville.

Keyser H H, Bohlool B B, Hu T S and Weber D F 1982 Fast-growing rhizobia isolated from root nodules of soybean. Science 215, 1631–1632.

Kirchner O 1895 Die Wurzelknöllchen der Sojabohne. Beitr. Biol. Pflanzen 7, 213–223.

Krishnan H B and Pueppke S G 1991a *nolC*, a *Rhizobium fredii* gene involved in cultivar-specific nodulation of soybean, shares homology with a heat-shock gene. Mol. Microbiol. 5, 737–745.

Krishnan H B and Pueppke S G 1991b Repetitive sequences with homology to *Bradyrhizobium japonicum* DNA and the T-DNA of *Agrobacterium rhizogenes* are closely linked to *nodABC* of *Rhizobium fredii* USDA257. Mol. Plant-Microbe Interact. 4, 521–529.

Krishnan H B and Pueppke S G 1991c Sequence and analysis of the *nodABC* region of *Rhizobium fredii* USDA257, a nitrogen-fixing symbiont of soybean and other legumes. Mol. Plant-Microbe Interact. 4, 512–520.

Krishnan H B and Pueppke S G 1993 Characterization of RFRS9, a second member of the *Rhizobium fredii* repetitive sequence family from the nitrogen fixing symbiont *R. fredii* USDA257. Appl. Environ. Microbiol. 59, 150–155.

Kuykendall L D, Saxena B, Devine T E and Udell S E 1992 Genetic diversity in *Bradyrhizobium japonicum* Jordan 1982 and a proposal for *Bradyrhizobium elkanii* sp. nov. Can. J. Microbiol. 38, 501–505.

Lewin A, Rosenberg C, Meyer H, Wong C H, Nelson L, Manen J-F, Stanley J, Dowling D N, Dénarie J and Broughton W J 1987 Multiple host-specificity loci of the broad host-range *Rhizobium* sp. NGR234, selected using the widely compatible legume *Vigna unguiculata*. Plant Molec. Biol. 8, 447–459.

Martínez E, Romero D and Palacios R 1990 The *Rhizobium* genome. CRC Crit. Rev. Plant Sci. 9, 59–93.

Meinhardt L W, Krishnan H B, Balatti P A and Pueppke S G 1993 Molecular cloning and characterization of sym plasmid locus that regulates cultivar-specific nodulation of soybean by *Rhizobium fredii* USDA257. Mol. Microbiol. 7, 17–29.

Morrison N A, Trinick M J and Rolfe B G 1986 Comparison of the host range of fast-growing *R. japonicum* strains with a fast-growing isolate from lablab. Plant and Soil 92, 313–317.

Otten L, Canaday J, Gérard J-C, Fournier P, Crouzet P and Paulus F 1992 Evolution of agrobacteria and their Ti plasmids–a review. Mol. Plant-Microbe Interact. 4, 279–287.

Perret X, Broughton W J and Brenner S 1991 Canonical ordered cosmid library of the symbiotic plasmid of *Rhizobium* species NGR234. Proc. Natl. Acad. Sci USA 88, 1923–1927.

Rastogi V K, Bromfield E S P, Whitwill S T and Barran L R 1992. A cryptic plasmid of indigenous *Rhizobium meliloti* possesses reiterated *nodC* and *nifE* genes and undergoes DNA rearrangement. Can J. Microbiol. 38, 563–568.

Rodriguez-Quinones F, Banfalvi Z, Murphy P and Kondorosi A 1987 Interspecies homology of nodulation genes in *Rhizobium*. Plant Molec. Biol. 8, 61–75.

Romero D, Brom S, Martinez-Salazar J, de Lourdes Girard M, Palacios R and Dávila G 1991 Amplification and deletion of a *nod-nif* region in the symbiotic plasmid of *Rhizobium phaseoli*. J. Bacteriol. 173, 2435–2441.

Sadowsky M J, Bohlool B B and Keyser H H 1987 Serological relatedness of *Rhizobium fredii* to other rhizobia and to bradyrhizobia. Appl. Environ. Microbiol. 53, 1785–1789.

Scholla M H and Elkan G H 1984 *Rhizobium fredii* sp. nov., a fast-growing species that effectively nodulates soybean. Int. J. Syst. Bacteriol. 34, 484–486.

Scholla M H, Moorefield J A and Elkan G H 1984 Deoxyribonucleic acid homology between fast-growing soybean-nodulating bacteria and other rhizobia. Int. J. Syst. Bacteriol. 34, 283–286.

Stanley J and Cervantes E 1991 Biology and genetics of the broad host range *Rhizobium* sp. NGR234. J. Appl. Bacteriol. 70, 9–19.

Stowers M D and Eaglesham A R J 1984 Physiological and symbiotic characteristics of fast-growing *Rhizobium japonicum*. Plant and Soil 77, 3–14.

Trinick M J 1980 Relationships amongst the fast-growing rhizobia of *Lablab purpureus*, *Leucaena leucocephala*, *Mimosa* spp., *Acacia farnesiana* and *Sesbania grandiflora* and their affinities with other rhizobial groups. J. Appl. Bacteriol. 49, 39–53.

Vincent J M 1970 A Manual for the Practical Study of Root-nodule Bacteria. Blackwell Scientific Publications, Oxford.

Yanagi M and Yamasato K 1993 Phylogenetic analysis of the family Rhizobiaceae and related bacteria by sequencing of 16S rRNA gene using PCR and DNA sequencer. FEMS Microbiol. Lett. 107, 115–120.

Zdor R E and Pueppke S G 1991 Nodulation competitiveness of Tn5-induced mutants of *Rhizobium fredii* USDA208 that are altered in motility and extracellular polysaccharide production. Can. J. Microbiol. 37, 52–58.

Zhang X, Harper R, Karsisto M and Lindström K 1991 Diversity of *Rhizobium* bacteria isolated from the root nodules of leguminous trees. Int. J. Syst. Bacteriol. 41, 104–113.

Plant and Soil **161**: 31–41, 1994.
© 1994 *Kluwer Academic Publishers.*

Rapid identification of *Rhizobium* species based on cellular fatty acid analysis

B. D. W. Jarvis and S. W. Tighe
Department of Microbiology and Genetics, Massey University, Palmerston North, New Zealand and Analytical Services Inc., P O Box 626, Essex Junction, VT 05453, USA

Key words: FAME, fatty acid analysis, rapid identification, *Rhizobium*

Abstract

As understanding of the evolutionary relationships between strains and species of root nodule bacteria increases the need for a rapid identification method that correlates well with phylogenetic relationships is clear. We have examined 123 strains of *Rhizobium: R. fredii* (19), *R. galegae* (20), *R. leguminosarum* (22), *R. loti* (17), *R. meliloti* (21), and *R. tropici* (18) and six unknowns. All strains were grown on modified tryptone yeast-extract (TY) agar, as log phase cultures, scraped from the agar, lysed, and the released fatty acids derivatized to their corresponding methyl esters. The methyl esters were analysed by gas-chromatography using the MIDI/Hewlett-Packard Microbial Identification System. All species studied contained 16:0, 17:0, 18:0 and $19cyclo_w9C$ fatty acids but only *R loti* and *R tropici* produced 12:0 3 OH, 13:0 iso 3 OH, $18:1_w9C$ and 15:0 iso 3 OH, 17:0 iso 3 OH and $20:2_w6,9C$ fatty acids respectively. Principal component analysis was used to show that strains could be divided into clusters corresponding to the six species. Fatty acid profiles for each species were developed and these correctly identified at least 95% of the strains belonging to each species. A dendrogram is presented showing the relationships between *Rhizobium* species based on fatty acid composition. The data base was used to identify unknown soil isolates as strains of *Rhizobium* lacking a symbiotic plasmid and a bacterium capable of expressing a symbiotic plasmid from *R. leguminosarum* as *Sphingobacterium spiritovorum.*

Abbreviations: 16SrRNA – 16S ribosomal ribonucleic acid, MIS – Microbial identification system, FAME – Fatty acid methyl ester, FID – Flame ionization detector, TY – Tryptone yeast extract, ECL – Equivalent carbon length, ED – Euclidean distance and UPGMA – Unweighted pair group method with arithmetic averages.

Introduction

The application of modern taxonomic methods to the classification of root nodule bacteria has led to the recognition of a rapidly increasing number of new species and genera. Among *Rhizobium* these include: *R. leguminosarum* (Frank 1879) *R. meliloti* (Dangeard 1926), *R. loti* (Jarvis et al. 1982), *R. fredii* (Scholla and Elkan 1984), *R. galegae* (Lindström 1989). *R. tropici* (Martinez-Romero et al. 1991), *R. huakuii* (Chen et al. 1991) and *R. etli* (Segovia et al. 1993). The methods used are discussed by Graham et al. (1991) and include: numerical taxonomy, multi-locus enzyme electrophoresis, DNA restriction fragment patterns, DNA:DNA relatedness, rRNA:DNA hybridization and 16SrRNA sequence analysis. These methods provide the best available indication of the natural relationship between strains, species and genera, especially when several methods are used in concert, but their use is usually restricted to specialist laboratories and requires considerable time and effort. They are not appropriate as tools for *Rhizobium* identification. Identification requires routine access to the developing phylogenetic classification. An identification method must correlate well with phylogenetic methods and be rapid, simple and inexpensive so that it can be used with confidence to examine the large numbers of samples generated in ecological and agronomic studies.

The aim of this investigation was to determine whether the automated, computer-based Hewlett-Packard Microbial Identification System (MIS), based on fatty acid methyl-ester (FAME) analysis of whole bacterial cells, could be used to identify the recognised species of *Rhizobium*.

Materials and methods

Bacterial strains

The strains and species of *Rhizobium* used in this investigation and the collections from which they were obtained are listed in Table 1.

Cultivation techniques

All strains were sub-cultured three times on a modified TY (Beringer 1974) agar containing calcium chloride and 0.1% D Mannitol. The last sub-culture was streaked in quadrants on medium which was 48h. old in order to ensure a reproducible moisture content in the medium. Cells for analysis were incubated at $28 \pm 1°C$ for $48 \pm 2h$.

Harvesting and extraction of cellular fatty acids

Approximately 50 mg of cell mass was harvested from the 2nd and 3rd quadrant of a streak plate using a platinum loop, and transferred to a Teflon-capped 13 mm glass tube. Fatty acids were extracted by saponifying the cells in 1 mL of a 1.125M Sodium hydroxide aqueous methanol solution at 100°C for 30 minutes, then methylated by adding 2.0 mL 6.0 M Hydrochloric acid in

methanol followed by incubation at 80° for 10 minutes. The methylated fatty acids were transferred to a solvent phase (1:1 hexane and methyl tertiary butyl ether) and cleaned up by washing the extract with 0.3 M NaOH. A portion of the washed sample extract was then transferred to a gas-chromatograph vial.

Analysis of cellular fatty acids

Fatty acids were analysed with a Hewlett-Packard 5890 Series II gas chromatograph fitted with a 25 m 0.2 mm methyl phenyl silicone capillary column and a flame ionization detector. Hydrogen was the carrier gas. The injector temperature was 250 °C and the column temperature was programmed to change from 170°C to 270°C at a rate of 5°C/min. The FID temperature was 300°C. Sample ($2.0\mu L$) was injected into a 100:1 split ratio injection port. Peaks were integrated with a Hewlett-Packard 3392A integrator. Cellular fatty acids were quantified by a computer equipped with Microbial Identification System (MIS) software (Microbial I D, Inc. Newark DE). Identifications, dendrograms and 2D principal component plots were generated using the MIS software.

Results

The fatty acids and related compounds for each *Rhizobium* strain were identified by determining their equivalent carbon length (ECL) and comparing these with a known standards using MIS software. The quantity of each acid present was expressed as a percentage of the total named fatty acids present in concentrations of more than 0.2%.

Table 2 shows the fatty acid profile for each species. Nineteen different fatty acids occur as more than 0.2% of the total fatty acids present in a species. All species contained 16:0,17:0,18:0 and 19:0 cyclo$_w$ 9C fatty acids and only *R. tropici* and *R. loti* lacked 16: 1w7C and 16:0 3OH and 20:3w6,9,12C, respectively. On the other hand the following fatty acids were found only in the species named; 15:0 2OH *R. leguminosarum*,

Table 1. *Rhizobium* strains used in this study

Identification	Strain number	Source[a]
Rhizobium fredii		
as	USDA 193	J P W Young
ba	USDA 201	J P W Young
au	USDA 205	J P W Young
bc	USDA 206	J P W Young
cz	USDA 214	J P W Young
bf	CCBAU 103	W X Chen
be	CCBAU 105	W X Chen
az,eb	CCBAU 110	W X Chen
bd	CCBAU 112	W X Chen
dy	CCBAU 114	W X Chen
av	CCBAU 115	W X Chen
ck	CCBAU 116	W X Chen
aw, er	CCBAU 169	W X Chen
at	USDA 257 (MU646)	M Fenton
ad	USDA 217 (MU640)	M Fenton
ea	USDA 298 (MU638)	M Fenton
dz	USDA 194 (MU637)	M Fenton
cj	USDA 191 (MU636)	M Fenton
da	USDA 192 (MU643)	MFenton
Rhizobium galegae		
de	1141/3	A K Lindström
ei	1144/2	A K Lindström
dp	1174/2	A K Lindström
dg	1189	A K Lindström
dn	1184/2	A K Lindström
df	490/2	A K Lindström
am	1145/2	A K Lindström
ao	1143/2	A K Lindström
al	1428	A K Lindström
dh	1147/2	A K Lindström
dm	1151/2	A K Lindström
dq	1122/2	A K Lindström
dt	503	A K Lindström
dv	1185/2	A K Lindström
do	1183/3	A K Lindström
ds	1186	A K Lindström
ak	54013	A K Lindström
dr	1461	A K Lindström
aj	1460	A K Lindström
an	1146/2	A K Lindström
Rhizobium leguminosarum biovar *phaseoli*		
bp	MU266 (NZP5097)	M Fenton
cf	MU273 (NZP5459)	M Fenton

Table 1. Continued

Identification	Strain number	Source[a]
bq	MU362 (F300)	M Fenton
ac	MU367(F310)	M Fenton
dx	MU632(NZP5479)	M Fenton
cg	M 1928	not known
ch	B 248	not known
cy	SPI8	not known

Rhizobium leguminosarum biovar *trifolii*

ap, aq	MU224(NZP1/6)	M Fenton
dw	MU 231(NZP514)	M Fenton
du	MU 237(TA1)	M Fenton
ar	MU 252(NZP540 ICMP2666)	M Fenton
ay	MU 258(K8)	M Fenton
cs	MU 269(NZP540, SU 391)	M Fenton
es	ATCC 14479	American Type Culture Collection
ax	ATCC 14480	American Type Culture Collection

Rhizobium leguminosarum biovar *viceae*

dd	MU267(NZP5225)	M Fenton
dc	MU268(NZP 5230,TA101)	M Fenton
ci	MU 272(NZP 5262,CB 596)	M Fenton
db	MU 363(NZP5487,A1)	M Fenton
ab	MU 424(NZP5421,92 AA1)	M Fenton
aa	MU 520(NZP5488,F2)	M Fenton

Rhizobium loti

bg	ATCC 33669	American Type Culture Collection
cc	NZP 2298(L72M103c)	M Fenton
cm	NZP 2146	M Fenton
ee	NZP 2195	M Fenton
–	NZP 2238(LC26SD a)	M Fenton
ef	NZP 2014	M Fenton
af	NZP 2079	M Fenton
cl	NZP 2235(cc812a)	M Fenton
–	NZP 2203	M Fenton
ed	NZP 2241	M Fenton
ae,bi	NZP 2048	M Fenton
cb	NZP 2227(461)	M Fenton
ec	NZP 2300L72M115c	M Fenton
cn	NZP 2234(cc811)	M Fenton
ce	NZP 2230(cc809a)	M Fenton
bh	NZP 2196(SU 343)	M Fenton
cd	NZP 2037	M Fenton

Table 1. Continued

Identification	Strain number	Source[a]
Rhizobium meliloti		
cx	MU 443(CB 1170)	M Fenton
eo	MU 663(2011)	M Fenton
bn	MU 442(CC2017)	M Fenton
ep	MU 441(SU 47)	M Fenton
by	MU 464(Balsac)	M Fenton
ca	MU 463(300a13)	M Fenton
bm	MU 462(U45)	M Fenton
ct	MU 461(PDDCC 1322)	M Fenton
bz	CC 2003	B D Eardly
ej	15AG	B D Eardly
eq	U102	B D Eardly
cv	74B12	B D Eardly
et	M1	B D Eardly
ek	CC 2013	B D Eardly
bo	M161	B D Eardly
cw	M275	B D Eardly
-	M3	B D Eardly
cu	102 FSI	B D Eardly
el	M 205	B D Eardly
en	ATCC 9930	American Type Culture Collection
em	ATCC 10310	American Type Culture Collection
R. tropici		
bt	BR 842	E Martínez-Romero
cr	CIAT 899	E Martínez-Romero
cp	BR 857	E Martínez-Romero
di	C-05-I	E Martínez-Romero
bv	CFN 299	E Martínez-Romero
dj	BR 852	E Martínez-Romero
bw	BR 863	E Martínez-Romero
bu	BR 864	E Martínez-Romero
cq	BR 846	E Martínez-Romero
co	BR 862	E Martínez-Romero
br	BR 836	E Martínez-Romero
dl	BR 853	E Martínez-Romero
bx	BR 10042	E Martínez-Romero
ah	BR 10043	E Martínez-Romero
ag	BR 859	E Martínez-Romero
ai	BR 845	E Martínez-Romero
bs	BR 847	E Martínez-Romero
dk	BR 850	E Martínez-Romero

Table 1. Continued

Identification	Strain number	Source[a]
Rhizobium sp. *(Parasponia)*		
bb	NGR 234	J P W Young
Unknowns		
eg	NZP 2668	M Fenton
eh	NZP2163	M Fenton
bj	NR40	M Fenton
bk	NR64	M Fenton
bl	NR42	M Fenton
Sphingobacterium multivorum		
	ATCC29837	National Health Institute

[a]American Type Culture Collection, 12301 Parklawn Drive, Rockville, Maryland 20852, USA. W X Chen, Department of Microbiology, Beijing Agricultural University. Beijing, Peoples Republic of China. B D Eardly. Pennsylvania State University, Berks Campus P O Box 7009 Reading PA 19610–6009, USA. M Fenton, Department of Microbiology and Genetics, Massey University, Palmerston North, New Zealand. A K Lindström, Department of Applied Chemistry and Microbiology, Faculty of Agriculture and Forestry, University of Helsinki, SF-00710 Helsinki, Finland. M E Martinez-Romero, Centro de Investigacion sobre. Fijacion de Nitrogeno, Universidad Nacional Autonoma de Mexico, Apartado Postal 565-A, Cuernavaca, Morelos, Mexico. J P W Young Department of Biology, University of York, Heslington, Yorkshire YO1 SDD. United Kingdom. Curator, National Health Institute, Kenepuru Drive, Porirua, New Zealand.

12:0 3OH 13:0iso3OH and $18:1_w$ 9C *R. loti*, 15:0 iso 3OH 17:0 iso 3OH and $20:2_w$ 6,9C *R. tropici*. Thus *R. loti* and *R. tropici* have a qualitatively different fatty acid composition from the other four species considered and *R. fredii*, *R. galegae*, *R. leguminosarum* and *R. meliloti* show quantitative differences in fatty acid composition. Mean fatty acid profiles were computed for each species and a computerized library database was constructed using the major fatty acid components from each taxon. Minor fatty acids which occurred on only one occasion were omitted from the library database.

The ability of this database to discriminate between 123 strains representing six species and six unknown strains on the basis of their fatty acid content was examined using principal component analysis in which the sample data for each strain was transformed into a set of three principal component scores and plotted in two dimensions to show the separation between strains. Figure 1 shows a plot of principal components 1 and 3. The strains form clusters which correspond with their species designations. This indicates that the differences in fatty acid composition between these *Rhizobium* species is sufficient to serve as a reliable identification method. Plots of principal components 1 and 2 or 1 and 3 were less successful but separate clusters of *R loti* in the component 1 and 2 plot and *R. galegae* and *R loti* in the components 2 and 3 plot were observed.

Dendrograms were obtained to show the relationship between the 123 strains. For this pur-

Table 2. Percent total fatty acid compositon of six *Rhizobium* species (>0.2%)[a]

Fatty acid	R. fredii	R. galagae	R. leguminosarum			R. loti	R. meliloti	R. tropici
			bv. phaseoli	bv. trifolii	bv. viceae			
12:0 3OH	ND[b]	ND	ND	ND	ND	0.40 ± 0.40	ND	ND
13:0 ISO 3OH	ND	ND	ND	ND	ND	1.39 ± 0.41	ND	ND
16:1 w7Cis	0.79 ± 0.77	0.54 ± 0.34	0.63 ± 0.34	0.37 ± 0.36	0.51 ± 0.26	0.45 ± 0.57	0.34 ± 0.21	<0.2
16:0	5.52 ± 1.88	9.49 ± 1.78	4.74 ± 1.98	4.31 ± 1.31	3.37 ± 1.05	16.38 ± 2.21	4.94 ± 1.14	7.72 ± 2.26
15:0 ISO 3OH	ND	ND	ND	ND	ND	ND	ND	3.48 ± 0.50
15:0 2OH	ND	ND	0.95 ± 0.53	0.83 ± 0.59	0.81 ± 0.59	ND	ND	ND
15:0 3OH	ND	ND	<0.2	ND	0.20 ± 0.49	ND	ND	ND
17:0 ISO	ND	ND	ND	ND	ND	9.11 ± 2.27	ND	1.00 ± 0.27
17:0 CYCLO	0.27 ± 0.48	<0.2	<0.2	0.24 ± 0.36	<0.2	0.82 ± 0.54	0.75 ± 0.31	0.75 ± 0.50
17:0	0.66 ± 0.47	0.44 ± 0.40	0.51 ± 0.65	0.31 ± 0.38	0.79 ± 1.13	0.59 ± 0.65	0.54 ± 0.24	<0.2
16:0 3OH	0.50 ± 0.38	3.18 ± 0.63	1.23 ± 0.19	1.39 ± 0.22	1.46 ± 0.68	ND	0.68 ± 0.12	3.97 ± 0.26
18:1 w9Cis	<0.2	<0.2	<0.2	<0.2	0.28 ± 0.50	0.38 ± 0.49	<0.2	ND
18:0	4.72 ± 1.23	1.29 ± 0.52	8.26 ± 2.51	8.61 ± 2.04	8.19 ± 2.50	5.01 ± 1.40	2.70 ± 0.72	2.66 ± 1.21
17:0 ISO 3OH	ND	ND	ND	ND	ND	ND	ND	0.31 ± 0.24
19:0 aCYCLO	4.68 ± 3.86	31.27 ± 9.12	14.29 ± 6.04	18.25 ± 6.43	13.48 ± 5.17	25.9 ± 9.19	22.56 ± 8.25	52.74 ± 9.63
18:1 2OH	ND	ND	<0.2	ND	ND	0.50 ± 0.71	ND	0.98 ± 0.74
19:0 10METHYL	0.51 ± 0.51	2.24 ± 0.49	<0.2	ND	<0.2	1.37 ± 0.51	0.85 ± 0.26	ND
20:3 w6,9,12Cis	3.34 ± 1.01	1.34 ± 0.43	2.32 ± 0.99	2.59 ± 0.35	2.27 ± 0.75	ND	2.94 ± 0.33	2.09 ± 0.64
20:2 w6,9Cis	ND	<0.2	ND	ND	ND	<0.2	<0.2	0.50 ± 0.12
Summed feature 3[c]	7.96 ± 2.63	4.88 ± 0.46	6.50 ± 1.25	6.56 ± 0.49	6.25 ± 0.82	ND	6.45 ± 0.70	2.48 ± 0.68
Summed feature 7[d]	70.6 ± 5.53	44.91 ± 10.45	59.66 ± 6.20	56.45 ± 8.44	61.84 ± 6.38	37.56 ± 7.08	57.01 ± 8.27	21.13 ± 10.53

[a]Mean and Standard Deviation

[b]ND is not detectable

[c]Summed feature 3 is composed of 12:0 aldehyde (?), 16:1 ISO, 14:0 3OH fatty acids and an unknown compound at an ECL value of 10.928.

[d]Summed feature 7 is composed of 18:1w7Cis/w9Trans/w12Trans/,18:1w9Cis/w12Trans/w7Cis and 18:1wTrans/w9Trans/w7Cis fatty acids.

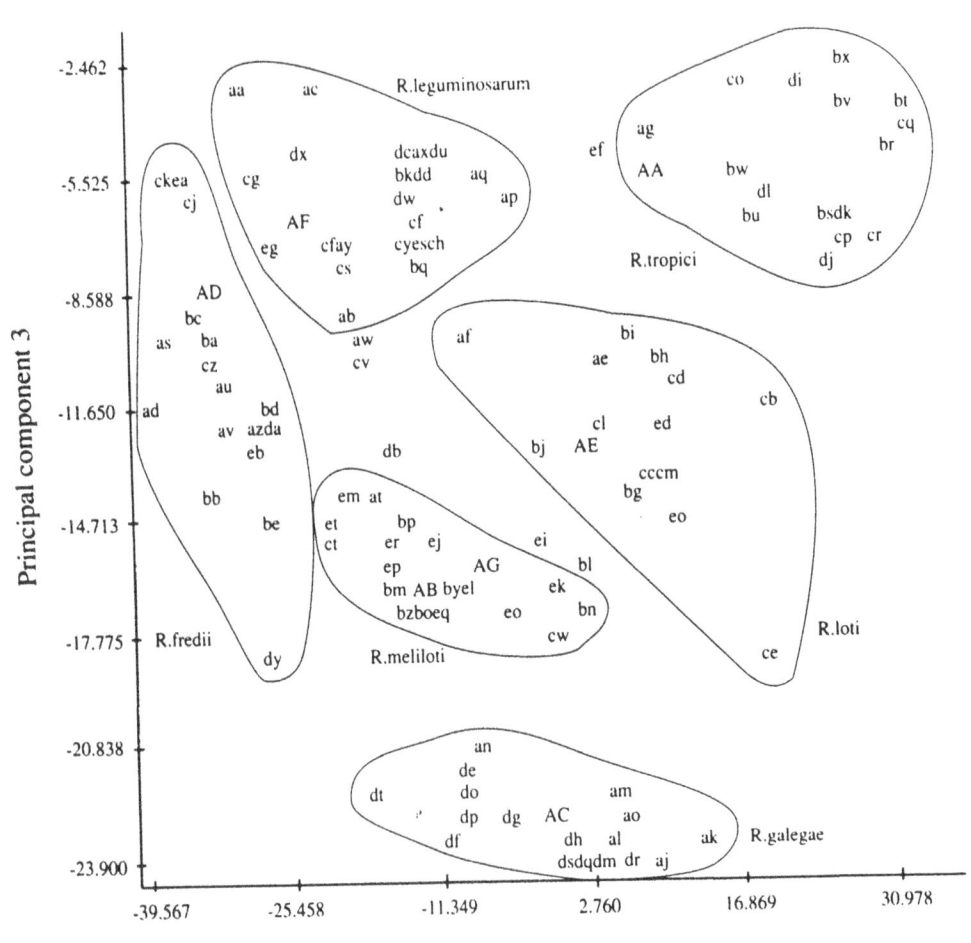

Fig. 1. Two dimensional plot of principal component analysis for 123 strains of *Rhizobium*. For individual strain identifications see Table 1. Samples grouped together are as follows: AA = ai + ah; AB = cx + ca; AC = dv + dn; AD = dz+bf; AE=ec+cn; AF=eh+ar; AG=en+cu.

pose the library data was transformed using the [mapped] feature available in the MIS software, Euclidean distances (ED) and the unweighted pair group method with arithmetic averages (UPGMA) clustering method. A simplified form of this dendrogram is shown in Figure 2. In this figure *R. loti* and *R. tropici* are clearly distinguished from the remaining species, *R. fredii* forms a distinct cluster but strains of *R. galegae*, *R. leguminosarum* and *R. meliloti* each form more than one cluster at ED values of 10–20 and these can only be joined at ED 27.0. This suggests a close relationship between these three species and *R. fredii*

The ability of the MIS system to identify individual strains and distinguish between similar species was determined by comparing each strain with the species fatty acid profiles in the data base. The results are presented in Table 3 and show that at least 95% of each species were correctly identified. Where *R. fredii* is miss-identified as *R. meliloti* or vice versa this can be justified on the basis that *R. fredii* and *R. meliloti* show 40% DNA:DNA relatedness (Wedlock and Jarvis 1986) and other evidence of phylogenetic relationship (Jarvis et al 1986; Jarvis et al 1992). Similarly *R. galegae* exhibits low level DNA relatedness with *R. meliloti* (Wedlock and Jarvis 1986).

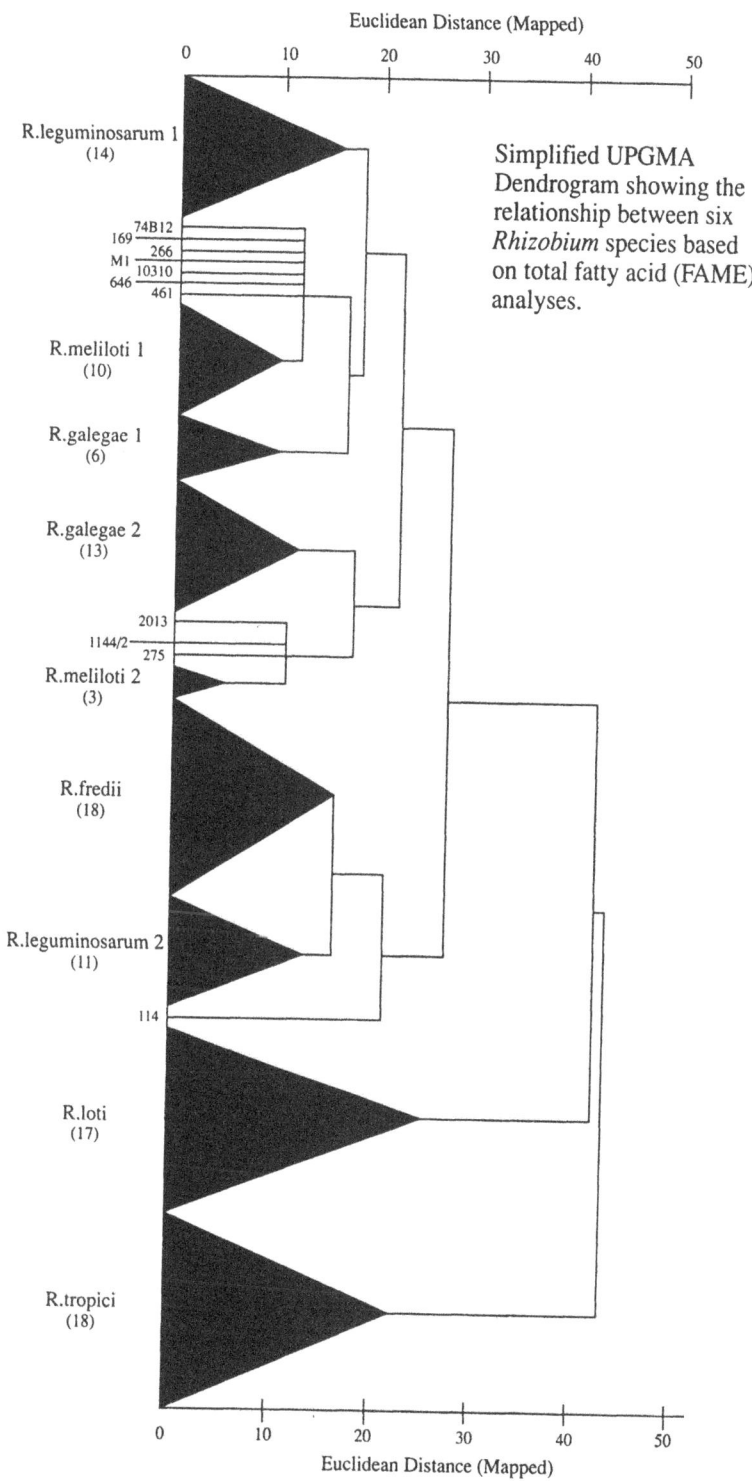

Fig. 2. Simplified UPGMA dendrogram showing the relationship between six *Rhizobium* species based on total fatty acid (FAME) analyses. For individual strain numbers see Table 1.

Table 3. Identification of *Rhizobium* strains (%) from their fatty acid profiles

MIS Identification	*Rhizobium* species					
	fredii	*galegae*	*leguminosarum*	*loti*	*meliloti*	*tropici*
fredii	95.5	–	–	–	5.0	–
galegae	–	95.0	–	–	–	–
leguminosarum	–	–	100.0	–	–	–
loti	–	–	–	100.0	–	–
meliloti	4.5	5.0	–	–	95.0	–
tropici	–	–	–	–	–	100.0
Total number of strains	22.0	20.0	24.0	16.0	21.0	18.0

Several unknowns were included among the strains submitted for examination. Two of these strains were clover inoculant strains from New Zealand, ICMP2668 and ICM2163, and these strains were successfully identified as *R. leguminosarum*. *Rhizobium* sp. NGR234 nodulates the non-legume *Parasponia*. It has been shown to be closely related to *R. meliloti* and *R.fredii* (Jarvis et al 1992). MIS identified it as *R. fredii* and placed it in the *R. fredii* cluster on the dendrogram. A strain of *Sphingobacterium* (*Flavobacterium*) *multivorum* (ATCC 29837) was introduced as an unknown, rejected as a *Rhizobium* and subsequently identified as *Sphingobacterium spiritivorum* by reference to another MIS library.

Discussion

This study demonstrates that *Rhizobium* species can be identified with a high degree of precision on the basis of their fatty acid composition. It is essential, however, to carefully standardize the conditions under which cells are grown, as medium, medium age, inoculum activity, inoculum age and incubation temperature can all influence cellular fatty acid composition. We found that the six species of *Rhizobium* examined contained 19 fatty acids in quantities greater than 0.2% of the total. All species contained 16:0, 17:0, 18:0 and 19:0 cyclo$_w$ 9C fatty acids and *R. fredii, R galegae,*

R. leguimosarum and *R. meliloti* were differentiated mainly on the proportions of the different fatty acids they contained. *R. loti* and *R tropici* were qualitatively different in fatty acid composition. These results can be compared with those of Sawada et al (1992) for the closely related genus *Agrobacterium* (Willems and Collins 1993). They found that 66 strains could be classified in three groups corresponding to biovars 1, 2 and 3 based on fatty acid composition. There appears to be sufficient difference in fatty acid composition to permit differentiation between members of these genera and we are currently exploring this possibility.

Principal component analysis enabled us to separate the 123 strains into six groups corresponding to their species designations. The dendrogram (Fig. 2) shows the relationship between these strains based on their overall similarities expressed as pairwise Euclidean distance coefficients clustered by the UPGMA method. This dendrogram can be compared with recent phylogenetic trees based on 16SrRNA sequence data. Willems and Collins (1993) present trees obtained with the Fitch and Margoliash algorithm and parsimony analysis. In both these trees *R. meliloti* and *R. fredii* are closely related as are *R. leguminosarum* and *R. tropici*, but *R. galegae* and *R. loti* show less relationship to these species and to each other. It can be concluded that whilst fatty acid analysis is and excellent basis for recognising strains belonging to the same species, and thus an effective identification tool, it is unlikely to be reliable as a source of detailed phylogenetic information. This apparent anomaly is probably due to the different algorithms used by the MIS software for identification and for the computation of overall similarity and relationship. FAME analysis may be useful for generating hypotheses to be tested by sequence-based methods and can be relied upon to detect broader phylogenetic differences such as that between *Rhizobium* and *Bradyrhizobium* (P.H. Graham pers. comm.).

The advantages of the MIS system of identification rest on its ability to recognise unique and characteristic features of bacteria, the rapidity and reproducibility with which results can be obtained and the low operating costs involved. Disadvantages include; the requirement for pure cultures, the need for careful standardization of media and methods and the capital cost of the analytical apparatus required. These disadvantages may be off-set if access to an analytical service is available.

The value of fatty acid analysis for identification purposes and the ease with which such data can be obtained dictate that the description of a new species of bacteria should include its fatty acid composition. This would require the examination of at least 20 strains so that a reliable, reproducible fatty acid profile for the species could be included in its description. General availability of fatty acid profiles has the potential to considerably simplify species identification and thus aid ecological and agronomic studies.

Acknowledgements

Provision of *Rhizobium* strains by the persons listed in Table 1 is gratefully acknowledged. This work was supported by a grant from the Massey University Research Fund

References

Beringer J E 1974 R-factor transfer in *Rhizobium leguminosarum*. J. Gen. Microbiol. 84, 188–198.

Chen W X, Li G S, Qi Y L, Wang E T, Yuan H L and Li J L 1991 *Rhizobium huakuii* sp. nov. isolated from the root nodules of *Astragalus sinicus*. Int. J. Syst Bacteriol 41, 275–280.

Dangeard P A 1926 Recherches sur les tubercles radicaux dex legumineuses Botaniste (Paris) 16, 1–275.

Frank B 1879 Ueber die parasiten in den wurzelanschwillungen der Papilionaceen. Ber. Deut. Bot. Ges. 37, 376–387, 394–399.

Graham P H, Sadowsky M J, Keyser H H, Barnet Y M, Bradley R S, Cooper J E, DeLey J, Jarvis B D W, Roslycky E B, Strijdon B W and Young J P W 1991 Proposed minimal standards for the description of new genera and species of root-and stem- nodulating bacteria. Int. J. Syst.Bacteriol. 41, 582–587.

Jarvis B D W, Downer H L and Young J P W 1992 Phylogeny of fast- growing soybean-nodulating rhizobia supports synonymy of *Sinorhizobium* and *Rhizobium* and assignment to *Rhizobium fredii*. Int. J. Syst. Bacteriol. 42, 93–96.

Jarvis B D W, Gillis M and DeLey J 1986 Intra- and intergeneric similarities between the ribosomal ribonucleic acid cistrons of *Rhizobium* and *Bradyrhizobium* species and some related bacteria. Int. J. Syst. Bacteriol 36, 129–138.

Jarvis B D W, Pankhurst C E and Patel J J 1982 *Rhizobium loti* a new species of legume root nodule bacteria. Int. J. Syst. Bacteriol 32, 378–380.

Lindström K 1989 *Rhizobium galegae* a new species of legume root nodule bacteria. Int.. J. Syst. Bacteriol. 39, 365–367.

Martínez-Romero E, Segovia L, Martius-Mercante F, Franco A A, Graham P and Pardo M A 1991 *Rhizobium tropici* a novel species nodulating *Phaselus vulgaris* L. beans and *Leuceana* sp trees. Int. J. Syst. Bacteriol. 41, 417–426.

Sawada H, Takikawa Y and Ieki H 1992 Fatty acid methyl ester profiles of genus *Agrobacterium*. Ann. Phytopath. Soc. Japan. 58, 46–51.

Scholla M H and Elkan G H 1984 *Rhizobium fredii* sp. nov., a fast- growing species that effectively nodulates soybeans. Int. J. Syst. Bacteriol. 34, 484–486.

Segovia L, Young J P W and Martinez-Romero E 1993 Reclassification of American *Rhizobium leguminosarum* biovar phaseoli type I strains as *Rhizobium etli* sp. nov. Int. J. Syst. Bacteriol. 43, 374–377.

Wedlock D N and Jarvis B D W 1986 DNA homologies between *Rhizobium fredii*, rhizobia that nodulate *Galega* sp. and other *Rhizobium* and *Bradyrhizobium* species. Int. J. Syst. Bacteriol. 36, 550–558.

Willems A and Collins M D 1993 Phylogenetic analysis of rhizobia and agrobacteria based on 16SrRNA gene sequences. Int. J. Syst. Bacteriol. 43, 305–313.

Plant and Soil **161**: 43–57, 1994.
© 1994 *Kluwer Academic Publishers.*

Ammonium sensing in nitrogen fixing bacteria: Functions of the *glnB* and *glnD* gene products

Christina Kennedy[1], Natalie Doetsch[1], Dietmar Meletzus[1], Eduardo Patriarca[2], Mohamad Amar[2] and Maurizio Iaccarino[2]

[1]*Department of Plant Pathology, University of Arizona, Tucson, AZ85721, USA and* [2]*International Institute of Genetics and Biophysics, Via Marconi 10, 80125 Naples, Italy*

Key words: Azospirillum, Azotobacter, Klebsiella, nitrogen fixation, Rhizobium

Abstract

A plentiful supply of fixed nitrogen as ammonium (or other compounds such as nitrate or amino acids) inhibits nitrogen fixation in free-living bacteria by preventing nitrogenase synthesis and/or activity. Ammonium and nitrate have variable effects on the ability of *Rhizobiaceae* (*Rhizobium*, *Bradyrhizobium* and *Azorhizobium*) species to nodulate legume hosts and on nitrogen fixation capacity in bacteroid cells contained in nodules or in plant-free bacterial cultures. In addition to effects on nitrogen fixation, excess ammonium can inhibit activity or expression of other pathways for utilization of nitrogenous compounds such as nitrate (through nitrate and nitrite reductase), or glutamine synthetase (GS) for assimilation of ammonium. This paper describes the roles of two key genes *glnB* and *glnD*, whose gene products sense levels of fixed nitrogen and initiate a cascade of reactions in response to nitrogen status. While work on *Escherichia coli* and other enteric bacteria provides the model system, *glnB* and, to a lesser extent, *glnD* have been studied in several nitrogen fixing bacteria. Such reports will be reviewed here. Recent results on the identity and function of the *glnB* and *glnD* gene products in *Azotobacter vinelandii* (a free-living soil diazotroph) and in *Rhizobium leguminosarum* biovar *viciae*, hereinafter designated *R.l. viciae* will be presented. New data suggests that *Azotobacter vinelandii* probably contains a *glnB*-like gene and this organism may have two *glnD*-like genes (one of which was recently identified and named *nfrX*). In addition, evidence for uridylylation of the *glnB* gene product (the PII protein) of *R. l. viciae* in response to fixed nitrogen deficiency is presented. Also, a *glnB* mutant of *R. l. viciae* has been isolated; its characteristics with respect to expression of nitrogen regulated genes is described.

Abbreviations: KD – kiloDaltons, kb – kilobase, UMP – uridine monophosphate

The *Escherichia coli* model: Roles for the *glnB* and *glnD* gene products

Transduction of the environmental signal of ammonium status has been best studied in enteric organisms by a combination of biochemical characterization of key proteins and genetic analysis of the genes encoding them. The laboratories and co-workers of Drs. Boris Magasanik, Sue Goo Rhee, and Earl Stadtman were particularly important in elucidating the model of the nitrogen control cascade in *E. coli* (and in *Klebsiella aerogenes*) (See Fig. 1). Fundamental information about the nitrogen cascade in *Salmonella typhimurium* obtained in Dr. Sydney Kustu's laboratory also contributed to this model. In *E. coli*, the cellular ratio of glutamine to α-ketoglutarate is probably the primary indicator of intracellular fixed N status (Garcia and Rhee, 1983). The glutamine:α-ketoglutarate ratio is sensed by the product of the *glnD* gene, a large 95 kD uridyl transfer/uridyl removing (UT/UR) enzyme (Reitzer and Magasanik, 1987). Under ammonium limitation (low N), the product of the *glnB* gene, known as the PII protein (a tetramer with identical 11.5 kD subunits), is uridylylated by the UT activity at a specific tyrosine residue, Tyr-51, while under ammonium excess (high N), UMP is removed from PII by UR (see Fig. 1a) (Adler et al., 1975; Chock et al. 1980; Garcia and Rhee, 1983; Rhee et al., 1985a; Son and Rhee, 1987).

The uridylylation state of PII has major physiological consequences in *E. coli*: it determines whether GS enzyme is adenylylated or not and also the rate at which the transcriptional activator NtrC is phosphorylated (the active form) or dephosphorylated (the inactive form) (Keener and Kustu, 1988; Ninfa and Magasanik, 1986; Rhee et al., l985b).

Deuridylylated PII (high N) stimulates the activity of adenylyl transferase (ATase or ATA, encoded by *glnE*), leading to the adenylylation (inactivation) of GS. Therefore during conditions of nitrogen excess, GS activity is greatly diminished (Kustu et al. 1984). Under these conditions, glutamate dehydrogenase (GDH), which has lower affinity for NH_4^+ than does GS, is important for incorporation of ammonium into cellular constituents. When PII is uridylylated (low N), The *glnE* gene product removes adenylyl from GS (AT_D activity, see Fig 1B). In *glnB* mutants lacking PII protein, steady state levels of adenylylated GS are similar to those in wild type grown under the same conditions. However, in *glnB* mutants, adenylylation/deadenylylation occurs more slowly than in wild type in response to a change from high N to low N conditions (Foor et al. 1980). Deadenylylated GS has very much higher activity for glutamine biosynthesis than does adenylylated GS. Thus, in low N, GS is highly active. With respect to GS adenylylation/deadenylylation, PII

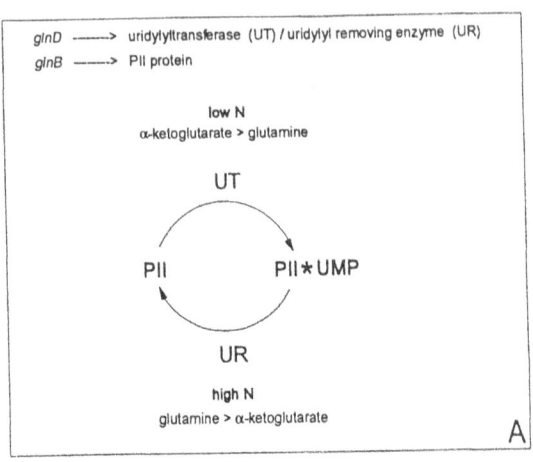

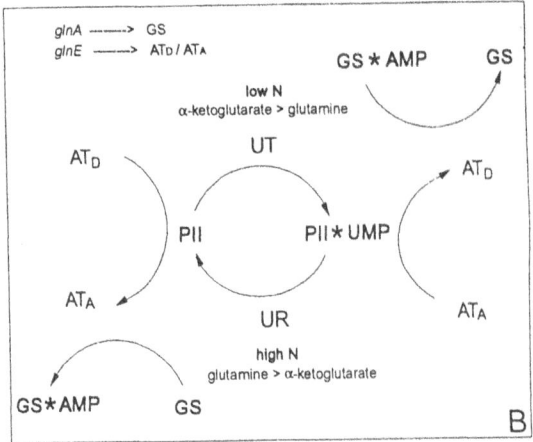

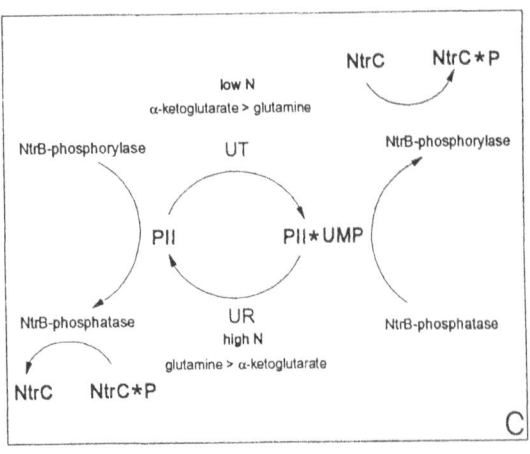

Fig. 1. Model of the nitrogen-sensing cascade in *E. coli* and other enteric bacteria. 1A, uridylylation/deuridylylation of PII protein (*glnB*) by the *glnD* gene product. 1B, PII stimulates adenylylation of glutamine synthetase (GS). 1C PII stimulates the phosphatase activity of NtrB to dephosphorylate NtrC

is active while PII-UMP is inactive. This is indicated by the fact that *glnD* insertion mutants of *E. coli* are glutamine auxotrophs (GS is inactive) while both *glnB* and *glnBglnD* double insertion (or deletion) mutants have high GS activity in both low N and high N conditions (Bueno et.al., 1985; Streicher et al., 1977).

NtrC is a transcriptional activator of certain operons encoding nitrogen utilization pathways. It is dephosphorylated (high N) or phosphorylated (low N) by NtrB (Keener and Kustu, 1988; Ninfa and Magasanik, 1986; Weiss and Magasanik, 1988). The *ntrBC* genes are adjacent and immediately downstream of the *glnA* gene encoding GS. NtrB and NtrC belong to a family of two component regulatory proteins, one of which is a sensor of some environmental stimulus (in this case low N) that phosphorylates the second protein, a regulator of transcription of some subset of genes (Nixon et al., 1986 Stock et al., 1989). Among the genes directly activated by NtrC are those in the p1p2*glnA*p*ntrBC* regulon and also, in the related *K. aerogenes*, *nac*, which encodes another transcriptional activator necessary for expression of histidine utilization *hut*, proline utilization *put*, and urease *ure* genes (Bender, 1991). Phosphorylated NtrC activates expression at the p2*glnA* promoter (which is recognized by the alternative δ^{54} factor-containing RNA polymerase). NtrC-P binds strongly to two specific sequences located 100 to 147 bp upstream from the start point of transcription (Reitzer and Magasanik, 1986). This binding facilitates NtrC-P interaction with RNA polymerase bound at a downstream δ^{54} recognition site (TGGC-N8-CGCT). This protein-protein interaction, along with ATP hydrolysis by NtrC, changes the RNA polymerase-promoter complex from the closed to the open configuration which is active for transcription (Popham et al. 1989). Both NtrC and Nac proteins are also repressors of transcription: NtrC during nitrogen excess at the p1*glnA* and at *pntrB* (Magasanik, 1988) and Nac during nitrogen limitation at *gdh* (encoding glutamate dehydrogenase, GDH) (Bender, 1991). Unlike NtrC, Nac is not a member of a two component regulator system nor are the promoters it activates recognized by δ^{54} (Bender, 1991).

The NtrB protein, like other histidine kinase proteins, is autophosphorylated at a conserved histidine residue in its C-terminal domain. It catalyzes the transfer of phosphate to a conserved aspartate residue in NtrC (for review, see Stock et al. 1989). In addition to this phosphorylation activity, NtrB has phosphatase activity that removes phosphate from NtrC. The phosphatase activity is stimulated by (deuridylylated) PII (high N) (Keener and Kustu, 1988) (Fig.1C). Thus in high N conditions, NtrC-P is not available as a transcriptional activator. In low N, when PII is uridylylated, the phosphotransferase activity of NtrB dominates and NtrC-P is available for transcriptional activation. Although GS activity has been extensively measured in *E. coli glnB* and *glnD* mutants, little data is available on the expression of genes directly activated by NtrC in these *E. coli* mutant backgrounds. Therefore whether PII-UMP is completely inactive with respect to NtrC phosphorylation/dephosphorylation by NtrB is not known.

Keener and Kustu (1988) reported that PII-UMP did not enhance the rate of NtrC phosphorylation by NtrB in *S. typhimurium* in vitro. Although a role for PII-UMP in NtrC phosphorylation in *E. coli* is uncertain, the major effect of PII on NtrC activity occurs through its stimulation of NtrB to dephosphorylate NtrC.

The *glnD* gene maps at 3.8 minutes on the *E. coli* chromosome while *glnB* is located at 55 minutes (Magasanik, 1982; Van Heeswijk et al. 1992). The *glnB* sequence was reported in 1987 by Son and Rhee. It is now known to be downstream from an ORF (named *orfXB*) which has been partially sequenced and shows homology to the central and C-terminal domains of NtrC (Van Hesswijk et al., 1993). Two transcripts are synthesized from this region, one from *orfXBglnB* and the other from *glnB* alone. The latter transcript may also be under negative control by PurR, a repressor of genes involved in purine biosynthesis (He et al., 1993). This regulation could reflect the requirement for the amido group of glutamine as a precursor in purine biosynthesis. The *E. coli glnD* and *glnE* gene sequences and transcriptional organization were recently reported (Van Hesswijk et al., 1993). The *glnD* gene is transcribed from its own promoter which is not regulated by fixed N supply. The *glnD* gene is downstream from the *map* gene encoding methionine aminopeptidase but is not cotranscribed with it. The *glnE* gene is downstream and cotranscribed with an ORF (*orfXE*) whose function is unknown. Importantly, Van Hesswijk et al.(1993) show that none of these three genes, *glnB*, *glnD*, *glnE*, involved in the nitrogen regulatory cascade in *E. coli*, is regulated by levels of fixed nitrogen nor do they require the *rpoN*-encoded δ^{54} factor for expression. Thus, the activities but not the expression of PII, UT/UR, and ATase are controlled by levels of fixed N.

glnB and *glnD* in nitrogen fixing bacteria

A *glnB* gene encoding PII has been identified in a number of nitrogen fixing bacteria, including the eubacteria *K. pneumoniae*, *Rhodobacter capsulatus*, *Azospirillum brasilense*, *Bradyrhizobium japonicum*, and *R.l. viciae*, and also in two archaebacterial genera, *Methanobacterium* and *Methanosarcina* (Sibold et al., 1991). A *glnD*-like gene named *nfrX* was recently characterized in *A. vinelandii*. Both *glnB* and *glnD* have not yet been characterized in any single nitrogen fixing bacterium.

glnB *in* K. pneumoniae

In *K. pneumoniae*, phosphorylated NtrC (NtrC-P) activates expression of the *nifLA* operon from the *nifL* promoter in addition to activating expression from p2*glnA* as in *E. coli* (Austin et al., 1987; Minchin et al., 1988) Expression of *nifLA* is the first step towards derepression of the 7 other *nif* operons leading to the expression of nitrogenase activity. NifA is a protein related by structural and functional homology to NtrC and is the transcriptional activator of the seven other *nif* operons whose products include the nitrogenase enzyme subunits (*nifHDK*) as well as ancillary proteins required for FeMocofactor biosynthesis (*nifBQ*, *nifEN*, *nifV*), maturation of nitrogenase subunits

(*nifUS, nifM*), and electron transfer to nitrogenase (*nifF, nifJ*) (for reviews, see Dean and Jacobson, 1992; Merrick, 1992). However, there is no evidence that NifA protein is activated by phosphorylation in the absence of fixed nitrogen as is NtrC. Indeed, The domain containing The conserved aspartate residue which is phosphorylated in other transcriptional activators of two-component regulatory systems, including NtrC as described above, is not present in NifA. Nevertheless, NifA activity is inhibited in ammonium due to the function of its partner regulatory protein NifL which may directly bind to NifA in the presence of ammonium (or oxygen) (for review see Merrick, 1992). Thus, NifL activity is modulated by fixed nitrogen (and oxygen). Therefore, in *K. pneumoniae*, excess fixed nitrogen not only inhibits NtrC and GS activities but also causes NifL to inhibit NifA function. Important questions are whether or not all of these effects are mediated by the *glnD* and *glnB* gene products in *K. pneumoniae* and whether or not *glnB* and *glnD* structure and function are similar to those in *E. coli*.

Work of Dr. Mike Merrick and co-workers provides much of what is now known about the function of *glnB* in *K. pneumoniae*. The primary structure of the PII protein deduced from the sequence of the *glnB* gene is very similar to that of *E. coli* and includes the conserved tyrosine 51, the site of uridylylation by the *glnD* gene product in low N culture conditions (Holtel and Merrick, 1988). Two classes of well-defined *K. pneumoniae glnB* mutants, similar to those characterized in other enteric organisms (Bueno et al., 1985; Foor et al., 1980; Leonardo and Goldberg, 1980) exist: one class consists of a single mutant with a Glu → Lys substitution at position 50 adjacent to Tyr-51, the other contains a Tn5 insertion in *glnB* (Holtel and Merrick, 1989). In the former, the Glu → 50 Lys substitution either prevents uridylylation of PII or prevents PII-UMP from taking the conformation of wild-type PII-UMP. Measurements in the mutant of GS activity, growth on N sources, nitrogenase activity, as well as expression of β-galactosidase from *lacZ* fusions inserted downstream of the two NtrC-activated promoters, *glnA* and *nifL*, led to the following observations and conclusions.

Firstly, the missense mutant was unable to express *nifLA* and to fix nitrogen (NifA is required for nitrogenase synthesis); it had little GS activity and required glutamine for growth; and it could not express *glnA* from the NtrC-P activated p2*glnA* promoter. These results were those expected from the model that unuridylylated PII stimulates adenylylation of GS and dephosphorylation of NtrC. In fact, introduction of a plasmid carrying the *ntrC* gene expressed from a constitutive promoter made no change in the mutant phenotypes. Instead, introduction of a plasmid carrying the *ntrB* gene expressed from a constitutive promoter restored GS activity, *nifL* expression and nitrogenase activity and normal expression of these activities by high N conditions. Thus, PII in *K. pneumoniae* stimulates NtrB phosphatase activity; excess NtrB can apparently titrate the effect of PII. In addition, introduction of a plasmid carrying *glnB* itself into the missense mutant restored a wild-type phenotype. The dominance of wild type over mutant *glnB* might indicate that mixing of mutant and wild type PII subunits in the tetrameric protein results in a partially uridylylated protein which fails to stimulate the phosphatase activity of NtrB. These results suggest that PII-UMP may play a physiologically significant role in blocking the phosphatase activity of NtrB, and ensure a rapid response to changes in nitrogen status.

The *glnB* null mutant (*glnB*::Tn5) had high GS and nitrogenase activities and expressed both the p2*glnA* and *nifL* promoters. Nitrogenase activity and *nifL* expression were completely repressed by high N conditions while GS activity and expression of *glnA* were repressed less in the mutant than in the wild type strain. Therefore, the ability of NifL to inactivate NifA in the presence of ammonium is not regulated by PII and a PII-independent mechanism must exist for transmitting the signal of fixed N excess in *K. pneumoniae*. A PII-independent N-sensing mechanism is also indicated because expression of *glnA* and *nifLA* were at least partially repressed by high fixed N in the *glnB*::Tn5 strain.

In summary, in *K. pneumoniae*, PII stimulates the dephosphorylation of NtrC by NtrB in high N conditions, is not absolutely required for nitrogen regulation of NtrC-activated promoters, and is not involved in NifL-mediated inactivation of NifA. The latter point is discussed further in relation to NifL in *A. vinelandii* (see below).

glnB *in* Rhodobacter capsulatus

The *glnB* gene of *R. capsulatus* was identified and characterized in Dr. Robert Kranz's laboratory. In this case, the phenotype was known before the gene was identified. This phenotype was constitutive expression of the *nifHDK* operon fused to *lacZ* in high N conditions. A number of such mutants had been isolated (Kranz and Haselkorn, 1985) and a DNA fragment that restored normal nitrogenase repression was cloned and sequenced (Kranz et al., 1990). The translation product of an open reading frame (ORF) of 339 bp showed a high degree of homology to the *glnB* gene product from *E. coli* and from *R.l. viciae* including the conserved Tyr-51, the amino acid uridylylated in enteric organisms. Sequence of an incomplete ORF starting 83 bp downstream from the *R. capsulatus glnB* gene terminus indicated homology to *glnA*. A *glnBglnA* operon structure was indicated from the fact that a Tn5 insertion in *glnB* could give viable genomic *glnB* substitution mutants in *R. capsulatus* only when glutamine was provided in the selection medium used to isolate *glnB*::Tn5 recombinants. Therefore *glnB* and *glnA* are cotranscribed in *R. capsulatus*.

The *glnB*::Tn5 mutant strain synthesized nitrogenase Fe protein (the *nifH* gene product) when grown anaerobically in the presence of repressive amounts of glutamine, but not under aerobic conditions (as the wild-type strain). Therefore PII mediates fixed N control of nitrogenase expression but not oxygen repression. In *R. capsulatus*, nitrogenase gene operons are activated by NifA as in *K. pneumoniae* and all other Gram-negative diazotrophs examined. No gene equivalent to NifL has been identified, and NifA activity is not inhibited by fixed nitrogen. NifA activity is inhibited by oxygen, as in the *Rhizobiaceae*, but this inactivation does not apparently involve another *nif* gene product as in *K. pneumoniae* and *A. vinelandii*. A unique feature of *R. capsulatus nif* gene regulation is that there are two copies of *nifA*, and both must be mutated in order to prevent nitrogenase synthesis (Klipp et al.,1988). Expression of both *nifA* genes requires NtrC (also called NIFR1 in this organism) presumably as NtrC-P and is repressed by ammonium (Foster-Hartnett and Kranz, 1992). As in *K. pneumoniae*, *ntrB* is present (called NIFR2) (Kranz and Foster-Hartnett, 1990). The constitutive expression of *nif* genes in the *glnB* mutants is most likely explained according to the *E. coli* model: in the absence of PII, NtrB phosphorylates NtrC and NtrB phosphatase activity is absent in high N growth conditions. Therefore NtrC is probably fully phosphorylated and NifA expression is not prevented in high N conditions. Because NifA activity is not inhibited by a NifL-like gene product, as in *K. pneumoniae*, expression of nitrogenase genes is constitutive in high N. Whether there is any residual fixed N repression of *nifA* expression, as occurs in *K. pneumoniae glnB* mutants, could not be deduced from this work because the amount of nitrogenase Fe protein expressed in high N was not compared to that produced under low N conditions. This was presumably not possible because of the glutamine requirement of the *glnB*::Tn5 mutant.

Another aspect of nitrogen regulation in *Rhodospirillum rubrum*, *R. capsulatus* and *A. brasilense* is that nitrogenase activity is inhibited in high N conditions by covalent modification of the nitrogenase Fe protein. This occurs by a reversible ADP-ribosylation of the protein mediated by the DRAT/DRAG proteins (DRAT catalyzes the addition of ADP-ribose, DRAG removes it) (for review, see Roberts and Ludden, 1992). Space does not permit a full description of this system, but an interesting question is whether the GlnD/GlnB nitrogen sensing system mediates the activites of the DRAT/DRAG proteins. A recent paper (Hallenbeck, 1992) reported experiments with two of the original *R. capsulatus* Nif constitutive mutants (Kranz and Hasel-

korn, 1988), one of which was later shown to map in *glnB*. Hallenbeck (1992) transferred a plasmid containing the *nifHDK* genes into the mutant strains. This was necessary because the original *glnB* mutants contained a *nifH-lacZ* fusion and was Nif. Nitrogenase activity and levels of ADP-ribosylation of the Fe protein were measured before and after addition of ammonium. While nitrogenase activity was higher in the two mutants both before and after addition of ammonium, a significant fraction of Fe protein in each was modified by ADP-ribosylation after ammonium addition. The two constitutive mutants apparently had more total nitrogenase than the wild type strain from the outset, and so contained more residual active enzyme after ammonium addition than wild type. Nevertheless, because a significant fraction of Fe protein was ADP-ribosylated in the *glnB* mutant, PII probably does not control the activity of the DRAT/DRAG system in *R. capsulatus*.

So far no information is available concerning GS activities in the *R. capsulatus glnB* mutant, nor has the identification of *glnD* gene been reported.

glnB *in Azospirillum brasilense*

The *glnB* gene of *A. brasilense* was identified in Dr. Claudine Elmerich's laboratory on the basis of sequence analysis of DNA upstream of *glnA* (De Zamaroczy et al., 1990), as previously indicated by Colonna-Romano et al. 1987. The homology of the *A. brasilense glnB* gene product to other PII proteins is striking and includes the conserved Tyr-51, the site of uridylylation. This group recently reported that a mutant generated by insertion of a kanamycin (Km) resistance gene cartridge into the *glnB* gene was NiF$^-$ (Liang et al., 1992). The resulting mutant did not require glutamine and growth was normal in all media except in the absence of fixed N. *glnB* and *glnA* are co-transcribed under low N conditions from two promoters (p1 and p2) upstream of *glnB*; p2 is a δ^{54} (*rpoN*-dependent) recognized promoter but it does not require NtrC for expression. *glnA* is transcribed under high N conditions from a promoter just upstream from this gene.

As in the other organisms described above, NifA is present and required for expression of the other *nif* genes. Analysis of expression of a *nifA-lacZ* fusion construct introduced into wild-type and different regulatory mutant backgrounds shows that *nifA* expression is controlled differently from *nifH* and *nifB* expression (Liang et al. 1992). *nifH* and *nifB* are expressed under low N and microaerobic (0.5% O_2) conditions, but not if ammonium is present or in aerobically grown cultures. Expression of *nifA*, on the other hand, while maximum in low N and microaerobic conditions, also occurs if ammonium is present or in air (reduced to about 50% and 20%, respectively, in the latter two conditions). Therefore, the activity of *A. brasilense* NifA protein is inhibited by high N or oxygen, as in *K. pneumoniae* (and *A. vinelandii*, see below), but the mechanism of this inhibition is probably different because no *nifL* gene has been identified upstream of *nifA* or elsewhere in this organism. While NifA is probably directly inactivated by oxygen (its sequence contains the characteristic inter-domain linker associated with NifA oxygen sensitivity in Rhizobium and other species), its activity in the absence of ammonium almost certainly involves PII. This is indicated by the fact that *nifA-lacZ* expression is normal in the *glnB* mutant strain but neither *nifH*- nor *nifB-lacZ* fusions are expressed in this background (de Zamaroczy et al., 1993). PII in *A. brasilense* may stimulate a post-translational modification of NifA which is essential for its activity. Whether PII is uridylylated or not in response to low N or high N conditions must be determined. So far, a *glnD*-like gene has not been identified in this organism.

The nfrX *gene of* Azotobacter vinelandii: *structural and functional homology to* glnD

In *A. vinelandii* ammonium neither inhibits nitrogenase activity nor prevents expression of *nifA* (Bali et al., 1992). The only level of ammonium control of nitrogen fixation is via NifL, which, as in *K. pneumoniae*, prevents expression of other *nif* genes in cultures exposed to ammonium or excess oxygen (Blanco et al., 1993). Although

ntrBC genes are present in *A. vinelandii*, they do not influence nitrogen fixation (Toukdarian and Kennedy, 1986). Unlike in *K. pneumoniae*, the *A. vinelandii* *nifLA* operon is expressed from δ^{54}-independent promoter (Blanco et al.,1993). A *nifL-lacZ* fusion construct was inserted into the chromosome of *rpoN$^-$* (o^{54-}) and *ntrC$^-$* mutants; both strains had similar levels of β-galactosidase as wild type and expression was the same in ammonium-grown cultures (Blanco et al., 1993). An interesting consequence of *A. vinelandii* having only one target of ammonium control is that nitrogenase activity in *nifL* insertion or deletion mutants is completely derepressed and mutants excrete large amounts of ammonium (up to 10 mM) into the growth medium during nitrogen fixation (Bali et al., 1992; Blanco et al., l993).

A gene identified in *A. vinelandii* as *nfrX* was initially characterized in several Tn5 regulatory mutants unable to synthesize nitrogenase (Santero et al., 1988). Their regulatory nature was established because the mutants could be complemented to a Nif$^+$ phenotype by introduction of a *nifA* plasmid; however, the Tn5 inserts were not in *nifA* since the transposon was located in a genomic region separate and distinct from *nifA*. Sequencing of the *nfrX* wild type DNA revealed a large ORF encoding a 105 kD protein (Contreras et al., 1991). Homology was found to the N-terminal end of the *E. coli glnD* gene product, for which at that time only 175 amino acids were available. This discovery prompted further experiments to compare function of *nfrX* and *glnD*. Cloned *nfrX* complemented *E. coli* and *K. aerogenes glnD* mutants for growth on arginine and nitrate, respectively, and cloned *E. coli glnD* complemented *A. vinelandii nfrX* mutants for nitrogen fixation (growth on N_2). Therefore *nfrX* is structurally and functionally homologous to *glnD*. The homology has now been extended beyond the N-terminal regions. The full sequence of *E. coli glnD* was recently reported (Van Hesswijk et al., 1993); the full homology between *NfrX* and *GlnD* is shown in Fig 2.

Why are *nfrX* mutants Nif$^-$ if NtrC is not necessary for nitrogenase gene expression? An important clue comes from the observation that

```
E.c.    1 MNTLPEQYANTALPTLPGQPQNPCVWPRDELTVGGIKAHIDTFQRWLGDAFDNGISAEQLIEARTEFIDQLLQRLWIEAGFSQIADLALVAVGGYGRGEL 100
          :|:        | | :   | |  :     : : :       |:  |::|    ||| | :||:| |    ::| ||:|:|:||||||||||
A.v.    1 ...MPQ.....VDPDLFDPGQFQAELALKSSPIPAYKKALRCAREVLDARFQEGRDIRRLIEDRAWFVDQILALAWNRFDWSEDADIALIAVGGYGRGEL 92

E.c.  101 HPLSDVDLLILSRKKLPDDQAQKVGELLTLLWDVKLEVGHSVRTLEECMLEGLSDLTVATNLIESRLLIGDVALFLELQKHIFSEGFWPSDKFYAAKVEE 200
          || |:||||| |  ||::|  || :|| ||||||:   || || |||    ||| |||:| | : |:    |:| | |  ||| |:|:  :|
A.v.   93 HPYSDIDLLILMDGADHEVFREPIEGFLTLLWDIGLEVGQSVRSLAECAEEAQADLTVITNLMESRTIAGPEHLRQRMQEVTSAQRMWPSRAFFLAKRDE 192

E.c.  201 QNQRHQRYHGTSYNLEPDIKSSPGGLRDIHTLQWVARRHFGATSLDEMVGFGFLTSAERAELNECLHILWRIRFALHLVVSRYDNRLLFDRQLSVAQRLN 300
          |   ||  ||::| ||||||||::|||||||||:|| |:||||:|  | |||  ||: ::|||||| :|  :| | :||||||: : ::|||||  |
A.v.  193 QKTRHARYNDTEYNLEPNVKGSPGGLRDIQTLLWIARRQFGFTINLHAMVGQGFLLESEYTLLASSQEFLWKVRYALHMLAGRAEDRLLFDLQRQIAGLLG 292

E.c.  301 YSG.EGNEPVERMMKDYFRVTRRVSELNQMLLQLFDEAILALPADEKPRPIDDEFQLRGTLIDLRDETLFMRQPEAILRMFYTMVHNSAITGIYSTTLRQ 399
          |: :  :||||:  | :| |::| :|  |::::|  ||  | | :     ||::| :  :::    :|  | |:| :| :  :  :   | |:    |:|
A.v.  293 YEDSDAKLAVERFMQKYYRVVLGIAELTELVFQHFEEVILPGDAAGRVEPLNERFQVRDGYLEVTHAGVFQETPSALLEIFVLLARRPEIRGVRADTIRL 392

E.c.  400 LRHARRHLQQPLCNIPEARKLFLSILRHPGAVRRGLLPMHRHSVLGAYMPQWSHIVGQMQFDLFHAYTVDEHTIRVMLKLESFASEETRQRHPLCVDVWP 499
          ||   :::   |:  :   ||: | |   |:|  :||   |:| :|| ||||| |||| ||||||| :|:| | | :||  || |:| | :  || |
A.v.  393 LRDHRYLIDDAFRRDPHNTGLFIELFKSRQGIHRNLRRMNRYGILGRYLPEFGHIVGQMQHDLFHIYTVDAHTLNLIKNLRKLFWPELAEKYPLASKLIE 492

E.c.  500 RLPSTELIFIAALFHDIAKGRGGDHSILGAQDVVHFAELHGLNSRETQLVAWLVRQHLLMSVTAQRRDIQDPEVIKQFAEEVQTENRLRYLVCLTVADIC 599
          :| |: |||| :|:|:|||||||||| |||  :| ||::|  |     :||||||| |||:|||||: ||:||||| |||||| ||||:| ||||||| |
A.v.  493 KLPKPELIYLAGLYHDIGKGRGGDHSELGAADALAFCQRHDLPAMDTQLIVWLVRNHLLMSTTAQRKDLSDPQVIFDFAQKVRDQTYLDYLYVLTVADIN 592

E.c.  600 ATNETLWNSWKQSLLRELYFATEKQLRRGMQNTPDMRERVRHHQLQALALLRMDNIDEEALHQIWSRCRANYFVRHSPNQLAWHARHLLQHDLSKPLVLL 699
          |||  |||||:  ||| |:| ||:|| ||:|| | :  ||:| |:|  ||:  | |   :|:  :||| ::||:  ::|||  :||| :||| :  :|
A.v.  593 ATNPTLWNSWRASLLRQLYTETKHALRRGLEQPVGREEQIRQTQKAALDILVRSGTDPDDAEHLWTQLGDDYFLRHTSSDIAWHTEAILQHPSSGGPLVL 692

E.c.  700 SPQRTR....GGTEIFIWSPDRPYLFAAVCAELDRRNLSVHDAQIFTTRDGMAMDTFIVLEPDGNPLSAD..RHEVIRFGLEQVLTQSS.WQPPQPRRQP 792
          :|    | |:|:|||| ||    | :|:||||| |   || : | :   :  | |  |||  | :       :  | | |  | |||   :
A.v.  693 IKETTQREFEGATQIFIYAPDQHDFFAVTVAAMDQLNLSIHDARVITSTSQFTLDTYIVLDADGGSIGNNPARIQEIRQGLVEALRNPADYPTIIQRRVP 792

E.c.  793 AKLRHFTVETEVTFLPTHTDRKSFLELIALDQPGLLARVGKIFADLGISLHGARITTIGERVEDLFIIATADRRALNNELQQEVHQRLTEALNPNDKG*C 892
          |:|| | :||:   ||:|| |||||||| ||||||:| |:|  ||:|| |   :|:| |:|||||:|:|:| | :|  | | :|| ||  || |
A.v.  793 RQLKHFAFAPQVTIQNDALRPVTILEIIAPDRPGLLARIGKIFLDFDLSLQNAKIATLGERVEDVFFVTDAHNQPLS...DPELCARLQLAIAEQLADGD 889

E.c.  893 VY*YEKSLTMQQLQNIIETAFER 915
          |          :  :
A.v.  890 SYIQPSRISI*........... 900
```

Fig. 2. Alignment of amino acid sequence of *A. vinelandii* nfrX gene product and *E. coli* glnD product showing indentical (‖) or similar (:) residues (from Contreras et al., 1991 and van Hesswijk et al., 1993). The dark underline indicates the extent of NfrX and GlnD originally compared in Contrerase et al., 1991.

nfrX nifL double mutants are Nif⁺. In other words, a mutation in *nifL* suppresses the *nfrX* phenotype. One explanation is that the *nfrX* gene product, *NfrX*, either directly or indirectly modifies the NifL protein in low N conditions such that in this modified conformation, NifL is unable to inactivate NifA. Consistent with this hypothesis is the ability of overexpressed NifA to correct the Nif phenotype of *nfrX* mutants to Nif⁺; excess NifA probably titrates out the 'active' form of NifL. NifL and NifA form a complex in low N conditions in *K. pneumoniae*. Although this has not been directly demonstrated for *A. vinelandii*, it is likely that NifL and NifA interact in a similar way in this organism because of the high homology between the two NifL's and NifA's in both species (Bennett et al. 1988; Blanco et al. 1993).

Since GlnD of enteric bacteria catalyzes uridylylation/deuridylylation of PII, NfrX might reg-ulate NifL activity in the same way. We have searched the NifL protein sequence for tyrosine residues which are surrounded by amino acids identical or similar to those surrounding the conserved Tyr-51 in other PII proteins. Although 9 Tyr's are present in *A. vinelandii* NifL, none are in a region with particular homology to the PII Tyr-51 region. Experiments are in progress to determine how *A. vinelandii* NifL is modified in response to changing from high N to low N and vice-versa) growth conditions and whether such a modification occurs in *nfrX* mutants. On the other hand, NfrX may influence NifL activity only indi-rectly as GlnD influences NtrB activity via the PII protein. An interesting difference between NifL in *A. vinelandii* and *K. pneumoniae* is that the former has a sequence very homologous to the autophos-phorylation site, including the conserved histi-dine residue present in histidine protein kinases as described above. This sequence does not occur in *K. pneumoniae* NifL. Because neither NifA from the two organisms carries the conserved aspartate which is phosphorylated in the other transcriptional activators modified by histidine protein kinase sensory proteins, the physiological significance of the conserved His in *A. vinelandii* NifL cannot be known. Nevertheless, it is possible that phosphorylation of *A. vinelandii* NifL occurs in response to a signal transmitted by NfrX, possibly through a PII-like protein, and that either the phosphorylated or dephosphorylated form of NifL inhibits NifA activity. The absence of *nfrX* would cause one or the other to be constitutively present.

Although NtrC is not required for nitrogenase expression in *A. vinelandii*, *ntrC* mutants are unable to grow on nitrate or arginine (Luque et al., 1987; Toukdarian and Kennedy, 1986 and unpublished results). Both nitrate reductase and nitrite reductase activities require NtrC for expression and both activities are repressed in ammonium-grown cultures. Therefore, NtrB and NtrC respond to N status as in enteric bacteria, i.e. NtrC is active in low N conditions and inactive in high N. Preliminary sequencing of the cotranscribed *ntrBC* genes show that their gene products are very homologous to those of other bacteria, including conservation of the phosphorylation sites in these proteins (Saunders and Kennedy, unpublished results). Thus, NtrC activity is expected to be dependent on phosphorylation, mediated by NtrB activity. Also GS, encoded by the *glnA* gene which lies upstream of *ntrBC* in *A. vinelandii*, is adenylylated and deadenylylated in response to high N and low N conditions, respectively, as in other bacteria (see Kennedy and Toukdarian, 1987, for review). However, neither NtrC-dependent nitrate assimilation nor rates of adenylylation/deadenylylation of GS are affected in *nfrX* mutants (Contreras et al., 1991). Thus, another *glnD* gene, in addition to *nfrX*, might be present to mediate these functions. In order to test this possibility, the *glnD* gene from *K. pneumoniae* was hybridized to Southern blots of several restriction digests of *A. vinelandii* genomic DNA. In each apparently complete digest, 3 to 4 fragments ranging in size from 3 to 10kb hybridized

to the *glnD* probe, including fragments of the size expected for *nfrX* and others. This result is consistent with the idea that more than one *nfrX*-like gene is present in *A. vinelandii*.

We are also searching for a *glnB*-like gene in *A. vinelandii* to determine whether PII plays a part in signal transduction to NifL or to NtrB or to GS in response to changing N conditions. Southern blots of *A. vinelandii* genomic DNA digested with several different restriction enzymes were hybridized with an *E. coli glnB* gene probe. The fragment hybridizing most clearly was a 3 kb *EcoRI* fragment, which has now been cloned. If PII functions in *A. vinelandii* as it does in enteric bacteria, then *A. vinelandii glnB* mutants will have nitrate reductase activity in both low and high N conditions, and will adenylylate and/or deadenylylate GS at a reduced rate.

In one possible model for the nitrogen cascade in *A. vinelandii*, two *glnD*-like genes are present, one of which, *nfrX*, functions in nitrogen fixation to inactivate NifL, perhaps independently of any PII protein, and the other of which, in concert with PII, mediates NtrB phosphorylation/dephosphorylation of NtrC and also GS adenylylation/ deadenylylation.

glnB *and* glnD *in* Rhizobium leguminosarum *biovar* viciae

R.l.viciae nodulates peas, and, typically for fast-growing *Rhizobium* species, only expresses nitrogenase in the symbiotic bacteroid state within nodules. When increasing concentrations of fixed nitrogen are added to a plant inoculated with *Rhizobium*, the symbiosis is less and less efficient and at certain concentrations no nodules are observed (see Streeter, 1988; Dusha et al. 1989; Wang and Stacey, 1990 for a description and references). There is evidence (Pate and Dart, 1961) that in the symbiosis between *R. leguminosarum* and *Vicia atropurpurea* at least part of the regulation is at the bacterial level. In fact, addition of ammonium nitrate inhibited the rate of nitrogen fixation with different efficiency when two different *R. leguminosarum* strains were used. In another case, a mutant of *R. meliloti* was isolated capable of more efficient nodulation of alfalfa than the wild type strain in the presence of 2 mM ammonium sulfate (Dusha et al., 1989).

Two forms of glutamine synthetase, GSI and GSII, have been demonstrated in all species tested of the family *Rhizobiaceae* (Defez et al. 1990; Fuchs and Keister, 1980). GSI is similar to the single GS of enteric and other bacteria. It is a polymeric enzyme, relatively heat stable, made of 12 identical 50 kD subunits, and can be adenylylated (Darrow et al., 1981). In contrast, GSII is made up of smaller subunits, of 36 kD, which are heat labile and not know to be modified after translation. These proteins are products of different genes (Carlson et al., 1985; Colonna-Romano et al., 1987; Filser et al., 1986; Somerville and Kahn, 1983). A *Rhizobium* gene, *glnT*, coding for a third glutamine synthetase, GSIII, has been recently described (De Bruijn et al. 1989; Espin et al. 1990). GS activity expressed from this gene has been shown in an enteric bacterium (Chiurazzi et al. 1992), and recently in *R. meliloti* (Shatters et al., 1993).

The *glnB* gene has been located upstream of *glnA* in *R.l. viciae* and identified because of its similarity to the *glnB* sequence of *E. coli* and *K. pneumoniae* (Colonna-Romano et al., 1987; Hotel et al., 1989). It was able to compliment the *glnB* missense mutant of *K. pneumoniae* described above. A δ^{54}-recognized promoter adjacent to *glnB* requires NtrC for activation of *glnBglnA* expression while a second promoter adjacent to *glnA* is neither δ^{54}- nor NtrC-dependent. A similar alignment of the *glnB* and *glnA* genes and promoters was reported for *B. japonicum* (Martin et al., 1989). The concentration of the *glnBA* transcript in *R.l. viciae* was 2- to 3-fold higher in nitrate-grown cultures compared to ammonium-grown cultures. The *glnA*-only transcript is not regulated by fixed N sources (Chiurazzi and Iaccarino, 1990). NtrC is also required for full expression of *glnII* as a monocistronic transcript; this transcript is enriched 10-fold in nitrate-grown as compared to ammonium-grown cultures (Patriarca et al., 1992). The *ntrBC* genes of *R.leguminosarum* are downstream of and cotranscribed with an ORF showing homology to an ORF located upstream

of *ntrBC* in *R. capsulatus* (Foster-Hartnett et al., 1993) and to the ORF1 located upstream of the *fis* gene of *E. coli* (Ninneman et al., 1993). The *R.l. viciae* ORF-*ntrBC* operon is negatively regulated by NtrC (a consensus site for NtrC binding is present) but does not require NtrC for expression. Thus it is probable that, unlike in enteric bacteria, only changes in the activity state of NtrC and not in its intracellular concentration are important for activation of transcription from its target promoters at *glnB* and *glnII* (Patriarca et al., 1993).

The *R.l. viciae* and *E. coli* PII proteins show a high degree of homology which includes the uridylylation site, Tyr-51, suggesting that the *R.l. viciae* PII protein is also post-translationally modified. A recent paper shows that the PII protein of *R.l. viciae* is uridylylated in response to a change from high N (ammonium as N source) to low N (nitrate as N source) conditions (Colonna-Romano et al., 1993). This suggests that a *glnD*-like gene is present in *R.l. viciae*, also previously indicated by the discovery that a cosmid from a *R. leguminosarum* biovar *phaseoli* gene library complemented the Nif⁻ phenotype of an *A. vinelandii nfrX* mutant (Hawkins et al., 1991). We also report the phenotype of a *glnB* null mutant of *R.l. viciae*, isolated by insertion of a Km casette in the *glnB* gene, with respect to expression of the NtrC-activated *glnBglnA* and *glnII* transcriptional units.

A *glnB* mutant strain in which *glnB* is interrupted by the placement of a Km insertion cassette was constructed using a derivative of the suicide plasmid pSUP202. This derivative, pAM15, carried the *glnB*-Km fragment (Km inserted at the *SacII* site located 123 base pairs inside the coding region of *glnB*) along with a *sacB* gene inserted at the *PstI* site within pSUP202 itself, 3.2 kb distant from the *glnB*-Km fragment. (The *sacB* gene confers sensitivity to sucrose and transconjugants retaining *sacB* are unable to grow on sucrose). Introduction into *R.l. viciae* by conjugation followed by selection on rich medium containing sucrose and neomycin (Km) yielded 160 Cm^s colonies (Cm^r is carried on the pSUP202 DNA). Of 32 colonies tested for hybridization to a *glnB* probe, 2 had hybridizing fragments of the size predicted if the *glnB*-Km DNA had replaced the wild type *glnB* region in *R.l. viciae*.

One *glnB*-Km mutant strain, BS11, produced *glnA* but not *glnB* transcripts, showing that the *glnB* gene interruption had no polar effect on *glnA* expression and that expression of *glnA* from its own promoter was not affected. Immuno-blots of BS11 showed that the strain does not contain PII protein. The growth rate of BS11 was normal on rich medium with or without glutamine supplementation and on minimal medium containing ammonium, glutamine, glutamate, proline or arginine as N source. However, BS11 failed to grow on nitrate as N source although it grew as wild type on medium with ammonium + nitrate. When a plasmid carrying the *R.l. viciae glnB* gene was introduced into BS11, growth on nitrate was restored.

The two NtrC-activated promoters, *pglnB* and *pglnII*, are expressed at higher levels in low N conditions than in high N conditions in the wild type (Table 1). A *pglnB-lacZ* plasmid and a *pglnII-lacZ* plasmid were separately introduced into both the wild type strain and BS11, the *glnB*-Km mutant. In the wild type strain, β-galactosidase activity was higher in low N (glutamate) than in high N (ammonium) media. In BS11, the *glnB*-Km mutant, β-galactosidase activity was equally high in both high N and low N media. Expression from the NtrC-independent *glnA* and *ntrBC*

Table 1. Promoter expression in wild type and *glnB* strains of *R.l. viciae*

IacZ fusion	Wild type NH$_4^+$/glutamate	*glnB* mutant NH$_4^+$/glutamate
glnA-lacZ	93%	84%
ntrBC-lacZ	81%	83%
glnII-lacZ	19%	121%
glnB-lacZ	33%	135%

β-galactosidase activity was measured in glutamate-grown and and in NH$_4^+$-grown cultures. Activities in glutamate ranged from about 150 to 1500. The activity in NH$_4^+$-grown cultures is expressed as a percentage of that in the same strain grown with glutamate. The *lacZ* fusions were introduced into each strain on wide host range plasmids.

promoters was the same in wild type and BS11 on all media. Therefore the absence of PII in *R.l. viciae* results in fully active (and presumable fully phosphorylated) NtrC. It can be inferred that, as in *K. pneumoniae* and other enteric bacteria, PII in its deuridylylated form is required for the NtrC-P phosphatase activity of NtrB.

Why might PII be required for growth on nitrate in *R.l. viciae*? Because of the diversity of regulatory circuits controlled by NtrC in Gram-negative bacteria, we can speculate that in certain *Rhizobium* species PII is necessary for the expression of a gene(s) required for nitrate uptake or reduction. Since PII appears to influence the activity of the NtrB/NtrC system in a negative way, it is conceivable that it has an analogous but positive influence on a different and nitrate-specific regulator system controlling regulation of nitrate (and possibly nitrite) reductase. Results are not yet available on the nodulation and nitrogen fixation abilities of the *glnB* mutant of *R.l. viciae*.

The detection of PII protein in *R.l. viciae* was achieved immunologically (Colonna-Romano et al., 1993). An anti-serum to purified recombinant *R.l. viciae* PII protein was used to detect a 11.5 kD polypeptide after SDS-PAGE of crude extracts followed by western blotting. There was about a third the amount of immuno-reacting material present in ammonium-grown cultures compared to the amount in nitrate-grown cultures. Twenty-fold more immuno-reacting material was present in a strain carrying a second copy of *glnB* expressed from a constitutive Km promoter. The immuno-reacting band was absent in the *glnB* mutant strain described above.

Uridylylation of PII was detected by first labelling cultures grown under various conditions with α^{32}P-UTP. The labelling was achieved by first permeabilizing the cells with CETAB[1]; crude extracts were then run on SDS-PAGE as above. In addition to treatment of blots with anti-PII antiserum, they were also autoradiographed. Thus, the presence of PII protein at a location showing specific labelling by UMP is very strong evidence for the presence of uridylylated PII. In a nitrate-

---

[1] cetyldimethylethylammonium bromide

grown culture, a single 11.5 kD band of radioactivity was detected. This was not present if cells were subjected to a shock of ammonium addition before permeabilization. In cells treated with 18 mM α-ketoglutarate after permeabilization, the radioactive band was more intense, but disappeared if cells were incubated for a further time period in 18 mM glutamine. Finally, treatment of extracts with snake venom phosphodiesterase abolished the band of radioactivity while RNase had no effect (Colonna-Romano et al., 1993).

These results are important for several reasons: firstly they show that the *E. coli* model for uridylylation of PII applies to a *Rhizobium* species: it is controlled by the ratio of glutamine/α-ketoglutarate. They confirm that the amount and uridylylation state of PII protein reflects the amount of mRNA transcripts present in high N- or low N-grown cultures. Thirdly, the permeabilized cell labelling technique is probably applicable to the analysis of other bacterial enzymes thought to modify proteins without purification of the proteins involved (except for preparation of anti-serum for which proteins need not be active).

Although the *glnD* gene from *R.l. viciae* has not been isolated, a candidate gene was isolated from a *R. leguminosarum* bv. *phaseoli* cosmid library by its ability to complement an *A. vinelandii nfrX* mutant to a Nif$^+$ phenotype (Hawkins et al., 1991). This gene and its regulatory region, also present on the same cosmid, are worthy of further study. It would be particularly important to know the sequence of *glnD* cetyidimethylethylammonium bromide from a *Rhizobium* species to compare it to *E. coli glnD* and *A. vinelandii nfrX* and also to construct *glnD* mutants of *R.l. viciae*.

by *glnD*-like genes or others not yet identified, by transmission of this signal via *glnB*-encoded PII proteins which have been identified in several diazotrophs, and by PII-regulated proteins such as NtrB or ATase which control NtrC and GS activity, respectively. NtrC controls expression of different genes in different organisms, which may (as in *K. pneumoniae*) or may not (as in *A. vinelandii*) be directly involved in diazotrophic growth.

The study of ammonium sensing is not as well developed in symbiotic nitrogen fixing bacteria as it is in free-living diazotrophs. Results concerning the identity and function of *glnB*- and/or *glnD*-like gene products are up to now limited to two species of the *Rhizobiaceae* family, *R.l. viciae* and *B. japonicum*. It will be of great interest and importance to determine whether similar genes are present in other species of *Rhizobium*, *Bradyrhizobium*, and *Azorhizobium*. In the later, an N-responsive two component regulatory system comprised of NtrY and NtrX is present which modulates *nifA* expression; since expression of *ntrYX* is partially controlled by NtrC, the question of *glnB/glnD* involvement is pertinent. In addition, the control in certain *Rhizobiaceae* of *nod* and *nif* gene expression by ammonium may involve either the *glnB/glnD* system or other N-sensing pathways not yet identified.

Acknowledgements

Work in Tucson was supported by United States Department of Agriculture, National Research Initiative Competitive Grants Program grant 92-37305-7888. Work in Naples was partially supported by MAFDPGA, EEC CII-0408 and BIOT-0l66-C(EDB) and RAISA-CNR Subpr 1109.

Conclusions

Mechanisms for controlling nitrogen fixation and ammonium assimilation in diazotrophic bacteria are as diverse in detail as the number of organisms studied. Each species has its individual variations on a basic theme of sensing of N-status

References

Adler S P, Purich D and Stadtman E R 1975 Cascade control of *Escherichia coli* glutamine synthetase. Properties of the P$_{II}$ regulator protein and the uridylyltransferase-uridylylremoving enzyme. J.Biol.Chem.250, 6264–6272

Austin S, Henderson N and Dixon R 1987 Requirements for transcriptional activation in vitro of the nitrogen regulated *glnA* and *nifLA* promoters from *Klebsiella pneumoniae*: dependence on activator concentration. Mol.Microbiol.1, 92–100.

Bali A, Blanco G, Hill S and Kennedy C 1992 Excretion of ammonium by a *nifL* mutant of nitrogen fixing *Azotobacter vinelandii*. Appl.Environ.Microbiol 58, 1711–1718

Bender RA 1991 The role of the NAC Protein in the nitrogen regulation of *Kiebsiella* Appl.Environ.Microbiol. 58, 1711–1718.

Bennett L T, Cannon F C and Dean D 1988, Nucleotide sequence and mutagenesis of the *nifA* gene from *Azotobacter vinelandii*. Mol.Microbiol. 2, 315–321.

Blanco G, Drummond M D, Kennedy C and Woodley P 1993 Molecular analysis of the *nifL* gene of *Azotobacter vinelandii*. Mol.Microbiol. 9, 869–879.

Bueno R, Pahel G and Magasanlk B 1985 Role of *glnB* and *glnD* gene products in regulation of the *glnALG* operon of *Escherichia coli*. J.Bacteriol. 164, 816–822.

Carlson T A, Guerinot M L and Chelm B K 1985 Characterization of the gene encoding glutamine synthetase 1 (*glnA*) from *Bradyrhizobium japonicum*. J.Bacteriol 162, 698–703.

Chiurazzi M, and Lacarino M 1990 Transcriptional analysis of the *glnB-glnA* region of *Rhizobium leguminosarum* biovar *viciae*. Mol. Microbiol. 4, 1727–1735

Chiurazzi M, Meza R, Lara M, Lahm A, Defez R, Iaccarino M and Espim G 1992 The *Rhizobium leguminosarum* biovar *phaseoli glnT* gene, encoding glutamine synthetase 111. Gene 119, 1–8.

Chock P B, Rhee S G and Stadtman R R 1980 Interconvertible enzyme cascades in cellular recognition. Annu. Rev. Biochem. 49, 813–843.

Colonna-Romano S, Riccio A, Guida M, Defez R, Lamberti A, Iaccaiino M, Arnold W, Priefer U and Puhler A 1987 Tight linkage of *glnA* and a putative regulatory gene in *Rhizobium leguminosarum*. Nucl.Acids Res.15, 1951–1964

Colonna-Romano S, Patriarca E J, Amar M, Bernard P, Manco G, Lamberti A, Iaccarino M and Defez R 1993 Uridylylation of the PII protein in *Rhizobium leguminosarum*. Febs Letts. 330, 95–98.

Contreras C, Drummond M, Bali A, Blanco G, Garcia E, Bush K, Kennedy C and Merrick M 1991 The product of the nitrogen fixation regulatory gene *nfrX* of *Azotobacter vinelandii* is functionally and structurally homologous to the uridylyltransferase encode by *glnD* in enteric bacteria. J.Bacteriol. 24, 7741–7749.

Darrow R A, Crist D, Evans W R, Jones B L, Keister D L and Knotts R R 1981 Biochemical and physiological studies on the two glutamine synthetases of Rhizobium. *In* Current Perspectives in Nitrogen Fixation.Eds A H Gibson and W E Newton. Australian Academy of Sciences, Canberra.

De Bruijn F J, Rossbach S, Schneider M, Ratet P, Messmer S, Szeto W W, Ausubel F M and Schell J 1989 *Rhizobium meliloti* 1021 has three differentially regulated loci involved in glutamine biosynthesis, none of which is essential for symbiotic nitrogen fixation. J.Bacteriol. 171, 1673–1682.

De Zamaroczy M, Delorme F and Elmerich C 1990 Characterization of three different nitrogen-regulated promoter regions for the expression of *glnB* and *glnA* in *Azospirillum brasilense*. Mol.Gen.Genet.224, 421–430.

De Zamaroczy M, Paquelin A and Emerick C 1993 Functional organization of the *glnB-glnA* cluster of *Azospirilium brasilense*. J. Bacteriol. 175, 2507–2515.

Dean D and Jacobson M R 1992 Biochemical genetics of nitrogenase. *In* Biological Nitrogen Fixation. Eds. G Stacey, H J Evans adn R Burris. Chapman & Hall, New York, pp 763–834.

Defez R, Chiurazzi M, Manco G, Lamberti P, Riccio A, Lopes C, Colonna-Romnano S, Moreno S, Meza R, Espin G and Iaccarino M 1990 The glutamine synthetases of *Rhizobium leguminosarum* and their regulatory genes. *In* Nitrogen Fixation. Achievements and Objectives. Eds. P M Greshoff, J Roth, G Stacey and W E Newton. Chapman and Halll, New York, pp 715–716.

Dusha L, Bakos A, Kondorosi A, de Bruijn F J and Schell J 1989 The *Rhizobium meliloti* early nodulation genes (*nodABC*) are nitrogen- regulated: isolation of a mutant strain with efficient nodulation capacity on alfalfa in the presence of ammonium. Mol.Gen.Genet.21, 89–96

Espin G, Moreno S, Wild M, Meza R and Iaccarino M 1990 A previously unrecognized glutamine synthetase expressed in *Klebsiella pneumoniae* from the *glnT* locus of *Rhizobium leguminosarum*. Mol.Gen.Genet.223, 513–516.

Filser M, Moscatelli C, Lamberti A, Vincze E, Guida M, Salzano Gand Iaccarino M 1986 Characterization and cloning of two *Rhizobium leguminosarum* genes coding for glutamine synthetase activities. J.Gen.Microbiol.132, 2561–2569.

Foor F, Reuveny Z and Magasanik B 1980 Regulation of the synthesis of glutamine synthetase by the P_{II} protein in *Klebsiella aerogenes*. Proc.Acad.Sci. USA 77, 2636–2640.

Fostei-Hartnett D and Kranz R G 1992 Analysis of the promoters and upstream sequences of *nifA1* and *nifA2* in *Rhodobacter capsulatus* - Activation requires *ntrC* but not *rpoN*. Mol.Microbiol. 6, 1049–1060.

Foster-Harmett D, Cullen P J, Gabbert K K and Kranz R G 1993 Sequence, genetic, and *lacZ* fusion analyses of a *nifR3-ntrB-ntrC* operon in *Rhodobacter capsulatus*. Mol.Microbiol. 8, 903–914.

Fuchs R L and Keister D L 1980 Comparative properties of glutamine synthetase I and II in *Rhizobium* and *Agrobacterium* spp. J.Bacteriol. 144, 641–648.

Garcia E and Rhee S G 1983 Cascade control of *Escherichia coli* glutamine synthetase. J.Biol.Chem.258, 2246–2253

Hallenbeck P C 1992 Mutations affecting nitrogenase switch-off in *Rhodobacter capsulatus*. Biochim. Biophys. Acta 1118, 161–168.

Hawkins F K L, Kennedy C and Johnston A W B 1991 A *Rhizobium leguminosarum* gene required for symbiotic nitrogen fixation, melanin synthesis and normal growth on certain growth media. J.Gen.Microbiol.137, 1721–1728.

He B, Choi K Y and Zalkin H 1993 Regulation of *Escherichia coli glnB*, *prsA* and *speA* by the purine repressor. J.Bacteriol. 175, 3598–3606

Holtel H and Merrick M 1988 Identification of the *Klebsiella pneumoniae glnB* gene: nucleotide sequence of wild-type and mutant alleles. Mol.Gen.Genet.215, 134–138.

Holtel A and Merrick M J 1989 The *Klebsiella pneumoniae* PII protein (*glnB* gene product) is not absolutely required for nitrogen regulation and is not involved in NifL - mediated *nif* gene regulation. Mol.Gen.Genet.217, 474–480.

Keener J and Kustu S 1988 Protein kinase and phosphoprotein phosphatase activities of nitrogen regulatory proteins NTRB and NTRC of enteric bacteria: Roles of the conserved amino-terminal domain of NTRC. Proc.Natl.Acad.Sci.USA 85, 4976–4980.

Kennedy C and Toukdarian A 1987 Genetics of azotobacters: applications to nitrogen fixation and related aspects metabolism. Annu. Rev.Microbiol.41, 227–248.

Klipp W, Masepohl B and Puhler A 1988 Identification and mapping of nitrogen fixation genes of *Rhodobacter capsulatus*: Duplication of a *nifA-nifB* region. J.Bacteriol.17, 693–699.

Kranz R G, Pace W M and Caldicott I M 1990 Inactivation, sequence, and *lacZ* fusion analysis of a regulatory locus required for repression of nitrogen fixation genes in *Rhodobacter capsulatus*. J.Bacteriol.172, 53–62.

Kranz R G and Foster-Hartnett D 1990 Transcriptional regulatory cascade of nitrogen-fixation genes in anoxygenic photosynthetic bacteria: oxygen- and nitrogen-responsive factors. Mol.Microbiol.4, 1793–1800.

Kranz R G and Haselkorn R 1988 Ammonia-constitutive nitrogen fixation mutants of *Rhodobacter capsulatus*. Gene 71, 65–74.

Kustu S, Hirschman J, Burton D, Jelesko J and Meeks J C 1984 Covalent modification of bacterial glutamine synthetase: physiological significance. Mol.Gen.Genet .197, 309–317.

Leonardo J M and Goldberg R B 1980 Regulation of nitrogen metabolism in glutamine auxotrophs of *Klebsiella pneumoniae*. J. Bacteriol, 142, 99–110.

Liang Y Y, De Zamaroczy M, Arsene F, Paquelin A and Elmerich C 1992 Regulation of nitrogen fixation in *Azospirillum brasilense*Sp7: Involvement of *nifA, glnA* and *glnB* gene products. FEMS 100, 113–120

Luque F, Santero E, Medina J R and Tortolero M 1987 Mutants of *Azotobacter vinelandii* altered in the regulation of nitrate assimilation. Arch.Microbiol.148, 231–235.

Magasanik B 1982 Genetic control in nitrogen assimilation in bacteria. Annu. Rev. Genet. 16, 135–168.

Magasanik B 1988 Reversible phosphorylation of an enhancer binding protein regulates the transcription of bacterial nitrogen utilization genes. Trends Biochem.Sci.13, 475–479

Martin G B, Thomashow M F and Chelm B K 1989 *Bradyrhizobium japonicum glnB*, a putative nitrogen-regulatory gene, is regulated by NtrC at tandem promoters. J. Bacteriol 171, 5638–5645.

Merrick M J 1992 Regulation of nitrogen fixation genes in free-living and symbiotic bacteria. *In* Biological Nitrogen Fixation. Eds. G Stacey, H J Evans and R H Burris. Chapman and Hall, New York, pp 835–876.

Minchin S D, Austin S and Dixon R A 1988 The role of activator binding sites in transcriptional control of the divergently transcribed *nifF* and *nifLA* promoters from *Klebsiella pneumoniae*. Mol.Microbiol.2, 433–442

Ninfa A J and Magasanik B 1986 Covalent modification of the *glnG* product, NR$_I$, by the *glnL* product, NR$_{II}$, regulates the transcription of the *glnALG* operon in *Escherichia coli*. Proc.Natl.Acad.Sci.USA 83, 5909–5913.

Ninneman O 1992 The *E.coli fis* promoter is subject to stringent control and autoregulation. EMBO J. 11, 1075–1083.

Nixon B T, Ronson C W and Ausubel F M 1986 Two-component regulatory systems responsive to environmental stimuli share strongly conserved domains with the nitrogen assimilation regulatory genes *ntrB* and *ntrC*. Proc.Natl.Acad.Sci.USA 83, 7850–7854

Pate J S and Dart P J 1961 Nodulation studies in legumes. Plant and Soil 15, 329–345.

Patriarca E J, Chiurazzi M, Manco G, Riccio A, Lamberti A, De Paolis A, Rossi M, Defez R and Iaccarino M 1992 Activation of the *Rhizobium leguminosarum glnII* gene by NtrC is dependent on upstream DNA sequences. Mol.Gen.Gen. 234, 337–345.

Patriarca E J, Riccio A, Tate R, Colonna-Romano S, Iaccarino M amd Defez R 1994 The ntrBC genes of *Rhizobium leguminosarum* are part of a complex operon subject to negative autoregulation. Mol.Microbiol. (*In press*).

Pawlowski K, Klosse U and de Bruijn F J 1991 Characterization of a nove *Azorhizobium caulinodans* ORS571 two-component regulatory system, NtrY/NtrX involved in nitrogen fixation and metabolism. Mol.Gen.Genet 239, 124–138.A

Popham D L, Szeto D, Keener J and Kustu S 1989 Function of a bacterial activator protein that binds to transcriptional enhancers. Science 243, 629–635.

Reitzer L J and Magasanik B 1986 Transcription of *glnA* in *E. coli* is stimulated by activator bound to sites far from the promoter. Cell 45, 785–792.

Reitzer L J and Magasanik B 1987 Ammonia assimilation and the biosynthesis of glutamine, glutamate, aspartate, asparagine, L-alanine, and D-alanine. *In Escherichia coli* and *Salmonella typhimurium*. Cellular and Molecular Biology. Volume 1 Ed. F C Neidhardt. American Society for Microbiology, Washington DC. pp 302–320.

Rhee S G, Chock P B and Stadtman E R 1985 Nucleotidylations involved in the regulation of glutamine synthetase in *Escherichia coli*. *In* The Enzymology of Post-translational Modifications of Proteins. Volume 2. Eds. R B Freedman and H C Hawkins. Academic Press Inc., New York, pp 273.

Rhee S G, Park S C and Koo J H 1985b The role of adenylyltransferase and uridylyltransferase in the regulation of glutamine synthetase in *Escherichia coli*. Cur.Top.Cell.Reg. 27, 221–232.

Roberts G and Ludden P 1992 Nitrogen Fixation by Photosynthetic bacteria. *In* Nitrogen Fixation. Ed. G Stacey. Chapman and Hall, New York, pp 135–165.

Santero E, Toukdarian A, Humphrey R and Kennedy C 1988 Identification and characterization of two nitrogen fixation regulatory regions *nifA* and *nfrX* in *Azotobacter vinelandii* and *Azotobacter chroococcum*. Mol. Microbiol. 2, 303–314.

Shatters R G, Liu Y and Kahn M L 1993 Isolation and characterization of a novel glutamine synthetase from *Rhizobium leguminosarum*. J.Biol.Chem. 268, 469–475.

Sibold L, Henriquet M, Possot O and Aubert J.-P 1991 Nucleotide sequence of *nifH* regions from *Methanobacterium ivanovii* and *Methanosarcina barkeri* 227 and characterization of *glnB*-like genes. Res.Microbiol. 142, 5–12.

Somerville J E and Kahn D 1983 Cloning of the glutamine synthetase I gene from *Rhizobium meliloti*. J.Bacteriol. 156, 168–176.

Son H S and Rhee S G 1987 Cascade control of *Escherichia coli* glutamine synthetase. Purification and properties of P_{II} protein and nucleotide sequence of its structural gene. J.Biol.Chem. 262, 8609–8695.

Stock J B, Ninfa A J and Stock A M 1989 Protein phosphorylation and regulation of adaptive responses in bacteria. Microbiol.Rev.53, 450–49.

Streeter J 1988 Inhibition of legume nodule formation and nitrogen fixation by nitrate. Crit.Rev.Plant.Sci. 7, 1–23.

Streicher S L, Bloom F R, Foor F, Levin M and Tyler B 1977 *Klebsiella pneumoniae* and *Escherichia coli* mutants altered in nitrogen assimilation. Fed.Proc. 34, 300(Abstract).

Toukdarian A and Kennedy C 1986 Regulation of nitrogen metabolism in *Azotobacter vinelandii*: isolation of *ntr* and *glnA* genes and construction of *ntr* mutants. EMBO J.5, 399–407.

Van Heeswijk W, Kuppinger O, Merrick M and Kahn D 1992 Localization of the *glnD* gene on a revised map of the 200-kilobase region of the *Escherichia coli* chromosome. J.Bacteriol.174, 1702–1703.

Van Heeswijk W C, Rabenberg M, Westerhoff H V and Kahn D 1993 Genes of the glutamine synthetase adenylylation cascade are not regulated by nitrogen in *Escherichia coli*. Mol.Microbiol. 9.

Wang S P and Stacey G 1990 Ammonia regulation of *nod* genes in *Bradyrhizobium japonicum* Mol.Gen.Genet, 223, 329–331.

Weiss V and Magasanik B 1988 Phosphorylation of NR1 of *E coli*. Proc.Natl.Acad.Sci.USA 85, 8919–8923.

Plant and Soil **161**: 59–68, 1994.
© 1994 *Kluwer Academic Publishers.*

Regulation of nodulin gene expression

Frans J. De Bruijn[1,2,3], Rujin Chen[1,4], Susan Y. Fujimoto[1,3], Alexander Pinaev[1], David Silver[1,3] and Krzysztof Szczyglowski[1]

[1]*MSU-DOE Plant Research Laboratory,* [2]*Department of Microbiology,* [3]*Genetics Program and* [4]*Department of Biochemistry, Michigan State University, E. Lansing, MI 48824, USA*

Key words: cis-acting elements, *Enod2* gene, *gus* reporter genes, leghemoglobin (*lb*) gene, *Lotus corniculatus*, microbe-plant signalling, promoter analysis, *Sesbania rostrata*, transgenic plants

Abstract

The expression of plant genes specifically induced during rhizobial infection and the early stages of nodule ontogeny (early nodulin genes) and those induced in the mature, nitrogen-fixing nodule (late nodulin genes) is differentially regulated and tissue/cell specific. We have been interested in the signal transduction pathway responsible for symbiotic, temporal and spatial control of expression of an early (*Enod2*) and a late (Leghemoglobin; *lb*) nodulin gene from the stem-nodulated legume *Sesbania rostrata*, and in identifying the *cis*-acting elements and *trans*-acting factors involved in this process (De Bruijn and Schell, 1992). By introducing chimeric *S. rostrata lb* promoter-*gus* reporter gene fusions into transgenic *Lotus corniculatus* plants, we have been able to show that the *lb* promoter directs an infected-cell-specific expression pattern in *Lotus* nodules. We have been able to delimit the *cis*-acting element responsible for nodule-infected-cell-expression to a 78 bp region of the *lb* promoter (NICE Element) and have analyzed this element in detail by site-specific mutagenesis. We have studied the interaction of the NICE element, and further upstream *cis*-acting elements, with *trans*-acting factors of both plant- and rhizobial origin. We have obtained evidence for the involvement of rhizobial proteins in infected-cell-specific plant gene expression (Welters et al.,1993). We have purified one of the bacterial binding proteins from the *S. rostrata* symbiont *Azorhizobium caulinodans* (AcBBP1), and cloned and mutated the corresponding gene, in order to examine its symbiotic phenotype. We have also found that the *S. rostrata Enod2* gene is rapidly induced by physiologically significant concentrations of cytokinins, suggesting the role of cytokinin as a potential secondary signal involved in nodulation (Dehio and De Bruijn, 1992). We are examining whether the observed cytokinin induction, as well as the nodule-specific expression pattern, are modulated by the *SrEnod2* promoter.

Introduction

Biological nitrogen fixation is an extremely energy intensive process, utilizing up to 40 ATP molecules per dinitrogen molecule reduced. Since the rhizobia are aerobic bacteria, they generate the required ATP and low potential electrons via their respiratory chain(s), and therefore require considerable amounts of oxygen. However, the nitrogenase enzyme complex is extremely oxygen sensitive, with the polypeptides of the complex irreversibly denatured at oxygen concentrations of 15 n*M* or higher (Appleby, 1984). This presents a formidable problem for symbiotic nitrogen fixing systems, such as nodules on legume plants, and has been referred to as the "oxygen paradox".

We are interested in studying mechanisms, both in the nodule and in the rhizobial symbiont, that are involved in dealing with the oxygen paradox. Two of the mechanisms under study involve plant proteins that are specifically synthesized in developing and mature nodules (nodulins; Van Kammen, 1984). Nodulin genes are commonly divided in "early" and "late" groups, a division that reflects the rapidity of their expression. Early nodulins appear to be involved in the infection process and structural aspects of nodule ontogeny. Late nodulins are induced in the mature nodule, around the onset of nitrogen-fixation and several representatives of this group participate in aspects of nodule functioning, such as oxygen transport, nitrogen assimilation and carbon metabolism (see Nap and Bisseling, 1990). Nodulin genes are interesting plant genes to study, since they are not only developmentally regulated in response to signals coming directly or indirectly from the infecting rhizobia, but are also expressed in a tissue-(cell-) manner (see Nap and Bisseling, 1990; De Bruijn et al., 1990; De Bruijn and Downie, 1991; De Bruijn and Schell, 1992).

The first nodulin that is clearly involved in the oxygen paradox, leghemoglobin or Lb, is an oxygen transport protein that facilitates oxygen diffusion to the actively respiring, nitrogen-fixing bacteroids within the infected zone of the nodule (**the oxygen carrier**). The Lbs operate at an intracellular O_2 concentration of approximately 10 nM which is below the concentration known to irreversibly inactivate the nitrogenase enzyme complex (Appleby, 1984). Lbs may be truly symbiotic proteins, since in various symbiotic systems the heme component appears to be contributed by the rhizobia and the globin portion by the plant, but considerable controversy still exists about this hypothesis (see Sangwan and O'Brian, 1991; Pawlowski et al., 1993 and references therein). The second protein postulated to be involved in the oxygen paradox is Enod2, a proline-rich polypeptide consisting predominantly of two repeating pentapeptides (PPEYQ; PPHEK) and a putative signal peptide. It has been proposed to be a cell-wall protein, which may play a role in the creation of an oxygen diffusion barrier in the nodule (see Nap and Bisseling, 1990; **the oxygen barrier**). We have been studying the role of these proteins in symbiotic nitrogen fixation and are examining the structure and regulation of the genes encoding them, as well as the signal transduction pathway involved in their induction.

We have selected the *Azorhizobium caulinodans-Sesbania rostrata* symbiotic system for most of our studies on the *Enod2* and *lb* genes. This symbiosis is different from the more classical symbiotic systems in that nodules are not induced only on the roots but also on the stems of the plant (De Bruijn, 1989). Stem nodules contain bacteroid-filled (infected) cells immediately adjacent to cells harboring chloroplasts. The juxtaposition of an energy source (photosynthetic cells) and an energy sink (cells in which nitrogen fixation is taking place) has been postulated to be beneficial for the symbiosis; however, oxygen evolution associated with photosynthesis may put an additional strain on the oxygen sensitive nitrogenase enzyme (De Bruijn, 1989). Therefore, this symbiotic system has been of particular interest for studies on the oxygen paradox.

Our strategy for the *S. rostrata lb* (*Srglb3*) and *Enod2* (*SrEnod2*) gene regulation experiments has been to clone the genes, determine the DNA sequences of their 5' upstream regions, construct chimeric genes consisting of defined 5' upstream fragments fused to the glucuronidase (*uidA*; *gus*) reporter gene, introduce the chimeric constructs into *Lotus corniculatus*, via *Agrobacterium rhizogenes*- mediated transformation and to assay the expression of the reporter genes in nodules and other tissues of transgenic *Lotus* plants (for reviews see De Bruijn et al., 1990; De Bruijn and Schell, 1992). In addition, we have been examining the interaction of *cis*-acting elements in the *Srglb3* 5' upstream region with *trans*-acting factors (DNA binding proteins), using in vitro methods.

Here we will summarize our results on the characterization of a *cis*-acting element in the 5' upstream region of the *Srglb3* gene. This element confers Nodule-Infected-Cell-Expression in transgenic *Lotus* plants (NICE element; Szabados et al., 1990; De Bruijn and Schell, 1992; Szczyglowski et al., 1994) We will also discuss our findings that the NICE element, and a further upstream *cis*-acting element in the *Srglb3* 5′ upstream region (Bacterial Binding Site 1; BBS1), interact specifically with DNA binding proteins derived from the infecting rhizobia, suggesting a potential role for bacterial proteins in the symbiotic (infected-cell-specific) induction of the plant hemoglobin genes (Fujimoto et al., unpubl.; Welters et al., 1993). We will also present new data on the location of Matrix- (Scaffold-) Attachment Regions (MAR's; SAR's) in the 5′ and 3′ upstream regions of the *Srglb3* locus. We will then review our finding that *SrEnod2* gene expression is specifically and rapidly induced by cytokinin and discuss our attempts to show that the *SrEnod2* 5′ upstream region is involved in tissue- (nodule parenchyma-) specific- and cytokinin induced gene expression.

Results and discussion

Cis-acting elements involved in nodule specific expression of the S. rostrata *leghemoglobin (*glb3*) gene: Delimitation of the NICE element*

A functional analysis of the *S. rostrata* leghemoglobin *glb3* gene promoter region in transgenic *Lotus* plants revealed the presence of two enhancer-like, upstream, positive regulatory regions and an ATG-proximal regulatory element involved in nodule-specific gene expression (De Bruijn et al., 1990; De Bruijn and Schell, 1992; Szabados et al., 1990). By using a deletion and substitution analysis, DNA sequences in the *Sesbania rostrata* leghemoglobin *glb3* gene promoter region responsible for expression of chimeric *gus* (*uidA*) reporter genes in infected cells of nodules on transgenic *Lotus corniculatus* plants were delimited to a 78 bp *DraI-HinfI* fragment. This region, located between coordinates -194 to -116 relative to the start codon of the *Srglb3* gene, was named the **Nodule-Infected-Cell-Expression** (NICE) element (Szczyglowski

et al., 1994). The NICE element shares regions of DNA homology with the Organ Specific Element (OSE) and Negative Element (NE) of the soybean *lbc3* promoter, which have also been shown to be involved in nodule-specific expression (De Bruijn and Schell, 1992; see below, Stougaard et al., 1990)

Insertion of the NICE element into the truncated nopaline synthase (*nos*) promoter was found to confer a nodule-enhanced expression pattern on this normally root-enhanced promoter (Szczyglowski et al., 1994.) Similar results have been reported by Lauridsen et al. (1991) for the soybean *lbc3* gene. In this case, the OSE plus NE elements have been found to be able to enhance expression of the minimal CaMV 35S promoter, specifically in nodules.

Within the NICE element three distinct motifs [(A)AAAGAT, TTGTCTCTT and CACCC(T)] have been identified, that are highly conserved in the promoter regions of a variety of plant (leg)hemoglobin genes, including the OSE and NE elements of the soybean *lbc3* promoter (see De Bruijn and Schell, 1992; Sandal et al., 1987; Stougaard et al., 1990). In order to determine the possible role that these conserved motifs have in *lb* promoter activity, the *Srglb3* NICE element was subjected to extensive site-directed mutagenesis, and the expression pattern of 29 selected mutant promoter fragments was examined in transgenic plants. Mutations in the highly conserved (A)AAAGAT motif reduced *Srglb3* promoter activity to 25-50% of wild-type levels, while mutations in the TCTT portion of the TTGTCTCTT motif virtually abolished promoter activity, demonstrating the essential nature of these motifs for *Srglb3* gene expression. An A to T substitution in the CACCC(T) motif also abolished *Srglb3* promoter activity, while a C to T mutation at position four resulted in a reduction of promoter strength to 30% of the wild-type level, resembling the effect of analogous mutations in the conserved CACCC motif at approximately the same position in the promoter region of mammalian β-globin genes (Szczyglowski et al., 1994. see Figure 1).

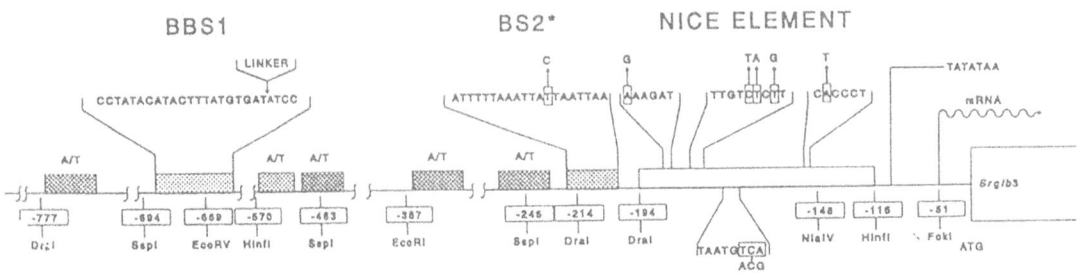

Fig. 1. **Structure of the** *S. rostrata glb3* **gene 5′ upstream region.** The positions of the BBS1 site (Welters et al., 1993), the BS2* site (Metz et al., 1988) and NICE element are indicated. The position of the conserved DNA motif AAAGAT, TTGTCTCTT and CACCCT within the NICE element are shown above the line, and the basepair mutations introduced in them are indicated by boxes plus vertical arrows. The positions of A/T rich motifs with high homology to BS2* are indicated by striped boxes labelled A/T (Welters et al., 1993). The restriction enzyme coordinates are indicated below the line in bp relative to the *glb3* translational start site (ATG). For further details, see text.

Extensive cytological Gus staining experiments revealed that expression of the *glb3* promoter could only be observed in the infected cells of the nodule (Szabados et al., 1990; Szczyglowski et al., 1994.) This finding suggests that a non-diffusible, *trans*-acting signal, possibly derived from the intracellular bacteria or bacteroids, may be required for *Srglb3* gene expression and possibly for late nodulin gene induction in general (see De Bruijn and Schell, 1992 and below).

Trans-acting factors interacting with DNA elements in the S. rostrata glb3 *5′ upstream region: The bacterial binding factors (BBFI and 2)*

Previously we have found a DNA sequence in the *glb3* 5′ upstream region (ATTTTTAAATTAT-TAATTAAA; Binding Site 2* or BS2*; Fig. 1), located immediately upstream of the NICE element, that specifically interacts with a proteinaceous DNA binding factor from nodule extracts (Metz et al., 1988). We also identified a region 570-700 bp upstream of the *glb3* start codon that did not contain BS2*-like sequences, but did interact specifically with a proteinaceous DNA binding factor from nodule extracts, but not from leaves or roots. The binding site for this factor was delimited by DNaseI footprinting. The DNA-binding activity of this factor was found to be heat-stable and dependent on divalent cations. Surprisingly, the factor was shown to be derived from the (infecting) *A. caulinodans* bacteria or

bacteroids (*A. caulinodans* Bacterial Binding Factor 1; AcBBFl; Welters et al., 1993).

A 9-10 kD protein was isolated from a free-living culture of *A. caulinodans* that co-purified with the DNA-binding activity (*A caulinodans* Bacterial Binding Protein 1; AcBBPl) and interacted specifically with its target (*S. rostrata* Bacterial Binding Site 1; SrBBSl; Fig. 1). The amino acid sequence of the N-terminal 27 residues of AcBBPl was determined and found to have 46% identity and 68% similarity with a domain of the herpes simplex virus major DNA binding protein ICP8. An insertion mutation in the SrBBSl was found to result in a substantial reduction of the expression of a *Srglb3-gus* reporter gene fusion in nodules of transgenic *L. corniculatus* plants, suggesting a role for this element in *Srglb3* promoter activity. Based on these results, we propose that (a) bacterial *trans*-acting factor(s) may play a role in (infected-cell-specific) expression of the symbiotically induced plant *lb* genes by migrating from the bacteroids into the nucleus and binding to a specific *cis*-acting element in the leghemoglobin promoter (Welters et al., 1993). This would constitute a unique example of prokaryote-eukaryote signalling.

The *A. caulinodans* gene encoding AcBBPl was cloned and characterized with the aid of oligonucleotides based on the N-terminal amino acid sequence of AcBBPl. The deduced AcBB-Pl amino acid sequence revealed the presence of a well defined helix-turn-helix motif at the C-terminal end, supporting the finding that AcBB-Pl efficiently binds DNA. A deletion/ substitution mutation was created in this locus by gene replacement. As determined by gel-shift assays, the resulting mutant strain lacked binding activity to the AcBBSl site in the *Srglb3* 5′ upstream region, confirming that it is a *bona fide* AcBBPl deficient mutant (Fujimoto et al., unpubl.). The mutant *A. caulinodans* strain was used to infect *S. rostrata* plants. Preliminary experiments revealed that the AcBBPl deficient strain was capable of inducing nitrogen fixing nodules on *S. rostrata* roots, but that nitrogen fixation may be reduced by up to 30% of the wild-type level (Fujimoto and De Bruijn, unpubl.).

These experiments clearly need to be repeated on a large scale, and the levels of *lb* transcripts and Lb proteins remain to be quantified, but the preliminary findings suggest that AcBBPl is not required for overall *lb* gene expression, but may play a role in the induction of specific *lb* genes (e.g. *Srglb3*) or play a role in optimizing *lb* gene expression in the nodule. This, in turn, would be consistent with the finding that an insertion mutation in the AcBBPl target site (AcBBSl) leads to a reduction, but not abolition, of *Srglb3* promoter activity in transgenic *Lotus* plants (Welters et al., 1993; see above). The latter transgenic *Lotus* plants, harboring a *Srglb3* promoter-*gus* fusion, are likely to be the most sensitive system to assay the effect of AcBBPl deficient mutations. Therefore, we are in the process of isolating such a mutation in the *Lotus* symbiont *Rhizobium loti*, which has been shown to produce a BBPI-like protein (Welters et al., 1993) and carries a DNA sequence with a high degree of homology to the cloned *A. caulinodans* BBPI gene, as determined by PCR analysis and DNA sequencing (Fujimoto and De Bruijn, unpubl.).

We also have examined the interaction of the NICE element with DNA-binding proteins using gel shift assays. The formation of two distinct complexes with extracts from root (Sr n) or stem (SrSn) nodules, but not from roots (SrR), was observed (Fig. 2). Two complexes of similar mobility were observed when extracts of free living *A. caulinodans* (Ac) were used for the gel shift assays. The DNA binding factor responsible for the observed complex formation was found both in luteolin-induced (Ac$^+$) and non-induced (Ac$^-$) cultures. This suggests that plant-derived nodulation (*nod*) gene inducers such as luteolin (see Hirsch, 1992) do not appear to affect the production of the DNA-binding factor. The *A. caulinodans* DNA-binding factor interacting with the NICE element was found to be different from the previously identified AcBBFl (Welters et al., 1993), in terms of heat stability and cation requirements, and was therefore designated AcBBF2. Again, the *in vivo* significance of this interaction has not been established and it is not clear which DNA sequences in the NICE element are responsible for the binding of this factor, or if there is any relationship between the putative binding site sequences and the conserved DNA sequence motifs shown to be essential for *Srglb3* promoter activity (see above, Fig. 1).

Chromatin structure of lb *loci and mapping of Matrix (Scaffold) Associated Regions (MAR's or SAR's)*

The experiments described above were designed to elucidate the molecular basis of nodule-specific gene expression and have been focused on the identification of short nucleotide sequences or even nucleotide base-pairs involved in *lb* gene expression and their interaction with DNA binding proteins of a putative regulatory nature. The relationship, however, between chromatin structure and nodule specific expression of genes, such as the *lb* loci of legumes, is not addressed in these types of experiments. We envision that chromatin structure analysis will constitute the next phase of our nodulin gene regulation studies and have initiated experiments to prepare the stage for such studies.

One of the important aspects of chromatin structure is the organization of chromosomes in loops and the role of scaffold-attachment- or matrix-attachment regions (SAR's; MAR's) in this process (see Slatter and Gray, 1991). These SAR's are not randomly distributed in chromosomal DNA and it has been suggested that highly

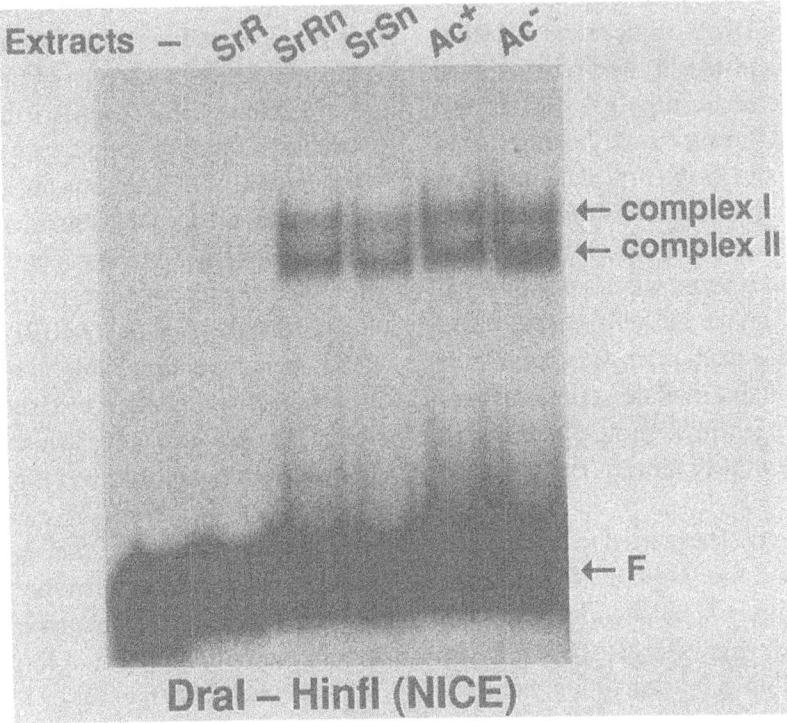

Fig. 2. **Interaction of Bacterial Binding Factor (BBF2) with the** *S. rostrata glb*3 **NICE element**. A photograph of the autoradiogram of a gel-shift experiment is shown. The experiment employed the radiolabelled *DraI-Hin*I fragment carrying the NICE element (see Fig. 1) and extracts from *S. rostrata* roots (SrR), root nodules (SrRn), stem nodules (SrSn), or *A. caulinodans* cell cultures induced with luteolin (Ac$^+$) and without luteolin (Ac$^-$). The gel shift assay was carried out as described (Welters et al., 1993). The position of the two complexes formed and the free fragment (F) are as indicated on the right.

transcribed genes tend to be located in smaller loops than are less frequently transcribed ones (Gasser, 1988; Slatter and Gray, 1991). Interestingly, SAR's have been found to contain A/T rich motifs (Gasser and Laemmli, 1986), that are very similar to the A/T motifs interacting with nuclear proteins that we and others have identified in the 5' upstream region of *lb* genes of both soybean and *S. rostrata* (De Bruijn et al., 1989, 1990; Jensen et al., 1988; Metz et al., 1988), the N23 gene of soybean (Jacobsen et al., 1990) and the *gln*γ gene of French bean (Forde et al., 1990). In fact, Jacobsen et al. (1990) have reported that DNA sequences in the soybean *lbc3* gene can bind to the *Drosophila* nuclear scaffold.

SAR's have also been found to be located in the 3' region of the pea plastocyanin gene (Slatter et al., 1991) and flanking several root- specific tobacco genes (Hall et al., 1991). SAR's often are located close to promoters and in the vicin-

ity of upstream enhancer sequences (Gasser and Laemmli, 1986; Slatter and Gray, 1991), suggesting that the location of DNA at the nuclear scaffold or matrix may be involved in transcriptional enhancement. Therefore, we have initiated experiments to map SAR's in the *S. rostrata glb*3 locus and correlate their positions with those of the protein-binding A/T sequences (BS2*-like elements; De Bruijn et al., 1989; De Bruijn and Schell, 1992), by using the "exogenous SAR assay described by Hall et al. (1991) and Cockerill and Garrard (1986).

For this purpose, the 17 kb insert of the original λ phage carrying the *Srglb*3 locus (λ*Srglb*3; Metz et al., 1988) was mapped and subcloned as overlapping restriction fragments (Fig. 3). The subfragments were radioisotope-labelled and employed for the exogenous SAR assay, using nuclear matrix prepared by DNaseI digestion and high salt extraction from *S. rostrata* leaf nuclei, as described by Cockerill and Garrard (1986).

Two distinct restriction fragments (a 2.65 kb *Hind*III-*Eco*RI fragment of pLB3EI/pLB3H2 and a 2.3 kb *Eco*RI fragment of pLB3H2; Fig. 3) containing the *Srglb3* gene and immediate 5' and 3' surrounding regions were found to interact specifically with the nuclear matrix, as evidenced by the recovery of specific labelled DNA fragments in the matrix pellet (P) versus the supernatant (S; Fig. 3). Other DNA fragments, such as the inserts of pLB3E3 and E4, did not interact with the nuclear matrix, as evidenced by the fact that all the input label led fragments (I) could be recovered in the supernatant (S; Fig. 3). These preliminary results suggest that there may be at least two distinct SAR like elements in close proximity to the *Srglb3* coding sequences (Pinaev and De Bruijn, unpubl.).

The position of SAR's probably topologically marks functional do mains of genes and therefore may give important information about the structure of the *glb3* locus. Moreover, it is important to map the positions of the SAR's relative to other *cis*-acting elements in the *glb3* promoter, in order to understand overall promoter architecture.

Cytokinin induction and tissue-specific expression of the S. rostrata *Enod2 gene (promoter)*

The *S. rostrata Enod2* gene (*SrEnod2*) is expressed around 8 days after infection and appears to be differentially expressed in stem-versus root nodules (Dehio and De Bruijn, 1992). The regulation of expression of early nodulin genes, such as the *SrEnod2* gene, appears to be quite different from the regulation of late nodulin genes, such as the *lb* genes De Bruijn et al., 1990; De Bruijn and Schell, 1992; see below). *Enod2* genes can be induced in the absence of infecting rhizobia, by an apparently diffusible signal, while *lb* gene activation appears to require the physical presence of bacteria (bacteroids) in the cytoplasm of infected cells (see De Bruijn and Schell, 1992; see above; Hirsch, 1992)

Hormones have been postulated to be the diffusible signals that activate early nodulin expres-sion (Hirsch, 1992; Verma, 1992). For example, auxin transport inhibitors (ATI's) have been found to induce nodule-like structures on alfalfa roots, in which the *Enod2* gene is expressed (Hirch, 1992). Therefore, we have previously examined the effect of plant hormones on *SrEnod2* expression. We were able to show that the *SrEnod2* gene is induced in the roots of *S. rostrata* seedlings treated with cytokinins (Dehio and De Bruijn, 1992). Similar results have been reported for the alfalfa *Enod2* gene by Hirsch et al. (1993). The cytokinin response appeared to be very specific, was observed with a variety of cytokinins and was time and concentration dependent (Dehio and De Bruijn, 1992).

In order to identify components of the signal transduction pathway responsible for nodule-specific and cytokinin-induced expression of the *SrEnod2* gene, we initiated studies on the *SrEnod2* promoter region using the same approach we previously used for the *Srglb3* promoter (De Bruijn et al., 1990; De Bruijn and Schnell, 1992; Szabados et al., 1990) Preliminary studies on the expression of chimeric *SrEnod2-gus* reporter gene fusions in transgenic *L. corniculatus* plants have thus far yielded inconclusive results. The majority of the constructs made, containing up to 3 kb of *SrEnod2* 5' upstream regions or deletion derivatives thereof, were poorly or non-expressed in transgenic *L. corniculatus* tissues. Only in a small subset of the independent transgenic plants harboring a particular construct with approximately 350 bp of *SrEnod2* 5' upstream region (AG738), nodule-specific Gus expression could be found, localized in the inner cortex of the nodule (nodule parenchyma; 4 Chen and De Bruijn, unpubl.), confirming the *in situ* RNA hybridization results of Van de Wiel et al. (1990).

In addition, when the roots of transgenic *L. corniculatus* plants, harboring a construct consisting of 3 kb of *SrEnod2* 5' upstream region fused to the *gus* gene, were treated with cytokinin, using the conditions described by Dehio and De Bruijn (1992), an induction effect was only observed in some of the independent transgenics (Chen and De Bruijn, unpubl.). These observations suggest that the *SrEnod2* promoter may be capable of

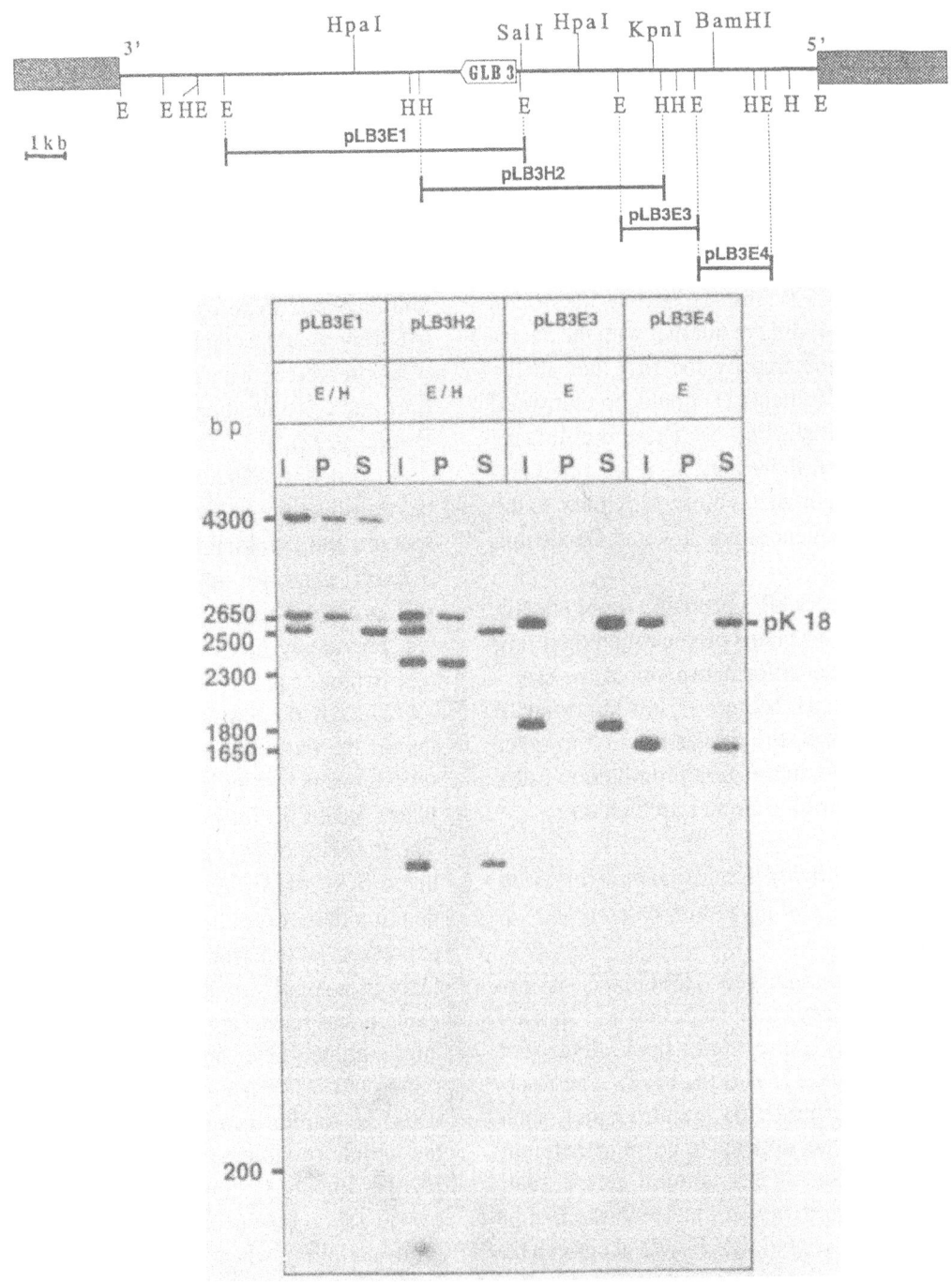

Fig. 3. **Exogenous SAR assay with fragments of the** *S. rostrata glb3* **locus.** The structure of the *Srglb3* locus, the position of restriction sites and the subcloned DNA fragments are shown in the top part of the panel. A computer image of an autoradiogram of a gel separating radioisotope-labelled *Srglb3* DNA fragments corresponding to the input DNA (I), DNA associated with the matrix pellet (P) or the supernatant (S) is shown in the bottom panel. The exogenous SAR assay was carried out as described by Cockerill and Garrard (1986). The length of the restriction fragments used is indicated on the left, in bp. The migration position of the pK8 vector fragment is indicated on the right. in the four cases shown, this vector fragment is always associated with the supernatant fraction (as expected) E indicates *EcoRl* and H, *Hin*dlll. For further details, see text.

conferring symbiotic induction and cell- (nodule parenchyma-) specific expression, in addition to being cytokinin induced, but that these are not very consistently observed phenomena in the experimental system used. The latter result appears to contrast with those reported for the soybean *Enod2* gene promoter, which has been found to be expressed consistently in the nodule parenchyma of both transgenic *L. corniculatus* (forming determinate nodules) and *Trifolium repens* (forming indeterminate nodules) plants (Lauridsen et al., 1993).

There are several possible explanations for the results we have obtained, including the generation of a mutation in the *SrEnod2* promoter region during the cloning/subcloning manipulations, limitations of the heterologous system used (a *S. rostrata Enod2* gene promoter in transgenic *L. corniculatus* plants) and the presence of important *cis*-acting elements further upstream of the 3 kb *SrEnod2* 5′ region used for our constructs. These different possibilities are presently being investigated. The question of whether rhizobial cytokinin production plays a direct role in *SrEnod2* gene induction is also currently under investigation.

Conclusions

It is clear from the published and unpublished experimental data reviewed above that although the nature of some of the *cis*-acting elements, *trans*-acting factors, and other signals involved in nodule(cell)-specific expression of early and late nodulin genes is beginning to be understood, there is still a tremendous amount of knowledge to be gathered about symbiotic control of plant gene expression. The final answers about the potential role of hormones (cytokinin) in early nodulin gene induction or the putative involvement of rhizobial DNA-binding proteins in late nodulin (*lb*) gene induction are still not known. Moreover the role of chromatin structure in nodule-specific gene expression has not even been seriously considered thus far. Nevertheless, we believe that the topic of regulation of nodulin gene expression is an exciting one due to the nature of the microbe-plant signalling involved, and particularly amenable to studies on the signal transduction pathways controlling plant gene expression, due to the availability of rhizobial mutants affected in the process.

Acknowledgements

We thank Uwe Rossbach, Dan Ragatz and Marlene Cameron for their help in preparing the figures. We also acknowledge financial support of the Department of Energy (DOE DE-FG02-91ER20021), the United States Department of Agriculture (USDA 92-37305-7735) and the National Science Foundation (NSF DCB 910592) for the studies described here.

References

Appleby C A 1984 Leghemoglobin and *Rhizobium* respiration. Annu. Rev. Plant Physiol. 35, 433–478.

Cockerill P N and Garrard W T 1986 Chromosomal loop anchorage of the kappa immunoglobulin gene occurs next to the enhancer in a region containing topoisomerase II sites. Cell 44, 273–282.

De Bruijn F J 1989 The unusual symbiosis between the diazotrophic, stem-nodulating bacterium *Azorhizobium caulinodans* ORS571 and its host, the tropical legume *Sesbania rostrata*. *In* Plant Microbe Interactions, Vol. 3. Eds. E Nester and T Kosuge. pp 457–493. McGraw-Hill, New York.

De Bruijn F J, Felix G, Grunenberg B, Hoffmann H-J, Metz B A, Ratet P, Simons-Schreier A, Szabados L, Welters P and Schell J 1989 Regulation of plant genes specifically induced in nitrogen-fixing nodules: *cis*-acting elements and *trans*-acting factors in leghemoglobin gene expression. Plant Mol. Biol. 13, 319–325.

De Bruijn F J , Szabados L and Schell J 1990 Chimeric genes and transgenic plants are used to study the regulation of plant genes involved in plant-microbe interactions (nodulin genes). Dev. Genet. 11, 182–196.

De Bruijn F J and Downie J A 1991 Biochemical and molecular studies of symbiotic nitrogen fixation. Curr. Opin. Biotech. 2, 184–192.

De Bruijn F J and Schell J 1992 Regulation of plant genes specifically induced in developing and mature nitrogen-fixing nodules: *Cis*-acting elements and *trans*-acting factors. *In* Control of Plant Gene Expression. Ed. D P S Verma. pp 241–258. CRC Press, Boca Raton, Florida.

Dehio C and De Bruijn F J 1992 The early nodulin gene *SrEnod2* from *Sesbania rostrata* is inducible by cytokinin. Plant J. 2, 117–128.

Forde B G, Freeman J, Oliver J E and Pineda M 1990 Nuclear factors interact with conserved A/T-rich elements upstream of a nodule enhanced glutamine synthetase gene from French bean. Plant Cell 2, 925–933.

Gasser S M 1988 Nuclear scaffold and the higher-order folding of eukaryotic DNA. *In* Architecture of eukaryotic genes. Ed. G Kahl. pp 461- 471. VHC Publishers, Weinheim, Germany.

Gasser S M and Laemmli U K 1986 Cohabitation of scaffold-binding regions with upstream enhancer elements of three developmentally regulated genes of *D. melanogaster.* Cell 46, 521–530.

Hall G, Allen G C, Loer D S, Thompson W F and Spiker S 1991 Nuclear scaffolds and scaffold-attachment regions in higher plants. Proc. Natl. Acad. USA 88, 9320–9324.

Hirsch A 1992 Developmental biology of legume nodulation. New Phytol. 122, 211–237.

Hirsch A M, Assad S, Fang Y, Wycoff K and Loebler M 1993 Molecular interactions during nodule development. *In* New Horizons in Nitrogen Fixation. Eds. R. Palacios, J Moira and W E Newton. pp 291–296. Kluwer Academic Publishers, Dordrecht,.

Jacobsen K, Laursen N B, Jensen E O, Marcker A, Poulsen C and Marcker K 1990 HMG 1-like proteins from leaf and nodule nuclei interact with different AT motifs in soybean nodulin promoters. Plant Cell 2, 85–94.

Jensen E O, Marcker K, Schell J and De Bruijn F J 1988 Interaction of a nodule-specific *trans*-acting factor with distinct DNA elements in the soybean leghemoglobin *lbc3* 5′ upstream region. EMBO J. 7, 1265–1271.

Lauridsen P, Sandal N, Kuehle A, Marcker K and Stougaard J 1991 Regulation of nodule specific genes. *In* Plant Molecular Biology 2. Eds. R G Herrmann and B Larkins. pp 131–137. Plenum Press, New York.

Lauridsen P, Franssen H, Stougaard J, Bisseling T and Marcker K 1993 Conserved regulation of the soybean early nodulin Enod2 gene promoter in determinate and indeterminate transgenic root nodules. Plant J. 3, 483–492.

Metz B A, Welters P, Hoffmann H-J, Jensen E O, Schell J and De Bruijn F J 1988 Primary structure and promoter analysis of leghemoglobin genes of the stem-nodulated tropical legume *Sesbania rostrata*: Conserved coding sequences, *cis*-elements and *trans*-acting factors. Mol. Gen. Genet. 214, 181–191.

Nap J P and Bisseling T 1990 Developmental biology of a plant-prokaryote symbiosis: The legume root nodule. Science 250, 948–954.

Pawlowski K, Gough S P, Kannangara C G and De Bruijn F J 1993 Characterization of a 5-aminolevulinic acid synthase mutant of *Azorhizobium caulinodans* ORS571. Mol. Plant-Micr. Interact. 6, 35–44.

Sandal N N, Bojsen K and Marcker K 1987 A small family of nodule-specific genes from soybean. Nucl. Acids Res. 15, 1507–1519.

Sangwan 1 and O'Brian M R 1991 Evidence for an interorganismal heme biosynthetic pathway in symbiotic soybean root nodules. Science 251 1220–1222.

Slatter R E and Gray J C 1991 Chromatin structure of plant genes. Oxford Surv. Plant Mol. Cell. Biol. 7, 115–142.

Slatter R E, Dupree P and Gray J C 1991 A scaffold-associated DNA region is located downstream of the pea plastocyanin gene. Plant Cell 3, 1239–1250.

Stougaard J, Joergenson J E, Christensen T, Kuhle A and Marcker K 1990 Interdependence and nodule specificity of *cis*-acting regulatory elements in the soybean leghemoglobin *lbc*3 and N23 promoters. Mol. Gen. Genet. 220, 353–360.

Szabados L, Ratet P, Grunenberg B and De Bruijn F J 1990 Functional analysis of the *Sesbania rostrata* leghemoglobin *glb*3 gene 5′ upstream region in transgenic *Lotus corniculatus* and *Nicotiana tabacum* plants. Plant Cell 2, 973–986.

Szczyglowski K, Szabados L, Fujimoto S Y, Silver D and De Bruijn F J 1994 Site-specific mutagenesis of the Nodule-Infected Cell Expression (NICE) element and the AT-Rich Element ATRE-BS2* of the *Sesbania rostrata* leg hemoglobin *glb*3 promoter. The Plant Cell 6 (*In press*).

Van de Wiel C, Scheres B, Franssen H, Van Lierop M-J, Van Lammeren A, van Kammen A and Bisseling T 1990 The early nodulin transcript *Enod2* is located in the nodule parenchyma (inner cortex) of pea and soybean nodules. EMBO J. 9, 1–7.

Van Kammen A 1984 Suggested nomenclature for plant genes involved in nodulation and symbiosis. Plant Mol. Biol. Rep. 2, 43–45.

Verma D P S 1992 Signals in root nodule organogenesis and endocytosis of *Rhizobium*. Plant Cell 4, 373–382.

Welters P, Metz B A, Felix G, Pal me K, Szczyglowski, K and De Bruijn F J 1993 Interaction of a rhizobial DNA binding protein with the promoter region of a plant leghemoglobin gene. Plant Physiol. 102, 1095–1107.

Plant and Soil **161**: 69–80, 1994.
© 1994 *Kluwer Academic Publishers.*

Synthesis, release, and transmission of alfalfa signals to rhizobial symbionts

D. A. Phillips[1], F. D. Dakora[2], E. Sande[1], C. M. Joseph[1] and J. Zoń[3]
[1]*University of California, Davis, USA,* [2]*University of Cape Town, Rondebosch 7700, South Africa, and*
[3]*Politechnika Wrocławska, Wrocław, Poland*

Key words: betaines, flavonoids, *Medicago sativa*, nodulation, *Rhizobium meliloti*

Abstract

In addition to the flavonoids exuded by many legumes as signals to their rhizobial symbionts, alfalfa (*Medicago sativa* L.) releases two betaines, trigonelline and stachydrine, that induce nodulation (*nod*) genes in *Rhizobium meliloti*. Experiments with ^{14}C-phenylalanine in the presence and absence of phenylalanine ammonia-lyase inhibitors show that exudation of flavonoid *nod*-gene inducers from alfalfa roots is linked closely to their concurrent synthesis. In contrast, flavonoid and betaine *nod*-gene inducers are already present on mature seeds before they are released during germination. Alfalfa seeds and roots release structurally different *nod*-gene-inducing signals in the absence of rhizobia. When *R. meliloti* is added to roots, medicarpin, a classical isoflavonoid phytoalexin normally elicited by pathogens, and a *nod*-gene-inducing compound, formononetin-7-*O*-(6"-*O*-malonylglycoside), are exuded. Carbon flow through the phenylpropanoid pathway and into the flavonoid pathway via chalcone synthase is controlled by complex *cis*-acting sequences and *trans*-acting factors which are not completely understood. Even less information is available on molecular regulation of the two other biosynthetic pathways that produce trigonelline and stachydrine. Presumably the three separate pathways for producing *nod*-gene inducers in some way protect the plant against fluctuations in the production or transmission of the two classes of signals. Factors influencing transmission of alfalfa *nod*-gene inducers through soil are poorly defined, but solubility differences between hydrophobic flavonoids and hydrophilic betaines suggest that the diffusional traits of these molecules are not similar. Knowledge derived from studies of how legumes regulate rhizobial symbionts with natural plant products offers a basis for defining new fundamental concepts of rhizosphere ecology.

Introduction

Initial discoveries that plant signals induce nodulation (*nod*) genes in *Rhizobium* and *Bradyrhizobium* (collectively referred to here as rhizobia) identified various flavonoids as the active molecules (Firmin et al., 1986; Kossalak et al., 1987; Peters et al., 1986; Redmond et al., 1986). Other investigators have since demonstrated that diverse groups of flavonoids are released as *nod*-gene inducers by some legumes. In addition to the flavonoid *nod*-gene inducers isolated during initial studies, chalcones (Maxwell et al.,1989; Recourt et al., 1991), anthocyanidins (Hungria et al., 1991), and conjugated isoflavonoids (Dakora et al., 1993) with *nod*-gene-inducing activities have subsequently been identified. The plant signals induce transcription of rhizobial *nod* genes by activating NodD proteins produced from regulatory *nodD* genes (Fisher and Long, 1992).

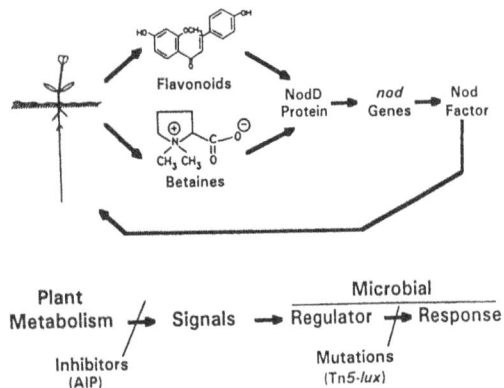

Fig. 1. Representative signals transmitted between alfalfa and *Rhizobium meliloti* bacteria. Inhibiting synthesis of the plant signal and blocking the induced response in the bacterium offer mechanisms for defining how these processes are integrated into the intact organisms. Structures of 4,4'-dihydroxy-2'-methoxychalcone from roots and stachydrine (*N,N*-dimethylproline) from seeds are shown as representative signals.

Some of these compounds also serve broader roles by inducing positive chemotaxis in rhizobia (Caetano-Anollés et al., 1988; Dharmatilake and Bauer, 1992).

Recent work in alfalfa (*Medicago sativa* L.) identified another chemical class of natural rhizobial *nod*-gene inducers in addition to flavonoids. Seeds of this plant species also release two betaines, stachydrine and trigonelline, which apparently induce transcription of *nodC-lacZ* fusions in *R. meliloti* by activating NodD2 protein (Phillips et al., 1992). Previous *nod*-gene inducers from alfalfa were identified by their capacity to activate NodD1 protein in *R. meliloti* (i.e. the indicator strains contained extra copies of *nodD1* genes) (Hartwig et al., 1990a; Maxwell et al., 1989; Peters et al., 1986). The structural diversity of *nod*-inducing signals from alfalfa, as well as the presence of multiple *nodD* genes in *R. meliloti* (Göttfert et al., 1986; Honma and Ausubel, 1987), makes the *R. meliloti*-alfalfa symbiosis an intriguing interaction for study and justifies that association as the focus of this review.

The *R. meliloti* response to alfalfa *nod*-gene inducer molecules has been well characterized (Dénarié et al., 1992), and Nod factors formed by products of the *nod* genes complete a signal-response cycle for *R. meliloti*-alfalfa interactions

(Fig. 1). The purpose of this review is to discuss major factors influencing the synthesis, release, and transmission of *nod*-gene-inducing signals from alfalfa to *R. meliloti*. Additional supporting details or contrasting responses that have been documented in other rhizobial-legume systems are noted.

Known *nod*-gene inducers released by alfalfa

Alfalfa seeds release four major *nod*-gene inducers (two flavones and two betaines) and at least four other flavonoids that fail to induce *nod*-genes in *R. meliloti* (Table 1). Methoxylation of the C-5 hydroxyl on luteolin eliminates the *nod*-gene-inducing activity. Two compounds with that modification are released in significant quantities by alfalfa seeds (3',5-dimethoxyluteolin and 5-methoxyluteolin), but no signaling function has been identified for them. Quercetin, which differs from luteolin by the presence of a C-3 hydroxyl group, has no *nod*-gene-inducing activity in *R. meliloti*, but it does increase growth rate of this organism on minimal media (Hartwig et al., 1991). This fact suggests that under natural conditions quercetin may help *R. meliloti* cells compete against other soil microbes for colonization of the seedling rhizosphere.

The active alfalfa seed flavones and betaines apparently induce *nodC-lacZ* fusions through different regulatory systems because multiple copies of *nodD1* genes are required in *R. meliloti* cells for activity of luteolin and chrysoeriol, while the betaines induce β-galactosidase only with multiple copies of *nodD2* genes. Whether the large difference in active concentrations of these two classes of molecules (nM for flavones vs. μM for betaines) is explained solely by their apparent activation of different NodD proteins is unknown. It is conceivable that the lipophilic flavonoids enter bacterial cells much more efficiently during the assay period than the hydrophilic betaines. Attempts to understand the mixture of compounds on alfalfa seeds by identifying interactions between the inactive *nod*-gene inducer quercetin and the two betaines have produced

Table 1. Rhizobial *nod*-gene inducers and related compounds released by alfalfa

Alfalfa compound	NodD activated	Amount released (pmol/plt h)$^{-1}$	Reference
Seed rinse			
Luteolin	D1[a]	70	1,2[b]
Chrysoeriol	D1	10	2
3',5-Dimethoxyluteolin	none	300	2
Luteolin-7-*O*-glucoside	none	800	2
5-Methoxyluteolin	none	825	2
Quercetin-3-*O*-galactoside	none	1,800	3
Trigonelline	D2	2,500	4
Stachydrine	D2	2,500	4
Root exudate			
4,4'-Dihydroxy-2'-methoxychalcone	D1&2	1	5
4',7-Dihydroxyflavone	D1	2	5
Liquiritigenin	D1	1	5
Formononetin	none	12	6
Formononetin-7-*O*-(6"-*O*-malonylglycoside)	D1&2	nd[c]	7
Medicarpin	none	174	8
Medicarpin-3-*O*-glycoside	none	349	8
Formononetin-7-*O*-glycoside	D1&2	nd	9

[a]Assayed in *R. meliloti* containing extra *nodD1* or *nodD2* (Hartwig et al., 1990b).
[b]References: 1, Peters et al., 1986; 2, Hartwig et al., 1990a; 3, Hartwig et al., 1991; 4, Phillips et al., 1992; 5, Maxwell et al., 1989; 6, Maxwell & Phillips, 1990; 7, Dakora et al., 1993; 8, Identified in Dakora et al., 1993 and quantified for this report; 9, León-Barrios et al., 1993;
[c]nd, not determined.

no new insights. One would like to know, for example, whether quercetin can activate NodD1 or NodD2 protein in the presence of charged betaines which might produce conformational changes in these proteins. However, the presence of trigonelline did not facilitate activation of NodD2 protein by quercetin, and quercetin did not allow trigonelline to activate NodD1 protein (Phillips et al., 1993).

Various flavonoid *nod*-gene inducers are released from alfalfa roots, but no trace of trigonelline or stachydrine has been found in root exudates of 72-hour-old seedlings. The 4,4'-dihydroxy-2'-methoxychalcone released by roots induces transcription of *nodC-lacZ* fusions in *R. meliloti* at a 10-fold lower concentration than lute-olin (Maxwell et al., 1989), and it gives an apparent activation of both NodD1 and NodD2 protein in strains containing pSym and extra copies of those genes. The 4',7- dihydroxyflavone and liquiritigenin (4',7-dihydroxyflavanone) induce *nodC-lacZ* fusions at concentrations similar to luteolin in the presence of extra *nodD1* genes. The isoflavone formononetin is inactive as a *nod*-gene inducer, but a formononetin-7-*O*-(6"-*O*-malonylglycoside) activated both NodD1 and NodD2 protein. Hydrolytic cleavage of either the malonyl or the sugar from the isoflavonoid, which was verified by ^{1}H-NMR and MS analyses, eliminated *nod*-inducing activity, and neither the sugar nor the malonyl moieties by themselves induced transcription of *nodC-lacZ* fusions (Dakora et al., 1993). It was therefore surprising that an undefined formononetin-7-*O*-glycoside isolated from alfalfa rhizosphere soil showed *nod*-gene-inducing activity without any ^{1}H-NMR or MS evidence for the labile malonyl group (León-Barrios et al. 1993). Because all results with isoflavonoids were obtained with samples purified in trace amounts from root exudates and soil extracts, none of the sugar conjugates was identified conclusively. Explanations for differences in *nod*-gene-inducing activity of the glycoside must await the availability of larger quantities for analysis.

Trigonelline occurs widely in plants (Tramontano et al., 1986), where it and stachydrine may serve as a cytoplasmic osmoticum associated with drought (Wyn Jones and Storey, 1981). Under osmotic stress, stachydrine is taken up by *R. meliloti* and increases growth rate dramatically (Bernard et al., 1986). Trigonelline, however, had a much smaller beneficial effect in the same experiments. An uptake system for stachydrine, which is inhibited by a 100-fold excess of trigonelline, has been characterized under salt stress in free-living *R. meliloti* cells (Gloux and Le Rudulier, 1989), and stachydrine is also transported into bacteroids (Fougère and Le Rudulier, 1990). *R. meliloti* contains genes on pSym for catabolism of trigonelline and stachydrine (Boivin et al., 1991; Goldmann et al., 1991), and both betaines are found in water-stressed alfalfa root nodules (Fougère and Le Rudulier, 1990; Jones et al., 1986). This latter point suggests that these compounds may affect bacteroid functions in the nodule, including possibly N_2 fixation. Early tests with trigonelline showed weak *nod*-inducing activity in *R. meliloti* (Schmidt et al., 1986), but its special role as a NodD2 activator was discovered only after it was identified as a natural component of alfalfa seed rinse (Phillips et al., 1992).

Synthesis of plant flavonoid signals to rhizobia

Flavonoid biosynthetic pathways have been documented in many plants (Stafford, 1990), and there

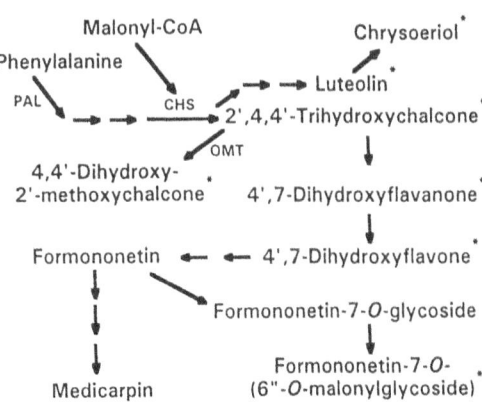

Fig. 2. Postulated pathways for biosynthesis of flavonoid *nod*-gene inducers released from alfalfa under controlled conditions. Compounds inducing *nod* genes in *R. meliloti* are marked with asterisks. Supporting evidence is discussed in the text. PAL, phenylalanine ammonia-lyase; CHS, chalcone synthase; OMT, *S*-adenosyl-L-methionine: isoliquiritigenin 2'-*O*-methyltransferase.

is no reason to postulate that alfalfa is unique. In all cases, flavonoids are produced through chalcone synthase (CHS) by condensation of 4-coumaroyl-CoA, which is derived from phenylalanine, and three malonyl-CoA molecules. As the first enzyme in the flavonoid pathway, CHS is an important regulatory point (see below). The initial molecule formed by CHS is hydroxylated at the C-4,4',6' positions of the chalcone or the corresponding C-4',7,5 positions in the flavone. Possible biosynthetic relationships among flavonoid *nod*-gene inducers reported from alfalfa (Fig. 2) reflect a dichotomy between the 5-hydroxy luteolin-chrysoeriol pathway and the 5-deoxy compounds found in root exudates. In *Glycyrrhiza* (Ayabe et al., 1988) and soybean (Welle and Grisebach, 1989) this difference is produced by an NAD(P)H reductase that functions with CHS to eliminate the 6'-hydroxyl on the resulting chalcone. A similar enzyme probably functions in alfalfa roots, but regulation of this enzyme is not known. Indeed, even the presence of such an enzyme in alfalfa has only been inferred by the presence of the 5-deoxy *nod*-gene inducers.

Although 2',4,4'-trihydroxychalcone (isoliquiritigenin) induces *nod* genes in *R. meliloti* (Maxwell et al. 1989), it has not been reported in alfalfa root exudates. Its methoxylated derivative, 4,4'-dihydroxy-2'-methoxychalcone, however, is present in root exudate and is the most active *nod*-gene inducer yet isolated in this system (I_{50} = 1 n*M*; Maxwell et al., 1989). The extremely active nature of the methoxychalcone suggests that the enzyme responsible for its synthesis from 2',4,4'-trihydroxychalcone may be very important for establishing an optimum symbiotic association with *Rhizobium*. Recent success in purifying the S-adenosyl-L-methionine: isoliquiritigenin 2'-*O*-methyltransferase (OMT) responsible for that methoxylation (Maxwell et al., 1992) has led to the cloning of the OMT structural gene from alfalfa (Maxwell et al., 1993). It will be interesting to examine the symbiotic properties of transgenic plants with elevated levels of this potentially important enzyme.

Regulating synthesis of plant flavonoid signals to rhizobia

The presence of *Rhizobium* cells is a major factor that clearly affects production of *nod*-gene-inducing molecules in several legumes. Rolfe et al. (1988) observed a *Rhizobium*-dependent increase in *nod*-gene-inducing activity in extracts of white clover, and recent results show that major flavonoid changes caused by rhizobia in *Lotus* are localized at the site of nodule formation (Cooper and Rao, 1992). Inoculating *Vicia* with *R. leguminosarum* biovar *viciae* increased exudation of *nod*-inducing chalcones and flavanones from the root (Recourt et al., 1991), but no isoflavonoids typical of those elicited by pathogens were observed. Comparable experiments with alfalfa showed that inoculation with symbiotic *R. meliloti* cells enhanced *nod*-gene-inducing activity in the root exudate and also elicited exudation of medicarpin, a classical phytoalexin that normally is produced by alfalfa in the presence of pathogens (Dakora et al., 1993). Carbon for medicarpin synthesis may come from stored formononetin (Maxwell and Phillips, 1990) or from new flavonoids formed by CHS. Evidence from cell cultures of alfalfa

(Dalkin et al., 1989) and *Canavalia* (Gustine et al., 1978) shows that medicarpin accumulation in the presence of microbial pathogens is associated with general increases in phenylpropanoid and flavonoid metabolism. Soybean root nodules contain the isoflavonoid phytoalexin glyceollin (Parniske et al., 1990), but sterile soybean seedlings, unlike alfalfa, exude isoflavonoid *nod*-gene inducers (Kosslak et al., 1987) and coumestrol (D'Arcy-Lameta 1986). Thus rhizobia are not required for release of isoflavonoids in soybean.

Several elicitors from rhizobia may initiate phytoalexin production. A role for Nod factors in eliciting additional flavonoid production from *Vicia* was cited by Spaink (1992). However, because similar responses are elicited by pathogens (Dixon and Lamb, 1990), rhizobial cell wall components present in many bacteria may also be involved in this process. Presumably the rhizobial signal is transduced through some mechanism, perhaps involving lectin, to promote the production and/or release of phytoalexins. Such observations suggest that rhizobia are perceived initially by alfalfa at the molecular level as a pathogen.

Detailed molecular studies indicate that external microbes regulate transcription of CHS in legumes. At least six functionally different copies of CHS are present in *P. vulgaris* (Ryder et al., 1987). Inoculating an infective *R.l.* bv. *viciae* strain on *vicia* increased CHS mRNA, but uninfective rhizobia had no effect (Recourt et al., 1992). Experiments in transgenic alfalfa cells with a chimeric gene containing a CHS promoter from common bean fused to a bacterial chloramphenicol acetyltransferase (CAT) reporter gene showed that the CAT gene was strongly induced by a fungal elicitor (Harrison et al., 1991a; Loake et al., 1991). The CHS promoter contained *cis*-acting sequences that functioned as both activators and silencers in the presence of *trans*-acting factors, and exogenous products from the phenylpropanoid pathway affected the regulation. When the *Phaseolus* CHS promoter was studied in soybean cells, a soybean nuclear protein that binds to the silencer region was purified and characterized (Harrison et al., 1991b). Evidence from soybean suggests that symbiotic and pathogenic microbes may affect different CHS and phenylalanine ammonia-lyase promoters (Estabrook and Sengupta-Gopalan, 1991). Such experiments support the concept that external biological factors, like symbionts and pathogens, may alter total flavonoid synthesis through molecular effects on key promoters. Presumably bacterial elicitors function parallel to, or even through the same regulatory systems as, the fungal elicitors.

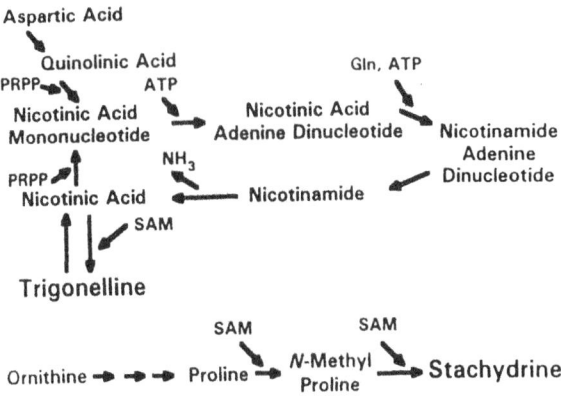

Fig. 3. Postulated pathways for biosynthesis of betaine *nod*-gene inducers released naturally from alfalfa. Supporting evidence is discussed in the text. PRPP, 5-phosphoribosyl-1-pyrophosphate; SAM, S-adenosylmethionine.

Synthesis of plant betaine signals to rhizobia

Betaines are a diverse group of molecules united primarily by the presence of both a positive and negative charge under physiological conditions. Trigonelline and stachydrine, the two betaine *nod*-gene inducers released by alfalfa seeds, are structurally related through N-methylations of nicotinic acid and proline, respectively. For this reason, one might postulate a competition for the methyl-transfer agent S-adenosylmethionine (SAM). However, the SAM: nicotinic acid-*N*-methyltransferase from soybean cells is very specific for nicotinate (Upmeier et al., 1988), and thus separate enzymes probably are responsible for methylating the two betaine *nod*-gene inducers. Many studies of stachydrine biosynthesis have used alfalfa as an experimental organism since it was first reported in that plant (Steenbock, 1918), but less is known about trigonelline formation in alfalfa. There is no published evidence that these betaines are synthesized in *R. meliloti*.

Data from biosynthesis and metabolism studies in *Lemna* indicate that trigonelline is formed directly from nicotinic acid and SAM by nicotinate methyltransferase (Taguchi et al., 1989a). Those workers outlined a pathway in which nicotinate is produced from quinolinic acid as part of a pyridine nucleotide cycle (Fig. 3). It was assumed that quinolinic acid was derived from aspartate. Subsequent metabolic experiments in *Lemna* supported the concept that trigonelline serves as a storage form of nicotinate (Taguchi et al., 1989b). Although no direct evidence for

a trigonelline demethylase was reported in *Lemna*, the enzyme has been found in many systems (Taguchi and Shimabayashi, 1983), and metabolism of [7-14C]-trigonelline by *Lemna* indicated the presence of a catabolic system.

Stachydrine can be extracted easily from alfalfa seedlings, and many attempts have been made to define a complete pathway for biosynthesis of this betaine. Regrettably, the postulated pathway (Fig. 3) is not completely supported by experimental data. After feeding 7-day-old plants with [2-14C]-ornithine, 14C-proline was isolated, but no label was found in stachydrine (Morgan and Marion, 1956). Using the same plant system, Wiehler and Marion (1958) found that label from [Me14C]-methionine was incorporated into stachydrine when plants were simultaneously fed folic acid. One might construe from those data that methionine molecules donate sequential methyl groups to proline and *N*-methylproline to form stachydrine (*N,N*-dimethylproline). However, later experiments showed that *N*-methylproline, but not proline alone, was converted to stachydrine (Robertson and Marion, 1960). Thus the logical conversion of proline to *N*-methylproline remains to be demonstrated. Assays with the more recently discovered, powerful methyl-transfer agent SAM might resolve that problem. Such studies should be conducted with shoot tissue because apparently only green tissue synthesizes stachydrine

Table 2. Effects of inhibiting phenylalanine ammonia-lyase on nodulation and exudation of *nod*-gene inducers from alfalfa

Chemical treatment	Nodulation (nodules/plant)	*nodC-lacZ* induction by root exudate (units β-galactosidase/plant)
None	2.54±0.29	81.9±9.4
30 μM AIP[a]	0.05±0.01	14.2±2.1
30 μM AIP + 1 μM 4',7-dihydroxyflavanone	0.12±0.07	Not determined

[a]AIP - 2-aminoindan-2-phosphonic acid. AIP, 4',7-dihydroxyflavanone, and *R. meliloti* strain 102F28 were supplied to hydroponic nutrient solution around roots of 1-day-old Moapa 69 seedlings. Root exudates in the nutrient solution were collected on day 4, and *nod*-inducing activity was assayed in *R. meliloti* 1021pRmM57. Nodules were counted on day 12.

(Sethi and Carew, 1974). No integrative studies on the molecular regulation of betaine synthesis by legumes are available.

Release of plant signals to rhizobia

Studies with alfalfa provide little evidence that the plant controls release of *nod*-gene inducers from seeds. Rinsing seeds with 50% methanol removed large quantities of flavonoids without producing normal hydration events associated with germination (Hartwig and Phillips, 1991). Thus it seems probable that flavonoids deposited on the seed coat during seed formation are rinsed off during imbibition without any regulatory controls that can be supplied through metabolic processes.

Roots, in contrast, control the storage and release of flavonoid *nod*-gene inducers. Alfalfa seedling roots, like many legume cells (Köster et al., 1983), contain conjugated forms of the isoflavonoid formononentin (Maxwell and Phillips, 1990), and exposing the plant to *R. meliloti* promotes release of a formononetin-7-*O*-(6"-*O*-malonylglycoside) (Dakora et al., 1993). Short-term (10-hour) experiments with the phenylalanine ammonia-lyase (PAL) inhibitor 2-aminooxy-3-phenylpropionic acid (AOPP) (Amrhein and Gödeke, 1977) showed that [U-^{14}C]-phenylalanine was rapid-

ly incorporated into flavonoid *nod*-gene inducers and exuded from roots in the absence of AOPP (Maxwell and Phillips, 1990). In the presence of AOPP, synthesis of the inducers was decreased by more than 95%, but exudation of 4',7-dihydroxyflavanone was inhibited by only 50%. The presence of so much unlabelled 4',7-dihydroxy-flavanone in root exudate from seedlings treated with AOPP indicated that this compound could be exuded from storage pools within the root. Exudation of the other *nod*-gene inducers, 4,4'-dihydroxy-2'-methoxychalcone and 4',7-dihydroxyflavone, was more tightly linked to concurrent synthesis.

Longer term experiments with the more powerful PAL inhibitor, 2-aminoindan-2-phosphonic acid (AIP) (Zoń and Amrhein, 1992) showed that inhibiting PAL activity for a longer period decreased root nodulation markedly (Table 2). If betaine *nod*-inducers play an important role in root exudates, then decreasing flavonoid exudation with a PAL inhibitor should have little effect on root nodulation. At this time, however, there is no evidence for either trigonelline or stachydrine in alfalfa root exudate. Although nicotinate has been reported in alfalfa root exudate on the basis of bioassays (Rovira and Harris, 1961), neither nicotinate nor proline induce transcription of a *nodC-lacZ* fusion in *R. meliloti* cells containing extra copies of *nodD2* (Phillips et al., 1993).

Unfortunately, the AIP inhibition of nodulation was not overcome by adding the normal *nod*-gene inducer 4',7-dihydroxyflavanone. The latter fact suggests that AIP may have disturbed other physiological processes that are required for normal root nodule development. Thus it would be inappropriate to conclude that the 98% inhibition of root nodulation associated with the AIP treatment necessarily reflects an absence of nonflavonoid *nod*-gene inducing compounds in root exudates.

Transmission of plant signals to rhizobia

Little is known about the fate of *nod*-gene-inducing compounds after they leave the plant. A more complete understanding of how plant signals are transmitted through soil would help relate facts about the genetic regulation of microbial genes by plant signals to the poorly understood ecological forces that control root colonization and nodulation. A recent review noted reports on the occurrence of simple phenolics and flavonoids in soils (Siqueira et al., 1991), and it was reassuring that *nod*-gene-inducing activity was extracted easily from soil around an alfalfa plant (León-Barrios et al., 1993). The decreasing concentrations of *nod*-gene-inducing activity extracted at greater distances from the rhizosphere in that study supported the concept that alfalfa roots were the original source of the compounds.

Structural differences between lipophilic flavonoids and amphoteric betaines suggest that these two classes of *nod*-gene inducers move differently in soils. Those diffusional patterns undoubtedly vary with chemical characteristics of the soil. Charged betaines are much more soluble than flavonoids in water, but they may be immobilized by ionic forces on some soil particles. Flavonoids, in contrast, are poorly soluble in water, but because they are biologically active at such low concentrations, that fact may not restrict their role as signals in the soil. Whether soil lipids facilitate flavonoid diffusion has not been addressed, but lipids comprise as much as 30% of soil organic matter and are found in most soils (Amblès et al. 1991).

Germinating alfalfa seeds release glycosidase activity that hydrolyzes luteolin-7-*O*-glucoside to luteolin (Hartwig and Phillips, 1991). Limited data from the same study support the concept that rhizobia can catalyze such reactions as a means of obtaining glucose for energy. A similar rationale suggests that any *nod*-gene-inducing compound can represent a source of energy to soil microbes. Rhizobia fit such a model in that they can degrade 100 μM to 10 mM concentrations of phenolic substances (Barz, 1970; Gajendiran and Mahadevan, 1990; Rao et al., 1991), but there are no published data showing disappearance of flavonoids at the much lower concentrations which can induce *nod* genes. Not all flavonoids are degraded easily in soil. For example, medicarpin can accumulate under alfalfa stands and inhibit alfalfa seed germination and seedling growth (Dornbos et al., 1990). The absence of detectable amounts of 4,4'-dihydroxy-2'-methoxychalcone, 4',7-dihydroxyflavanone, or 4',7-dihydroxyflavone in extracts from alfalfa rhizosphere soil (León-Barrios et al., 1993) offers indirect evidence for degradation of these alfalfa *nod*-gene inducers, but the identity of microbes that actually catabolize plant signals is unknown. Other ecological issues also are unresolved: How far do *nod*-gene inducers diffuse in soil? Do they produce other microbial responses in addition to *nod*-gene transcription? Are *nod*-gene inducers modified by microbes to serve secondary regulatory roles important for structuring microbial communities? Answers to these questions can only come from detailed measurements in soil.

Tracing signals from plants to microbes offers an important new molecular tool for defining ecological principles that influence plant-microbe interactions in soil (Phillips et al., 1993) . Although traditional genetic methods can find bacterial mutants with altered capacities for competition, root colonization, or root infection (e.g. Lam et al., 1991; Sanjuan and Olivares, 1989), the results of such studies are strongly affected by the assay system used to find mutants. One solution to this problem is to use a more general approach that identifies any bacterial genes which are regulated by plant signals. Accurate information on the signal molecules that are actually present in the soil environment can be used with a bank of mutant microbes generated by a promoterless reporter gene (e.g. Tn5-*luxAB*; Wolk et al., 1991). *R. meliloti* is a good test organism because it presumably contains many genes required for both general and specific rhizosphere competence. Using this approach, we currently are characterizing 22 Nod$^+$ *R. meliloti* 1021 mutants that produce a Lux$^+$ phenotype in the presence of stachydrine and/or trigonelline.

Obtaining rhizobial *nod*-gene inducers

Investigators should not be dissuaded from studying *nod*-gene inducers by the market prices charged for many of these compounds. Before commercial sources were available, many workers purified their own compounds from selected plant organs. That is still the only reasonable course available for obtaining many isoflavonoids, and bibliographical data for locating those sources have been collected by Ingham (1983). Guides to original literature describing the occurrence of other flavonoids are also available (e.g. Harborne, 1988).

Some *nod*-gene inducers can be obtained by direct chemical synthesis. Until recently, commercial companies supplied common alfalfa flavonoids, such as 4',7-dihydroxyflavanone, at affordable prices. Unfortunately that situation has changed. Investigators faced with this problem can find inexpensive methods for synthesizing chalcones and flavanones in the older literature (e.g. Geissman and Clinton, 1946). Biologists generally can be satisfied with the low yields (25-50%) characteristic of those reactions because of the large financial savings. Recent modifications of older methods are claimed to improve conversions of chalcones to flavanones (Harwood et al., 1990) and chalcones to flavones (Hans and Grover, 1993). Techniques for interconverting flavonoids were summarized by Seshadri (1962), and more complex syntheses of methoxylated flavones have been described recently (Srivastava and Srivastava, 1987). The success of some

sequential protocols depends on skills for separating desired intermediate products from complex mixtures. Biologists willing to develop those skills often can find physical traits of the intermediates in a standard reference for chemical syntheses (e.g. Beilstein, 1991). Trigonelline and stachydrine are both available commercially, and the price of the former is still reasonable. An extremely simple, efficient method for synthesizing stachydrine from proline (Musich and Rapoport, 1977) works well for biologists.

Future research

A clear understanding of factors that control the biosynthesis, accumulation, release, and transmission of plant-derived signal molecules may offer novel methods for manipulating *Rhizobium*-legume symbioses. Initial benefits may include promoting nodulation by particular rhizobial strains or possibly enhancing bacteroid N$_2$ fixation directly in nodules with plant signals. Future advantages could involve general methods that favor beneficial microbes over rhizosphere pathogens. All of these goals require defining which plant signals are perceived by soil microbes, identifying microbial genes that are affected, and clarifying how each gene alters organismic traits that influence population and community attributes. Techniques for achieving these objectives are available, and their application offers an opportunity for defining new principles of rhizosphere ecology.

Acknowledgements

This work was supported by U.S. Department of Agriculture National Research Initiative Competitive Grants Program award 91-373305-6513 and grant US-1884-90 from the U.S.-Israel Binational Agricultural Research and Development Fund (BARD).

References

Amblès A, Jacquesy J C, Jambu P, Joffre J and Maggi-Churin R 1991 Polar lipid fraction in soil: a kerogen-like matter. Organic Geochem. 17, 341–349.

Amrhein N and Gödeke K-H 1977 α-Aminooxy-β-phenylpropionic acid - a potent inhibitor of L-phenylalanine ammonia-lyase in vitro and in vivo. Plant Sci. Lett. 8, 313–317.

Ayabe S-I, Udagawa A and Furuya T 1988 NAD(P)H-dependent 6'-deoxychalcone synthase activity in *Glycyrrhiza* cells induced by yeast extract. Arch. Biochem. Biophys. 261, 458–462.

Barz W 1970 Isolation of rhizosphere bacterium capable of degrading flavonoids. Phytochemistry 9, 1745–1749.

Beilstein Handbook of Organic Chemistry. 1991. 4th edition. Springer-Verlag, Berlin.

Bernard T, Pocard J-A, Perroud B and Le Rudulier D 1986 Variations in the response of salt-stressed *Rhizobium* strains to betaines. Arch. Microbiol. 143, 359–364.

Boivin C, Barran L R, Malpica C A and Rosenberg C 1991 Genetic analysis of a region of the *Rhizobium meliloti* pSym plasmid specifying catabolism of trigonelline, a secondary metabolite present in legumes. J. Bacteriol. 173, 2809–2817.

Caetano-Anollés G, Crist-Estes D K and Bauer W D 1988 Chemotaxis of *Rhizobium meliloti* to the plant flavone luteolin requires functional nodulation genes. J. Bacteriol. 170, 3164–3169.

Cooper J E and Rao J R 1992 Localized changes in flavonoid biosynthesis in roots of *Lotus pedunculatus* after infection by *Rhizobium loti*. Plant Physiol. 100, 444–450.

Dakora F D, Joseph C M and Phillips D A 1993 Alfalfa root exudates contain isoflavonoids in the presence of *Rhizobium meliloti*. Plant Physiol. 101, 819–824.

Dalkin K, Edwards R, Edington B and Dixon R A 1990 Stress responses in alfalfa (*Medicago sativa* L.) I. Induction of phenylpropanoid biosynthesis and hydrolytic enzymes in elicitor-treated cell suspensions cultures. Plant Physiol. 92, 440–446.

D'Arcy-Lameta A 1986 Study of soybean and lentil root exudates II. Identification of some polyphenolic compounds, relation with plantlet physiology. Plant and Soil 92, 113–123.

Dénarié J, Debellé F and Rosenberg C 1992 Signaling and host range variation in nodulation. Annu. Rev. Microbiol. 46, 497–531.

Dharmatilake A J and Bauer W D 1992 Chemotaxis of *Rhizobium meliloti* towards nodulation gene-inducing compounds from alfalfa roots. Appl. Environ. Microbiol. 58, 1153–1158.

Dixon R A and Lamb C J 1990 Regulation of secondary metabolism at the biochemical and genetic levels. *In* Secondary Products from Plant Tissue Culture. Ed. B V Charlwood and M J C Rhodes. pp 103–118. Clarendon Press, Oxford.

Dornbos D L, Spencer G F and Miller R W 1990 Medicarpin delays alfalfa seed germination and seedling growth. Crop Sci. 30, 162–166.

Estabrook, E M and Sengupta-Gopalan C 1991 Differential expression of phenylalanine ammonia-lyase and chalcone synthase during soybean nodule development. Plant Cell 3, 299–308.

Firmin J L, Wilson K E, Rossen L and Johnston A W B 1986 Flavonoid activation of nodulation genes in *Rhizobium* reversed by other compounds present in plants. Nature 324, 90–92.

Fisher R F and Long S R 1992 *Rhizobium*-plant signal exchange. Nature 357, 655–660.

Fougère F and Le Rudulier D 1990 Uptake of glycine betaine and its analogues by bacteroids of *Rhizobium meliloti*. J. Gen. Microbiol. 136, 157–163.

Gajendiran N and Mahadevan A 1990 Utilization of phenolic substances by *Rhizobium* sp. Ind. J. Exp. Biol. 28, 1136–1140.

Geissman T A and Clinton R O 1946 Flavanones and related compounds. I. The preparation of polyhydroxychalcones and flavanones. J. Am. Chem. Soc. 68, 697–700. .

Gloux K and Le Rudulier D 1989 Transport and catabolism of proline betaine in salt stressed *Rhizobium meliloti*. Arch. Microbiol. 151, 143–148.

Goldmann A, Boivin C, Fleury V, Message B, Lecoeur L, Maille M and Tepfer D 1991 Betaine use by rhizosphere bacteria: genes essential for trigonelline, stachydrine, and carnitine catabolism in *Rhizobium meliloti* are located on pSym in the symbiotic region. Mol. Plant-Microbe Inter. 4, 571–578.

Göttfert M, Horvath B, Kondorosi E, Putnoky P, Rodriguez-Quinones F and Kondorosi A 1986 At least two different *nodD* genes are necessary for efficient nodulation on alfaifa by *Rhizobium meliloti*. J. Mol. Biol. 191, 411–420.

Gustine D L, Sherwood R T and Vance C P 1978 Regulation of phytoalexin synthesis in jackbean callus cultures. Plant Physiol. 61, 226–230.

Hans N and Grover S K 1993 An efficient conversion of 2'-hydroxychalcones to flavones. Synthetic Comm. 23, 1021–1023.

Harborne J B (Ed.) 1988 The Flavonoids, Advances in Research Since 1980. Chapman and Hall, London. 621 p.

Harrison M J, Choudhary A D, Dubery I, Lamb C J and Dixon R A 1991a Stress responses in alfalfa (*Medicago sativa* L.). 8. *Cis*-elements and *trans*-acting factors for the quantitative expression of a bean chalcone synthase gene promoter in electroporated alfalfa protoplasts. Plant Molec. Biol. 16, 877–890.

Harrison M J, Lawton M A, Lamb C J and Dixon R A 1991b Characterization of a nuclear protein that binds to three elements within the silencer region of a bean chalcone synthase gene promoter. Proc. Natl. Acad., Sci. USA 88, 2515–2519.

Hartwig U A, Joseph C M and Phillips D A 1991 Flavonoids released naturally from alfalfa seeds enhance growth rate of *Rhizobium meliloti*. Plant Physiol. 95, 797–803.

Hartwig U A, Maxwell C A, Joseph C M and Phillips D A 1990a Chrysoeriol and luteolin released from alfalfa seeds induce *nod* genes in *Rhizobium meliloti*. Plant Physiol. 92, 116–122.

Hartwig U A, Maxwell C A, Joseph C M and Phillips D A 1990b Effects of alfalfa *nod* gene-inducing flavonoids on *nodABC* transcription in *Rhizobium meliloti* strains containing different *nodD* genes. J. Bacteriol. 172, 2769–2773.

Hartwig U A and Phillips D A 1991 Release and modification of *nod*-gene-inducing flavonoids from alfalfa seeds. Plant Physiol. 95, 804–807.

Harwood L M, Loftus G C, Oxford A and Thomson C 1990 An improved procedure for cyclisation of chalcones to flavanones using celite supported potassium fluoride in methanol: total synthesis of bavachinin. Syn. Comm. 20, 649–657.

Honma M A and Ausubel F M 1987 *Rhizobium meliloti* has three functional copies of the *nodD* symbiotic regulatory gene. Proc. Natl. Acad. Sci., USA 84, 8558–8562.

Hungria M, Joseph C M and Phillips D A 1991 Anthocyanidins and flavonols, major nod-gene inducers from seeds of a black-seeded common bean (*Phaseolus vulgaris* L.). Plant Physiol. 97, 751–758.

Ingham J L 1983 Naturally occurring isoflavonoids (1855–1981). Fortschritte d. Chem. org. Naturst. 43, 1–266.

Jones G P, Naidu B P, Starr R K and Paleg L G 1986 Estimates of solutes accumulating in plants by ^{1}H nuclear magnetic resonance spectroscopy. Aust. J. Plant Physiol. 13, 649–658.

Kosslak R M, Bookland R, Barkei J, Paaren H E and Appelbaum E R 1987 Induction of *Bradyrhizobium japonicum* common *nod* gene by isoflavones isolated from *Glycine max*. Proc. Natl. Acad. Sci., USA 84, 7428–7432.

Köster J, Strack D and Barz W 1983 High performance liquid chromatographic separation of isoflavones and structural elucidation of isoflavone 7-*O*-glucoside 6"-malonates from *Cicer arietinum*. J. Med. Planta Res. 48, 131–135.

Lam S T, Ellis D M and Ligon J M 1991 Genetic approaches for studying rhizosphere colonization. *In* The Rhizosphere and Plant Growth. Ed. D L Keister and P B Cregan. pp 43–50. Kluwer Academic Publ, Dordrecht.

León-Barrios M, Dakora F D, Joseph C M and Phillips D A 1993 Isolation of *Rhizobium meliloti nod* gene inducers from alfalfa rhizosphere soil. Appl. Environ. Microbiol. 59, 636–639.

Loake G J, Choudhary A D, Harrison M J, Mavandad M, Lamb C J and Dixon R A 1991 Phenylpropanoid pathway intermediates regulate transient expression of a chalcone synthase gene promoter. Plant Cell 3, 829–840.

Maxwell C A, Edward R and Dixon R A 1992 Identification, purification and characterization of S-adenosyl-L-methonine: isoliquiritigenin 2'-*O*-methyltransferase from alfalfa (*Medicago sativa* L.). Arch. Biochem. Biophys. 293, 158–166.

Maxwell C A, Harrison M J and Dixon R A 1993 Molecular characterization and expression of alfalfa isoliquiritigenin 2'-*O*-methyltransferase, an enzyme specifically involved in the biosynthesis of a transcriptional activator of *Rhizobium meliloti* nodulation genes. Plant J. 4, 971–981.

Maxwell C A, Hartwig U A, Joseph C M and Phillips D A 1989 A chalcone and two related flavonoids released from alfalfa roots induce *nod* genes in *Rhizobium meliloti*. Plant Physiol. 91, 842–847.

Maxwell C A and Phillips D A 1990 Concurrent synthesis and release of *nod*-gene-inducing flavonoids from alfalfa roots. Plant Physiol. 93, 1552–1558.

Morgan A and Marion L 1956 The biogenesis of alkaloids XVII. Further study of the role of ornithine in the biogenesis of stachydrine. Can. J. Chem. 34, 1704–1708.

Musich J A and Rapoport H 1977 Reaction of O-methyl-*N,N'*-diisopropylisourea with amino acids and amines. J. Org. Chem. 42, 139–141.

Parniske M, Zimmermann C, Cregan P B and Werner D 1990 Hypersensitive reaction of nodule cells in the *Glycine* sp./*Bradyrhizobium japonicum*-symbiosis occurs at the genotype-specific level. Bot. Acta 103, 143–148.

Peters N K, Frost J W and Long S R 1986 plant flavone, luteolin, induces expression of *Rhizobium meliloti* nodulation genes. Science 233, 977–980.

Phillips D A 1992 Flavonoids: plant signals to soil microbes. *In* Phenolic Metabolism in Plants. Eds. H A Stafford and R K Ibrahim, Plenum Press, New York. Rec. Adv. Phytochem. 26, 201–231.

Phillips D A, Dakora F D, León-Barrios M, Sande E and Joseph C M 1993 Signals released from alfalfa regulate microbial activities in the rhizosphere. *In* New Horizons in Nitrogen Fixation. Ed. R Palacios, J Mora and W E Newton. pp 197–206. Nijhoff/Junk, Dordrecht.

Phillips D A, Joseph C M and Maxwell C A 1992 Trigonelline and stachydrine released from alfalfa seeds activate NodD2 protein in *Rhizobium meliloti*. Plant Physiol. 99, 1526–1531.

Rao J R, Sharma N D, Hamilton J T G, Boyd D R and Cooper J E 1991 Biotransformation of the pentahydroxy flavone quercetin by *Rhizobium loti* and *Bradyrhizobium* strains (*Lotus*). Appl. Environ. Microbiol. 57, 1563–1565.

Recourt K, Schripsema J, Kijne J W, Van Brussel A A N and Lugtenberg B J J 1991 Inoculation of *Vicia sativa* subsp. *nigra* roots with *Rhizobium leguminosarum* biovar *viciae* results in release of *nod* gene activating flavanones and chalcones. Plant Mol. Biol. 16, 841–852.

Recourt K, van Tunen A J, Mur L A, van Brussel A A N, Lugtenberg B J J and Kijne J W 1992 Activation of flavonoid biosynthesis in roots of *Vicia sativa* subsp. *nigra* by inoculation with *Rhizobium leguminosarum* biovar *viciae*. Plant Molec. Biol. 19: 411–420.

Redmond J W, Batley M, Djordjevic M A, Innes R W, Kuempel P L and Rolfe B G 1986 Flavones induce expression of nodulation genes in *Rhizobium*. Nature 323, 632–635.

Robertson A V and Marion L 1960 The biogenesis of alkaloids XXV. The role of hygric acid in the biogenesis of stachydrine. Can. J. Chem. 38, 396–398.

Rolfe B G, Batley M, Redmond J W, Richardson A E, Simpson R J, Bassam B J, Sargent C L, Weinman J J, Djordjevic M A, and Dazzo F B 1988 Phenolic compounds secreted by legumes. *In* Nitrogen Fixation: Hundred Years After. Ed. H Bothe, F J de Bruijn and W E Newton. pp 405–409. Gustav Fischer, Stuttgart.

Rovira A D and Harris J R 1961 Plant root excretions in relation to the rhizosphere effect V. The exudation of B-group vitamins. Plant and Soil 14, 199–214.

Ryder T B, Hedrick S A, Bell J N, Liang X, Clouse S D and Lamb C J 1987 Organization and differential activation of a gene family encoding the plant defense enzyme chalcone synthase. Molec. Gen. Genet. 210, 219–233.

Sanjuan J and Olivares J 1989 Implication of *nifA* in the regulation of genes located on a *Rhizobium meliloti* cryptic plasmid that affects nodulation efficiency. J. Bacteriol. 171, 4154–4161.

Schmidt J, John M, Wieneke U, Krüssmann H-D and Schell J 1986 Expression of the nodulation gene *nodA* in *Rhizobium meliloti* and localization of the gene product in the cytosol. Proc. Natl. Acad. Sci., USA 83, 9581–9585.

Seshadri T R 1962 Interconversions of flavonoid compounds. *In* The Chemistry of Flavonoid Compounds. Ed. T A Geissman. pp. 156–196. McMillan, NY.

Sethi J K and Carew D P 1974 Growth and betaine formation in *Medicago sativa* tissue cultures. Phytochemistry 13, 321–324.

Siqueira J O, Nair M G, Hammerschmidt R and Safir G R 1991 Significance of phenolic compounds in plant-soil-microbial systems. Crit. Rev. Plant Sci. 10, 63–121.

Spaink H P 1992 Rhizobial lipo-oligosaccharides: answers and questions. Plant Molec. Biol. 20, 977–986.

Srivastava S D and Srivastava S K 1987 Synthesis of a new flavone Ind. J. Chem. 26B, 257–58.

Stafford H A 1990 Flavonoid Metabolism. CRC Press, Boca Raton, Florida. 298 p.

Steenbock H 1918 Isolation and identification of stachydrin from alfalfa hay. J. Biol. Chem. 35, 1–13.

Taguchi H, Nishitani H, Okumura K, Shimabayashi Y and lwai K 1989a Biosynthesis and metabolism-of NAD in *Lemna paucicostata* 151. Agric. Biol. Chem. 53, 1543–1549.

Taguchi H, Nishitani H, Okumura K, Shimabayashi Y and Iwai K 1989b Biosynthesis and metabolism of trigonelline in *Lemna paucicostata* 151. Agric. Biol. Chem. 53, 2867–2871.

Taguchi H and Shimabayashi Y 1983 Findings of trigonelline demethylating enzyme activity in various organisms and some properties of the enzyme from hog liver. Biochem. Biophys. Res. Comm. 113, 569–574.

Tramontano W A, McGinley P A, Ciancaglini E F and Evans L S 1986 A survey of trigonelline concentrations in dry seeds of the Dicotyledoneae. Environ. Expt. Bot. 26, 197–205.

Upmeier B, Gross W, Köster S and Barz W 1988 Purification and properties of S-adenosyl-L-methionine: nicotinic acid-N-methyltransferase from cell suspension cultures of *Glycine max* L. Arch. Biochem. Biophys. 262, 445–454.

Welle R and Grisebach H 1989 Phytoalexin synthesis in soybean cells: elicitor induction of reductase involved in biosynthesis of 6'-deoxychalcone. Arch. Biochem. Biophy. 272, 97–102.

Wiehler G and Marion L 1958 The biogenesis of alkaloids XX. The induced biogenesis of stachydrine. J. Biol. Chem. 231, 799–805.

Wolk C P, Cai Y and Panoff J M 1991 Use of a transposon with luciferase as a reporter to identify environmentally responsive genes in a cyanobacterium. Proc. Natl. Acad. Sci., USA 88, 5355–5359.

Wyn Jones R G and Storey R 1981 Betaines *In* Physiology and Biochemistry of Drought Resistance in Plants. Ed. L G Paleg and D Aspinall. pp 171–204. Academic Press, Sydney.

Zoń J and Amrhein N 1992 Inhibitors of phenylalanine ammmonia-lyase: 2-aminoindan-2-phophonic acid and related compounds. Liebigs Ann. Chem. 1992, 625–628.

Plant and Soil **161**: 81–89, 1994.

Role of rhizobial lipo-oligosacharides in root nodule formation on leguminous plants

Otto Geiger[1], Tita Ritsema, Anton A. N. van Brussel, Teun Tak, André H. M. Wijfjes, Guido V. Bloemberg, Herman P. Spaink and Ben J. J. Lugtenberg
Institute of Molecular Plant Sciences, Leiden University, The Netherlands. [1]*Present address: Technische Universität Berlin, Institut für Biotechnologie, Fachgebiet Technische Biochemie, Seestrasse 13, 13353 Berlin, Germany*

Key words: fatty acid, *nod* genes, phospholipid

Abstract

During recent years signals leading to the early stages of nodulation of legumes by rhizobia have been identified. Plant flavonoids induce rhizobial *nod* genes that are essential for nodulation. Most of the *nod* gene products are involved in the biosynthesis of lipo-oligosaccharide molecules. The common *nodABC* genes are minimally required for the synthesis of all lipo-oligosaccharides. Host-specific *nod* gene products in a given *Rhizobium* species are responsible for synthesis or addition of various moieties to those basic lipo-oligosaccharide molecules. For example, in *R. leguminosarum*, the *nodFEL* operon is involved in the production of lipo-oligosaccharide signals that mediate host specificity. A *nodFE*-determined highly unsaturated fatty acid (*trans*-2, *trans*-4, *trans*-6, *cis*-11-octadecatetraenoic acid) is essential for inducing nodule meristems and pre-infection thread structures on the host plant *Vicia sativa*. Lipo-oligosaccharides also trigger autoregulation of nodulation in pea and, if applied in excessive amounts to a legume, can prevent nodulation and thereby might play a role in competition. During our studies on the biosynthesis of lipo-oligosaccharides, we discovered that, besides the lipo-oligosaccharides, other metabolites are synthesized *de novo* after induction of the *nod* genes. These novel metabolites appeared to be phospholipids, containing either one of the three fatty acids which are made by the action of NodFE in *R. leguminosarum*.

Introduction

Soil bacteria belonging to the genera *Rhizobium*, *Bradyrhizobium*, and *Azorhizobium*, collectively called rhizobia, invade the roots of their legume hosts and trigger the formation of a new organ, the nodule. In these root nodules, a differentiated form of the rhizobia, the bacteroid, is able to fix nitrogen into ammonium, which then can be utilized by the plant. The symbiosis occurs in a host-specific way, leading to the definition of cross-inoculation groups in which the bacterial species are classified according their ability to form nitrogen-fixing nodules on host plants. Examples of such cross-inoculation groups are *R. leguminosarum* biovar *viciae* with peas and vetch as hosts, *R. leguminosarum* bv. *trifolii* with clovers as host, *R. leguminosarum* bv. *phaseoli* with bean as host, and *R. meliloti* with alfalfa and sweet clovers as hosts. Several distinct steps of signal exchange between plant and bacterium have been recognized so far as being involved in the determination of host-specific nodulation. In the first step, plant substances (flavonoids and betaines) excreted from the legume roots induce the transcription of the bacterial nodulation (*nod* or *nol*) genes (Göttfert, 1993; Phillips et al., 1993). The bacterial NodD protein, a transcriptional regulatory protein that presumably interacts with the flavonoids/betaines, contributes to the host specificity of this first step. The molecular chaperone protein GroEL binds directly to NodD and is required for NodD-DNA binding (Long et al., 1994) in order to induce the *nod* or *nol* genes. In the second step, the bacterium, by the means of the activated *nod* or *nol* genes, produces metabolites (Nod metabolites) some of which can act as signals on respective host plants. In all instances characterized so far, these signals are lipo-oligosaccharides (Dénarié et al., 1992; Verma, 1992). Lipo-oligosaccharides are decorated with specific functional groups characteristic for the *Rhizobium* species involved in their synthesis (Fig. 1 and Table 1). Some of the *Rhizobium*-specific decorations have been shown to be essential for proper signaling of the lipo-oligosaccharides to the host plant, thereby initiating the early nodulation events. In this review we try to highlight some novel aspects on the role of lipo-oligosaccharides in root nodule formation.

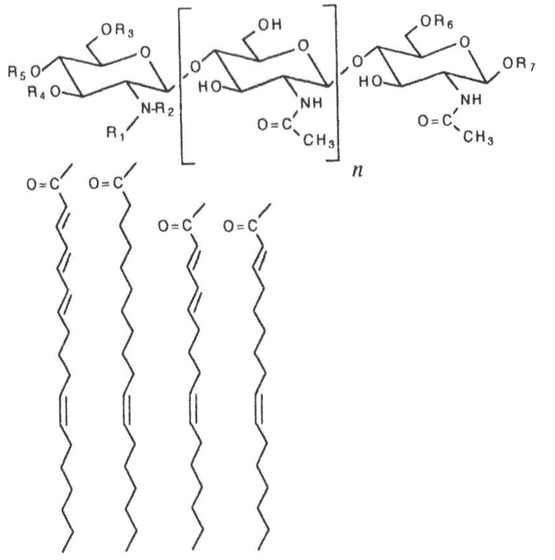

Structures of lipo-oligosaccharides

Since the initial discovery of lipo-oligosaccharides produced by a *R. meliloti* strain (Lerouge et al., 1990), the structures of lipo-oligosaccharides from a sufficient number of rhizobia have been reported to obtain a generalized picture (Bec-Ferté et al., 1993; Carlson et al., 1993; Martínez et al., 1993; Mergaert et al., 1993; Price et al., 1992; Sanjuan et al., 1992; Schultze et al., 1992; Spaink et al., 1991) (Fig. 1 and Table 1). The oligosaccharide backbone of β-1,4-linked *N*-acetyl-D-glucosamines varies in length between three and five sugar units. To the amino nitrogen of the non-reducing end sugar moiety, always a fatty acyl group, the structure of which can vary, is attached. In all rhizobia studied so far lipo-

Fig. 1. Basic structure of the rhizobial lipo-oligosaccharides. The most common C18:1 (*cis*-vaccenic acid) and the *nodFE*-derived fatty acyl residues are indicated. The nature of the substituents indicated by R1 to R7 is given in Table 1.

oligosaccharides which are substituted with *cis*-vaccenic acid are found. Such a substituent is not unusual because in rhizobia *cis*-vaccenic acid is the most abundant fatty acid. This is reflected in the fatty acid composition of the major fatty acid-containing pool, the phospholipids. Minor amounts of lipo-oligosaccharides from *R. meliloti* are substituted with ω-OH fatty acids (C16:0 to C26:0) (Demont et al., 1993). It is so far not understood why these presumptive biosynthetic precursors of the ω-OH C28:0 fatty acid, which is part of the lipopolysaccharides of rhizobia, are found in lipo-oligosaccharides of *R. meliloti*.

In all cases studied so far, rhizobial strains harbouring the genes *nodFE(G)* produce additional lipo-oligosaccharides substituted with specific multi-unsaturated fatty acids (Lerouge et al., 1990; Schultze et al., 1992; Spaink et al., 1991). The multiple unsaturation of the acyl residues is thought to be needed for the full biological activity of these specific lipo-oligosaccharide molecules on their respective hosts. In *R. leguminosarum nodFE* are essential for the production of lipo-oligosaccharides that cause mitogenic reactions on the host plant *V. sativa* (Spaink et al.,

Table 1. Lipo-oligosaccharide structures produced by various rhizobia

Strains	R_1	R_2	R_3	R_4/R_5	R_6	R_7	$n =$	Reference
R. leguminosarum bv viciae								
wild type	C18:4/C18:1	H	O-acetyl	H	H	H	2,3	Spaink et al.,1991
nodL−	C18:4/C18:1	H	H	H	H	H	2,3	Spaink et al., 1991
nodE−	C18:1	H	O-acetyl	H	H	H	2,3	Spaink et al.,1991
R. meliloti 2011 (pGMI149)								
wild type	C16:2	H	H/O-acetyl	H	SO₃H	H	2,3	Lerouge et al.,1990
nodPQ−	C16:2	H	H/O-acetyl	H	H/SO₃H	H	2,3	
nodH−	C16:2	H	H/O-acetyl	H	H	H	2,3	
AK41	C16:3	H	H/O-acetyl	H	SO₃H	H	1–3	Schultze et al., 1992
R. tropici CFN299	C18:1	CH₃	H	H	H/SO₃H	H	3	Martínez et al., 1993
Rhizobium NGR234	C16:0/C18:1	CH₃	H/carb	H/carb	2-0-CH₃-Fuc/ 2-0-CH₃-3-0-acetyl-Fuc/ 2-0-CH₃-4-0-sulfatyl-Fuc	H	3	Price et al., 1992
R. fredii USDA257	C18:1	H	H	H	Fuc/ 2-0-CH₃-Fuc	H	1–3	Bec-Ferté et al., 1993
B. japonicum								
Type I								
USDA 110	C18:1	H	H	H	2-0-CH₃-Fuc	H	3	Sanjuan et al., 1992
USDA 135	C16:0/C16:1/C18:1	H	H/O-acetyl	H	2-0-CH₃-Fuc	H	3	Carlson et al., 1993
Type II								
USDA 61	C18:1	H/CH₃	H/O-acetyl	H/carb	2-0-CH₃-Fuc/ Fuc	glycerol	2,3	Carlson et al., 1993
A. caulinodans ORS571	C18:0/C18:1	CH₃	H/carb	H	H/arabinosyl	H	2,3	Mergaert et al., 1993

carb: carbamoyl; Fuc: fucosyl

1991; Van Brussel et al., 1992). Common to all *nodFE*-derived fatty acyl residues found in lipo-oligosaccharides so far is an alpha-beta unsaturation in conjugation to the carbonyl group. From a chemical point of view such a configuration means that the delocalization of the π-electrons from the conjugated C=C double bond reduces the positive charge of the carbonyl-C, thereby reducing the possibility of a nucleophilic attack by hydrolytic enzymes. One therefore can predict that the amide bond of an alpha-beta unsaturated fatty acid is much more stable than that of a saturated one. The amide bond of *nodFE*-derived lipo-oligosaccharides therefore should show some protection against possible hydrolytic degradation by the plant. Consistent with this predicted stability is the fact that all those rhizobia containing *nodFE(G)* and therefore producing lipo-oligosaccharides with alpha-beta unsaturated fatty acids (*R. meliloti* and *R. leguminosarum* bv. *viciae*) induce nodule primordia at a distance from the root surface in the uninfected inner layers of the root cortex leading to the formation of indeterminate nodules. In contrast, rhizobia producing only lipo-oligosaccharides without alpha-beta unsaturated fatty acids induce cortical cell divisions just beneath the epidermis giving rise to determinate nodules. The presence of other substitutions depends on the rhizobial strain and are shown in Table 1.

Biological activities of lipo-oligosaccharides

External application of lipo-oligosaccharides, in concentrations varying between 10^{-8} and 10^{-12} M, can elicit various effects on root hairs and on cells of the outer and inner cortex of the host plant roots. All lipo-oligosaccharides tested so far cause root hair deformation (HAD) and thick and short roots (TSR) on *Vicia*, but the biological meaning of these effects is not clear (Spaink et al., 1991). Purified lipo-oligosaccharides carrying the respective host-specific decorations of *R. meliloti* or *R. leguminosarum* bv. *viciae* act as mitogens and are able to induce nodule primordia in the inner cortex of the roots

of *Medicago sativa* (Truchet et al., 1991) and *Vicia sativa* (Spaink et al., 1991), respectively. These primordia are indistinguishable from the nodule primordia in the first stage of normal nodule organogenesis. In the case of *Medicago* the nodule primordia can even develop into full-grown nodules, albeit free of bacteria (Truchet et al., 1991). In the outer cortex of *Vicia*, mitogenic lipo-oligosaccharides induce the formation of cytoplasmic bridges which are radially aligned, a phenomenon named pre-infection thread (PIT) structure (Van Brussel et al., 1992). Also, the formation of root hair-like structures is stimulated by mitogenic lipo-oligosaccharides (Van Brussel et al., 1992).

Another effect of mitogenic lipo-oligosaccharides, observable only when roots are not shielded from light, is the induction of flavonoid synthesis. The induction of new flavonoids can be monitored by an increase in *nod* gene-inducing activity (INI) of root exudates (Van Brussel et al., 1990). If lipo-oligosaccharides are applied earlier than *Rhizobium* or are applied in large amounts together with *Rhizobium* to the host plant, they can block nodulation (JAN = jamming of nodulation). This interesting finding indicates that lipo-oligosaccharides may play a role in competition (Van Brussel et al., 1993). Like *Rhizobium* bacteria mitogenic lipo-oligosaccharides inactivate a root nodulation factor in pea and it is speculated that this might be the mechanism involved in autoregulation of nodulation (Smit et al., 1993).

Mitogenic lipo-oligosaccharides also induce the early nodulin genes ENOD5 and ENOD12 which are related to the infection process in cells of the root epidermis (Nap and Bisseling, 1990). In the inner cortex ENOD40 is expressed preferentially opposite the proto-xylem poles and ENOD12 only in the cells of the primordium after induction with mitogenic lipo-oligosaccharide. The spatial pattern of ENOD12 and ENOD40 expression, induced by mitogenic lipo-oligosaccharides, corresponds to the pattern after *Rhizobium* infection (see references in Vijn et al., 1993).

Biosynthesis of lipo-oligosaccharides

Proteins encoded by the *nod* genes play an essential role in the biosynthesis of the lipo-oligosaccharides. The NodA, NodB, and NodC proteins are called common Nod proteins because they are functionally interchangeable. They are present in all rhizobia and they are sufficient for the production of a basic lipo-oligosaccharide molecule (Spaink et al., 1991). NodC is homologous to chitin synthases (see references in Spaink et al., 1993b) and it is involved in chitooligomer synthesis from UDP-*N*-acetylglucosamine (Spaink et al., 1993a). Therefore, NodC is thought to assemble the sugar backbone of the lipo-oligosaccharide. Interestingly, the protein most similar to rhizobial NodCs found so far is the FbfA protein which is needed for fruiting body formation by the myxobacterium *Stigmatella aurantiaca* (Schairer, pers. commun.). This is an indication that chitooligomer-derived signal molecules might be involved in morphogenetic processes of bacterial systems as well. In vitro experiments performed with purified NodB show that it can remove the *N*-acetyl group from the non-reducing terminus of chitooligosaccharides (John et al., 1993). In a final step, a fatty acyl residue must be attached to the free amine of the *N*-deacetylated chito-oligosaccharide to obtain a complete lipo-oligosaccharide molecule. The NodA protein may be involved in this acylation step (Spaink et al., 1993a).

Other host-specific Nod proteins are involved in the synthesis or addition of various structural modifications as indicated in Table 1. In *R. meliloti*, the NodPQ proteins function as ATP sulfurylase and APS kinase, respectively, leading to the production of the activated sulfate donor PAPS (Schwedock and Long, 1992). The NodH protein is involved in transferring the sulfate moiety from PAPS to the 6- position of the reducing end of a lipo-oligosaccharide acceptor (Atkinson et al., 1992; Dénarié and Roche, 1992). The NodL protein, which is produced by a number of rhizobial species, is an acetyltransferase involved in the addition of an *O*-acetyl residue at the non-reducing end sugar (Spaink et al., 1993a). In

Rhizobium leguminosarum bv. *viciae* the NodF and NodE proteins are involved in the production of a lipo-oligosaccharide which carries a highly unsaturated fatty acyl moiety. NodF presumably functions as an acyl carrier protein (Geiger et al., 1991; Shearman et al., 1986) NodE is homologous to β-ketoacyl synthases (Shearman et al., 1986). β-ketoacyl synthase and acyl carrier protein are known to function together in the condensing reaction step of fatty acid biosynthesis.

Are lipo-oligosaccharides the only nod gene-related signals?

Besides the production and secretion of lipo-oligosaccharide signal molecules at least some of the rhizobial Nod proteins have other functions or other functions in addition. The NodO protein is secreted by *R. leguminosarum* bv. *viciae* and is able to form pores in artificial lipid bilayers (Sutton et al., 1993). In *R. trifolii*, a *nodABCIJT*-dependent diglycosyl diacylglyceride is formed which can induce cortical cell division in roots of white clover (Orgambide et al., 1993). Also, some Nod proteins involved in the biosynthesis of lipo-oligosaccharides are involved in the synthesis of intermediates which might represent other potential signals. The presence of a NodC protein is sufficient to allow the synthesis of chitooligomers (Spaink et al., 1993a). Chitooligomers are known to function as signal molecules in their own right on plants during the attack by some phytopathogenic fungi (Ren and West, 1992). The proteins NodFE in a *R. leguminosarum* background are sufficient to allow the synthesis of the *trans*-2, *trans*-4, *trans*-6, *cis*-11-octadecatetraenoic acid. This *nodFE*-derived fatty acid is incorporated into the membrane phospholipids as detailed below (Geiger et al., 1993, 1994).

NodFE-dependent phospholipids

We have shown, that after the induction of the *nodFE* genes, even in the absence of *nodABC* genes, the *trans*-2, *trans*-4, *trans*-6, *cis*-11-octadecatetraenoic acid, which has an absorbance maximum of 303 nm, is still synthesized, suggesting that the biosynthesis of the unusual fatty acid is completed before it is linked to the sugar backbone of the lipo-oligosaccharide. We also found that the unusual C18:4 fatty acid is linked to the *sn*-2 position of the phospholipids (Geiger et al., 1994). In addition, the phospholipids contain other *nodFE*-derived fatty acids, a *trans*-4, *trans*-6, *cis*-11-octadecatrienoic acid (C18:3) which has an absorption maximum at 225 nm, and a octadecadienoic acid (C18:2) (Geiger et al., 1993). Even when lipo-oligosaccharide signals are produced in a wild type-*Rhizobium* cell, a fraction of all those unusual fatty acids is still bound to all major phospholipids (Geiger et al., 1994). Neither the C18:3 nor the C18:2 fatty acid have been observed so far in lipo-oligosaccharides, suggesting that a still unknown acyl transferase (possibly NodA) responsible for the assembly of the fatty acyl chain to the sugar backbone of the lipo-oligosaccharides, does not transfer all fatty acids synthesized by the action of NodFE to the lipo-oligosaccharides. Rather, it selects for alpha-beta unsaturated fatty acids during transfer. These findings offer interesting possibilities: 1) The phospholipids might be biosynthetic intermediates for the synthesis of lipo-oligosaccharide signals. 2) Phospholipids, containing one of the three different types of *nodFE*-derived fatty acids, might have a signal function on their own. 3) Phospholipids might be a dump for an excess of *nodFE*-derived fatty acids. 4) Phospholipids, containing multi-unsaturated fatty acids might have another, hitherto unexpected function.

Effect of nodFE overexpression on growth rate

Organisms normally adapt to lower temperatures by changing the composition of their membranes (Murata et al., 1992). In various prokaryotic and eukaryotic systems, the relative amount of unsaturated fatty acids in the membranes is increased at lower temperatures. It is thought that the loss of membrane fluidity associated with a decrease

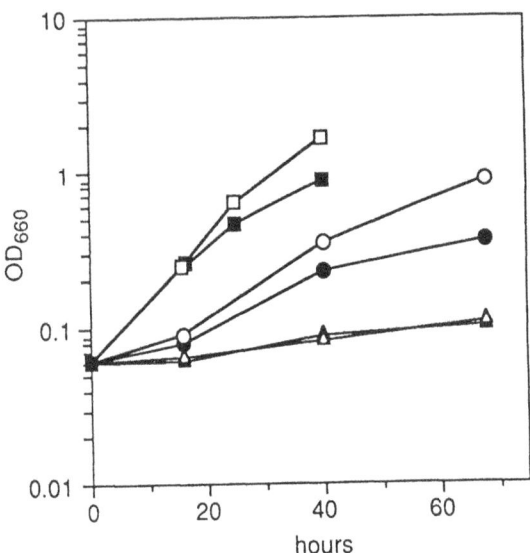

Fig. 2.   Growth of *Rhizobium leguminosarum* RBL5560.pMP1255 cultures after flavonoid induction. LPR5045 is a Sym plasmid-cured *Rhizobium* strain. RBL5560 is LPR5045 containing the *R. leguminosarum* bv. *viciae* Sym plasmid pRL1JI *mep*::Tn5 (Zaat et al., 1987). pMP1255 harbours cloned *nodFE* of pRL1JI in an IncQ plasmid (Geiger et al., 1991). Cells of strain RBL5560.pMP1255 were grown in B⁻ medium in the absence (□ ○ △) or presence (■ ● ▲) of the inducer naringenin (1.5 uM) at 20°C (□■), 16°C (○ ●) or at 4°C (△▲).

in temperature might be, at least partially, compensated by a fluidity increase due to the presence of more unsaturated fatty acids. Expression of the genes *nodFE* leads to increased amounts of unsaturated fatty acids found in the phospholipids of *R. leguminosarum* membranes (Geiger et al., 1993; Geiger et al., 1994). We noticed that *R. leguminosarum* cultures, in which *nodFE* are overproduced, show longer generation times at normal (29°C) growth temperature than control cultures which had not been induced (Fig. 2). Induced cultures of RBL5560.pMP190 that contain only one copy of *nodFE* on the Sym plasmid but an empty IncQ plasmid, without *nodFE*, show normal short generation times (data not shown).

The reduced growth of *nodFE*-overproducing *R. leguminosarum* cultures could be explained by an increased membrane fluidity which would be too high for optimal growth rates. If true, one would expect that at reduced temperatures in the NodFE-overexpressed situation, the contents of unsaturated fatty acids in the membranes causes a higher fluidity allowing better growth than that shown by cultures that do not express NodFE. Such a strategy might lead to the construction of *Rhizobium* strains with increased cold tolerance. We analyzed growth rates of non-induced and *nodFE*-induced cultures at normal and reduced growth temperatures and found that also at lower temperatures (16°C), the growth in the *nodFE*-induced culture was always significantly slower than in the uninduced control cultures (Fig. 2). At 4°C hardly any growth was noticed in non-induced or *nodFE*-induced cultures. We therefore think that the reduced growth rates observed after overproduction of NodFE are not due to a change in membrane fluidity. The most likely explanation is that such high levels of NodFE are interfering with normal fatty acid synthesis of the cell, thereby slowing down growth.

Biological activity of NodFE-dependent phospholipids

We tested NodFE-dependent phospholipids in various bioassays that are normally used to analyze lipo-oligosaccharides. We could not detect any induction of nodule meristems, preinfection threads or INI after incubation of *Vicia sativa* in the presence of NodFE-dependent phospholipids. We also tested the effects of fatty acids obtained after mild alkaline hydrolysis of the phospholipids. Such fatty acids obtained from *nodFE*-independent or *nodFE*-dependent phospholipids had no effect on HAD or TSR, if applied alone. In the presence of rhizobial strains, lacking *nodE*, and which normally make only non-mitogenic lipo-oligosaccharides (i.e. RBL5580.pMP1060) a stimulation of TSR and HAD is observed by addition of fatty acids derived from phospholipids that contain *nodFE*-determined fatty acids but not by monounsaturated or saturated fatty acids (data not shown). We could even show, that *nodFE*-determined fatty acids obtained by hydrolysis of phospholipids in combination with RBL5580.pMP1060 cause INI on *V. sativa* (Table 2). An explanation for these results is, that *nodFE*-determined fatty acids are taken up by rhizobia

Table 2. Induction of INI by *nodE*-deficient *Rhizobium* in the presence of NodFE-derived fatty acids

Hydrolyzed phospholipids from	Strain RBL5580 pMP1060	INI (units of β-galactosidase) 4 days	7 days
LPR5045 - nar pMP281	-	210	130
pMP1255	+	223	596
LPR5045 + nar pMP281	-	40	63
pMP1255	+	704	2400
LPR5045 + nar pMP281	-	30	36
	+	150	605
LPR5045 + nar pMP1255	-	1	47
	+	150	1559
RBL5580 + nar pMP1060	-	30	67
	+	110	430
-	+	70	681
-	RBL5601	3028	6591
-	-	40	50

INI was measured after 4 and 7 days of co-culture of *V. sativa* and *Rhizobium* (strain RBL5580.pMP1060 or RBL5601)(van Brussel et al., 1990) in the presence of phospholipid hydrolysates and 0.004% CHAPS. LPR5045 is a Sym plasmid-cured *Rhizobium* strain. RBL5601 is LPR5045 containing Sym plasmid pRL1JI. RBL5580 is LPR5045 containing pRL1JI::Tn1831 (deletion of *nodELMNO*). pMP281 (cloned *nodD* of pRL1JI in IncP). pMP1060 (cloned *nodL* of pRL1JI in IncP). pMP1255 (cloned *nodFE* of pRL1JI in IncQ). Hydrolyzed phospholipids used per INI assay were obtained from about 10^{11} of uninduced or naringenin-induced (nar) cells by a Bligh and Dyer extraction procedure. $CHCl_3$-soluble material was chromatographed on HPLC silica gel (Geiger et al., 1994) and allowed the separation of phospholipids from other lipids. Phospholipids were subsequently subjected to mild alkaline hydroysis (0.2 *M* NaOH in MeOH, 1 h, 20°C) and fatty acids were obtained in the $CHCl_3$ phase after another Bligh and Dyer partition.

and are attached to *N*-deacetylated chitooligomers during lipo-oligosaccharide biosynthesis to form mitogenic lipo-oligosaccharides which are able to cause INI. Experiments are under way to see whether lipo-oligosaccharides produced by RBL5580.pMP1060 (*nodE⁻*) in the presence of a chemically synthesized *trans*-2, *trans*-4, *trans*-6, *cis*-11-octadecatetraenoic acid (Verduyn et al., 1992) are indeed mitogenic lipo-oligosaccharides and give INI, NOI, and PIT.

References

Atkinson E M, Ehrhardt D W and Long S R 1992 In vitro activity of the *nodH* gene product from *Rhizobium meliloti*. *In* Sixth International Symposium on Plant-Microbe Interactions, Program and Abstracts. Abst.45.University of Washington, Seattle, WA, USA.

Bec-Ferté M P, Savagnac A, Pueppke S G and Promé J C 1993 Nod factors from *Rhizobium fredii* USDA257. *In* New Horizons in Nitrogen Fixation. Eds. R Palacios, J Mora and W E Newton. pp 157–158. Kluwer Academic Publishers, Dordrecht,

Carlson R W, Sanjuan J, Bhat U R, Glushka J, Spaink H P, Wijfjes A H M, Van Brussel A A N, Stokkermans T J W, Peters K and Stacey G 1993 The structure and biological activities of the lipo-oligosaccharide nodulation signals produced by type I and type II strains of Bradyrhizobium jaconicum. J. Biol. Chem. 268, 18372–18381.

Demont N, Debelle F, Aurelle H and Promé J C 1993 Structural studies of Nod factors from *Rhizobium meliloti*: Revisited structures and the role of *nodFE* genes in their biosynthesis. *In* New Horizons in Nitrogen Fixation. Eds. R Palacios, J Mora and W E Newton. p 224. Kluwer Academic Publishers, Dordrecht

Dénarié J and Roche P 1992 *Rhizobium* nodulation signals. *In* Molecular Signals in Plant-Microbe Communications. Ed. D P S Verma. pp 295–324. CRC Press, Boca Raton.

Dénarié J, Debellé F and Rosenberg C 1992 Signaling and host range variation in nodulation. Annu. Rev. Microbiol. 46, 497–531.

Geiger O, Spaink H P and Kennedy E P 1991 Isolation of *Rhizobium leguminosarum* NodF nodulation protein: NodF carries a 4'- phosphopantetheine prosthetic group. J. Bacteriol. 173, 2872–2878.

Geiger O, Spaink H P and Lugtenberg B J J 1993 Biosynthesis of lipo-oligosaccharides: Phospholipids of *Rhizobium* contain *nodE*-determined highly unsaturated fatty acid moieties. *In* New Horizons in Nitrogen Fixation. Eds. R Palacios, J Mora and W E Newton. p 233.Kluwer Academic Publishers, Dordrecht,

Geiger O, Thomas-Oates J E, Glushka J, Spaink H P and Lugtenberg B J J 1994 Phospholipids of *Rhizobium* contain *nodE*-determined highly unsaturated fatty acid moieties. J. Biol. Chem. 269 (*In press*).

Göttfert M 1993 Regulation and function of rhizobial nodulation genes. FEMS Microbiol. Rev. 104, 39–63.

John M, Röhrig H, Schmidt J, Wieneke U and Schell J 1993 *Rhizobium* NodB protein involved in nodulation signal synthesis is a chitooiigosaccharide deacetylase. Proc. Natl. Acad. Sci. USA 90, 625–629.

Lerouge P, Roche P, Faucher C, Maillet F, Truchet G, Promé J-C and Dénarié J 1990 Symbiotic host-specificity of *Rhi-zobium meliloti* is determined by a sulphated and acylated glucosamine oligosaccharide signal. Nature (London) 344, 781–784.

Long S R, Ehrhardt D W, Atkinson E M, Willits M, Ogawa J, Swanson J and Fisher R F 1994 Action of nodulation genes in *Rhizobium meliloti*. Plant and Soil, pages.

Martinez E, Poupot R, Promé J C, Pardo M A, Segovia L, Truchet G and Dénarié J 1993 Chemical signaling of *Rhizobium* nodulating bean. *In* New Horizons in Nitrogen Fixation. Eds. R Palacios, J Mora and W E Newton. pp 171–175. Kluwer Academic Publishers, Dordrecht,

Mergaert P, Van Montagu M, Promé J-C and Holsters M 1993 Three unusual modificadons, a D-arabinosyl, an N-methyl, and a carbamoyl group, are present on the Nod factors of *Azorhizobium caulinodans* strain ORS571. Proc. Natl. Acad. Sci. USA 90, 1551–1555.

Murata N, Ishizaki-Nishizawa O, Higashi S, Hayashi H, Tasaka Y and Nishida I 1992 Genetically engineered alteration in the chilling sensitivity of plants. Nature (London) 356, 710–713.

Nap J-P and Bisseling T 1990 Developmental biology of a plant-prokaryote symbiosis: The legume root nodule. Science 250, 948–954.

Orgambide G, Hollingsworth R and Dazzo F 1993 *Rhizobium trifolii* produces a diglycosyl diacylglyceride signal molecule which can elicit a-host-specific responses on white clover roots at subnanomolar concentrations. *In* New Horizons in Nitrogen Fixation. Eds. R Palacios, J Mora and W E Newton. p 248. Kluwer Academic Publishers, Dordrecht.

Phillips D A, Joseph C M and Maxwell C A 1992 Trigonelline and stachydrine released from alfalfa seeds activate NodD2 protein in *Rhizobium meliloti*. Plant Physiol. 99, 1526–1531.

Price N P J, Relíc B, Talmont F, Lewin A, Promé D, Pueppke S G, Maillet F, Dénarié J, Promé J-C and Broughton W J 1992 Broad-host-range *Rhizobium* species NGR234 secretes a family of carbamoylated, and fucosylated, nodulation signals that are O-acetylated or sulphated. Mol. Microbiol. 6, 3575–3584.

Ren Y and West C A 1992 Elicitation of diterpene biosynthesis in rice (*Oryza sativa* L.) by chitin. Plant Physiol. 99, 1169–1178.

Sanjuan J, Carlson R W, Spaink H P, Bhat U R, Barbour W M, Glushka J and Stacey G 1992 A O-methylfucose moiety is present in the lipo-oligosaccharide nodulation signal of *Bradyrhizobium japonicum*. Proc. Natl. Acad. Sci. USA 89, 8789–8793.

Schultze M, Quiclet-Sire B, Kondorosi E, Virelizier H, Glushka J, Endre G, Géro S D and Kondorosi A 1992 *Rhizobium meliloti* produces a family of sulfated lipo-oligosaccharides exhibiting different degrees of plant host specificity. Proc. Natl. Acad. Sci. USA 89, 192–196.

Schwedock J S and Long S R 1992 *Rhizobium meliloti* genes involved in sulfate activation: The two copies of *nodPQ* and a new locus, *saa*. Genetics 132, 899–909.

Shearman D H F, Rossen L, Johnston A W B and Downie J A 1986 The *Rhizobium leguminosarum* nodulation gene *nodF* encodes a polypeptide similar to acyl carrier protein and is regulated by *nodD* plus a factor in pea root exudate. EMBO J. 5, 647–652.

Smit G, Van Brussel A A N and Kijne J W 1993 Inactivation of a root factor by ineffective *Rhizobium*: A molecular key to autoregulation of nodulation in *Pisum sativum*. *In* New Horizons in Nitrogen Fixation. Eds. R Palacios, J Mora and W E Newton. p 371. Kluwer Academic Publishers, Dordrecht.

Spaink H P, Sheeley D M, Van Brussel A A N, Glushka J, York W S, Tak T, Geiger O, Kennedy E P, Reinhold V N and Lugtenberg B J J 1991 A novel highly unsaturated fatty acid moiety of lipo-oligosaccharide signals determines host specificity of *Rhizobium*. Nature (London) 354, 125–130.

Spaink H P, Wijfjes A H M, Geiger O, Bloemberg G V, Ritsema T and Lugtenberg B J J 1993 The function of the rhizobial *nodABC* and *nodFEL* operons in the biosynthesis of lipo-oligosaccharides. *In* New Horizons in Nitrogen Fixation. Eds. R Palacios, J Mora and W E Newton. pp 165-170. Kluwer Academic Publishers, Dordrecht.

Spaink H P, Wijfjes A H M, Van Vliet T B, Kijne J W and Lugtenberg B J J 1993 Rhizobial lipo-oligosaccharide signals and their role in plant morphogenesis; are analogous lipophilic chitin derivatives produced by the plant? Aust. J. Plant Physiol. 20, 381–392.

Sutton M J, Lea E J A, Crank S, Rivilla R, Economou A, Ghelani S, Johnston A W B and Downie J A 1993 NodO: A nodulation protein that formes pores in membranes. *In* Advances in Molecular Genetics of Plant-Microbe Interactions. Eds. E W Nester and D P S Vena. pp 163-167. Kluwer Academic Publishers, Dordrecht.

Truchet G, Roche P, Lerouge P, Vasse J, Camut S, de Billy F, Promé J-C and Dénarié J 1991 Sulphated lipo-oligosaccharide signals of *Rhizobium meliloti* elicit root nodule organogenesis in alfalfa. Nature (London) 351, 670–673.

Van Brussel A A N, Recourt K, Pees E, Spaink H P, Tak T, Wijffelman C A, Kijne J W and Lugtenberg B J J 1990 A biovar-specific signal of *Rhizobium leguminosarum* bv. *viciae* induces increased nodulation gene-inducing activity in root exudate of *Vicia sativa* subsp. *nigra*. J. Bacteriol. 172, 5394–5401.

Van Brussel A A N, Bakhuizen R, van Spronsen P, Spaink H P, Tak T, Lugtenberg B J J and Kijne J W 1992 Induction of pre-infection thread structures in the host plant by lipo-oligosaccharides of *Rhizobium*. Science 257, 70–72.

Van Brussel A A N, Tak T, Wijfjes A H M, Spaink H P, Smit G, Diaz C L and Kijne J W 1993 Jamming of nodulation of *Vicia sativa* ssp. *nigra* by Nod factors of *Rhizobium leguminosarum* bv. *viciae*. *In* New Horizons in Nitrogen Fixation. Eds. R Palacios, J Mora and W E Newton. p 261. Kluwer Academic Publishers, Dordrecht.

Verduyn R, Lagas R M, Dreef C E, van der Marel G A and van Boom J H 1992 Synthesis of a highly unsaturated fatty acid moiety of lipo-oligosaccharides determining host-specificity in *Rhizobium*. Recl. Trav. Chim. Pays-Bas 111, 367–368.

Verma D P S 1992 Signals in root nodule organogenesis and endocytosis of *Rhizobium*. Plant Cell 4, 373–382.

Vijn J, das Neves L, van Kammen A, Franssen H and Bisseling T 1993 nod factors and nodulation in plants. Science 260, 1764–1765.

Zaat S A J, Wijffelman C A, Spaink H P, van Brussel A A N, Okker R J H and Lugtenberg B J J 1987 Induction of the *nodA* promoter of *Rhizobium leguminosarum* Sym plasmid pRL1JI by plant flavanones and flavones. J. Bacteriol. 169, 198–204.

Plant and Soil **161**: 91–96, 1994.
© 1994 *Kluwer Academic Publishers.*

Sucrose transport and hydrolysis in *Rhizobium tropici*

Vasilly I. Romanov and Esperanza Martínez-Romero
A.N.Bach Institute of Biochemistry RAS, Moscow, Russia and CIFN-UNAM, Apdo. postal 565-A, Cuernavaca, Mor., México

Key words: catabolite, repression, *Rhizobium*, sucrose, transport

Abstract

The *Rhizobium tropici* strain CFN 299 was maintained on PY medium and was grown in minimal medium (MM) with sucrose, glucose, fructose and glutamate (or their combination) as carbon sources. Bacteria were able to simultaneously use different carbon sources and, with a combination sucrose and glutamate, the growth rate was faster than with either carbon source alone. Sucrose transport was induced by sucrose and partially repressed by glucose and glutamate if they were included in MM as additional carbon sources. The transport of sucrose was active because both an uncoupler (dinitrophenol, DNP) and inhibitors of terminal oxidation (KCN, NaN$_3$) severely reduced sucrose uptake. Sucrose transport was also sensitive to a functional sulfhydryl reagent but was much less sensitive to EDTA and arsenate. We obtained nonlinear Lineweaver-Burk plots for the uptake of sucrose (by sucrose-grown bacteria), and this implied the existence of at least two uptake mechanisms. Invertase (EC 3.2.1.26) is the main enzyme for sucrose hydrolysis in this organism. This enzyme was induced by sucrose and had high activity in mid-log phase cells when sucrose was the sole carbon source (0.2%). Invertase activity was not detected in growth medium. In general, the results obtained support the idea, that *R. tropici* is adapted to sucrose utilization and to multicarbon nutrition during its interaction with plants.

Introduction

In legume-*Rhizobium* interactions, the bacteria use carbon compounds supplied by host plants. Sucrose is a major component of the carbon material translocated in the phloem to the roots and nodules, and is a major carbon compound in nodules (Streeter, 1991). Sucrose synthase, the major enzyme for sucrose breakdown in nodules has been identified as nodulin 100 in soybean (Thummer and Verma, 1987). Whether sucrose is a potential carbon substrate for bacteria inside the host cell would be important to investigate.

A general dogma of symbiotic nitrogen fixation in legumes is that bacteroids use only C$_4$-dicarboxylates as carbon and energy sources, primarily malate and succinate (Day and Copeland, 1991; Streeter, 1991; Vance and Hiechel, 1991). However, some reports do not prove the concept of a exclusive role of C$_4$-dicarboxylates in the carbon nutrition of bacteroids (Duncan, 1981; Herrada et al., 1989; Romanov et al.,1985). *Rhizobium* may use sucrose as carbon source at least during the infection process and the early stages of nodule development. This process could be important in the formation of effective nodules, but information concerning sucrose metabolism in Rhizobia is very limited. This is why we decided to begin this project.

Rhizobium tropici CFN 299 is a broad host range *Rhizobium*, that nodulates *Phaseolus vulgaris* bean and *Leucaena* spp. (Martínez et al.,

1991). In the present paper we suggest that this organism possesses at least two active transport systems for sucrose uptake. These bacteria lack catabolite repression by glucose and are able to use glucose simultaneously with sucrose or glutamate.

Materials and methods

Chemicals

$(U^{14}C)$-sucrose (475 mCi/mmol) and $(U^{14}C)$-glucose (320 mCi/mmol) were from DuPont & Co. Inc.; L-$(U^{14}C)$-glutamate (262 mCi/mmol) and $(6,6(n)^3H)$-sucrose (14.4 Ci/mmol) were from Amersham. All other chemicals were reagent grade and were purchased from standard sources.

Cell Growth

One day old cultures of *R. tropici* CFN 299 grown on PY medium (Noel et al., 1984) agar plates were used to inoculate liquid MM. A basal MM of the following composition was used: K_2HPO_4 - 0.25 g; KH_2PO_4 - 1.0 g; $MgSO_4$ $7H_2O$ - 0.2 g; KNO_3 - 0.7 g; $CaCl_2$ - 0.2 g; $FeCl_3$ - 0.01 g; Biotin - 1 mg; H_3BO_3 - 3.17 mg; Na_2MoO_4 - 1.0 mg; $MnSO_4 4H_2O$ - 1.52 mg; $ZnSO_4 7H_2O$ - 0.25 mg; $CuSO_4$ $5H_2O$ - 0.087 mg; 1.0 liter demineralized water; the pH was adjusted to 6.8 with $1N$ NaOH. The basal MM was usually supplemented with 2 g of carbon source per liter. When two carbon sources were used, an equal mass of each carbon compound was added. In the MM containing glutamate, KNO_3, was omitted. Cultures were grown at 30°C (220 rpm). Growth rate was determined at 600 nm and/or by protein measurement.

Transport assays

A modified sugar uptake procedure (Wong, 1990) was used to analyze the transport kinetics of sucrose, glucose and glutamate. Cells were harvested at mid-log phase and washed two times with 0.05 M K-phosphate buffer, pH 6.6. Washed cells were resuspended in fresh buffer to 0.65–1.0

mg protein per mL. Suspensions were incubated 1 h with gentle stirring to deplete endogenous substrates. Twenty-four well tissue culture plates (Costar, Cambridge, Mass.) were used. Each duplicate well contained 0.8 mL carbon-free MM with radiolabeled sucrose, glucose, glutamate or its combinations; the concentrations of sucrose and glucose were 0.1 mM and glutamate was 0.05 mM. Specific radioactivity of each substrate was always kept at 0.1 mCi/mmol when either one or two substrates were used. The plates were taped on a standard shaker at 23–25°C, at 200 rpm. Transport experiments were started by adding 0.2 mL of cell suspension to each well. When the effect of inhibitors was studied, transport was started by adding (U^{14})-sucrose, inhibitors were added 1 min before the addition of radioactivity and uptake was measured for 5 min. Without inhibitors, after 1, 2, 3 or 5 min 0.15–0.3 mL cell samples from each well were transferred to 1 mL ice-cold 2% sucrose or glucose, or glutamate or its combination in MM. Cells were washed twice by centrifugation in a Microfuge at 4°C with the same MM, transferred to POPOP-PPO-toluene scintillator solution and counted in a Beckman LS 7000 scintillation counter at least two times. Data presented in the figures are the means calculated from four time points of duplicates if not specially mentioned. Boiled cells were used to determine the background counts for non specific binding after incubation with radiolabeled compounds during 5 min. This control usually shows very low radioactivity, less than 1% of the experiments.

Other assays

Washed cell suspensions in 0.05M K-phosphate buffer pH 6.6 were sonicated on ice, centrifuged for 10 min in a Microfuge in cold and the supernatant was used to determine invertase activity (Hoelzle and Streeter, 1990). Sucrose phosphorylase was assayed using the method of Mieyal (1972). The concentration of the sucrose, glucose, fructose and glutamate during the bacterria growth was measured in culture supernatants, using Boehringer kits. Protein was determined by the method of Bio-Rad laboratories.

Table 1. Invertase activity in *R. tropici* grown in the presence of different carbon sources[a]

Carbon sources	Invertase
Glucose	22.1 ± 1.4
Fructose	24.6 ± 1.6
Glutamate	32.0 ± 4.7
Glucose + Fructose	19.6 ± 0.9
Glucose + Sucrose	63.1 ± 2.1
Sucrose	81.9 ± 1.9

[a]Units are glucose liberated per minute per mg protein and are average of four replicate assays.

Results and discussion

There is a report in the literature that some of the fast-growing Rhizobia have a constitutive sucrose uptake system, while others have an inducible system (Glenn and Dilworth, 1981). In *R. tropici*, sucrose uptake is inducible (Fig. 1a) and an unexpected finding was that in glutamate-grown cells, sucrose transport was markedly higher than that of fructose- or glucose-grown cells. To obtain more information, cells were grown in two carbon sources and it was confirmed that glucose repressed sucrose transport more strongly than glutamate (Fig. 1b).

Invertase activity in *R. tropici* was found in cells grown on any of the carbon sources tested, but this enzyme was also strictly induced by sucrose (Table 1). Our attempts to show sucrose phosphorylase activity failed and this indicates that invertase is the main enzyme for sucrose hydrolysis in this organism. There was no invertase activity found in the culture supernatant. From the data presented in Figure 1 and Table 1, it may be speculated that in *R. tropici*, sucrose transport and invertase synthesis may be coordinately regulated.

The results from Figure 1 and Table 1 lead us to suggest that *R. tropici* is probably able to utilize different carbon substrates simultaneously, as it is known for *R. trifolii* (De Hollander and Stouthamer, 1979) and *R. leguminosarum* bv.

viciae (McKay et al., 1989). First, it was shown that during the growth of bacteria in the presence of two carbon sources, the concentrations of sucrose and glucose (Fig. 2b) or of sucrose and glutamate (Fig. 2c) in the culture supernatant decreased simultaneously. The same result was obtained for glucose + glutamate or glucose + fructose as carbon sources (data not shown). Other evidence for this phenomenon was obtained from the uptake experiments.

Sucrose + glucose -grown bacteria show uptake activity for both these carbon substrates measured either independently (Fig. 3a) or simultaneously (Fig. 3b). From the results of the co-transport experiment, it is clear that when present in equimolar concentrations (0. 1 mM), glucose does not influence sucrose transport, but sucrose sharply reduces glucose uptake. This was confirmed in experiments using (U^{14}C)-sucrose + glucose or with (U^{14}C)-glucose + sucrose. Glucose at a concentration 10 fold higher than sucrose reduced sucrose uptake by only 20% (data not shown).

Sucrose + glutamate grown bacteria show uptake activity of both substrates (Fig. 4a) and co-transport measurements demonstrate only weakly diminished glutamate uptake (Fig. 4b). It appears that these systems are probably working independently. It is important to note that with a combination of sucrose and glutamate, the bacteria growth rate was faster than with either of them independently (Fig. 2a and 2c). This could be important in the course of interaction of the bacteria with host legume.

A nonlinear (biphasic) Lineweaver-Burk plot was obtained for sucrose-grown cells when sucrose uptake was measured at different sucrose concentrations (Fig. 5). This suggested that a high-affinity (K_m = 0.04 mM) and a low-affinity (K_m = 1.0 mM) transport systems are present in *R. tropici*. The apparent K_m values for transport did not vary significantly in different experiments and high V_{max} values for both systems (56.8 ± 4.9 and 19.0 ± 1.6 nmol·min^{-1} mg protein^{-1}) were consistently obtained with sucrose-grown cells. This data indicates that *R. tropici* has a very effective system for sucrose uptake.

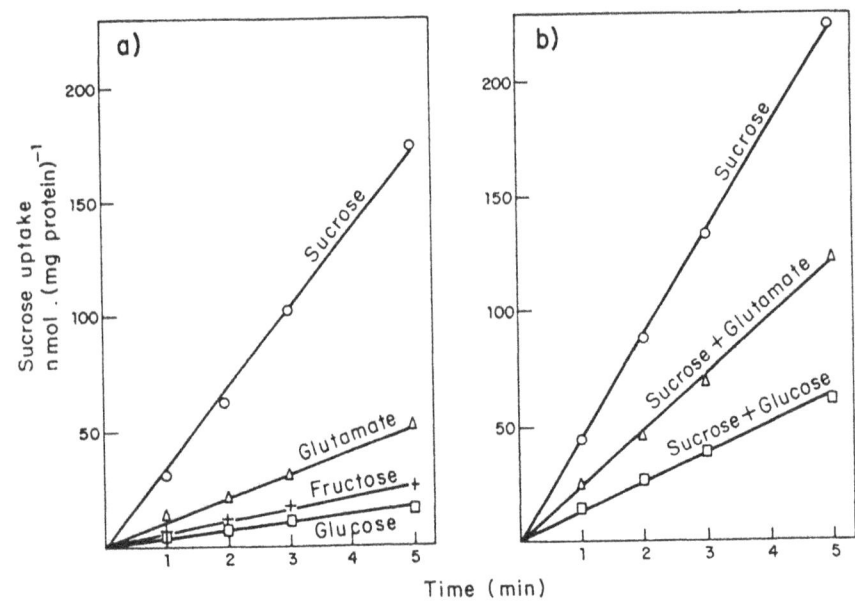

Fig. 1. Kinetics of (^{14}C)-sucrose uptake by *R. tropici*, grown on minimal medium plus **a** sucrose (O), glutamate (△), fructose (+), glucose (□), and **b** sucrose+glutamate (△) and sucrose+glucose (□). **a** - experiment 1; **b** - experiment 2.

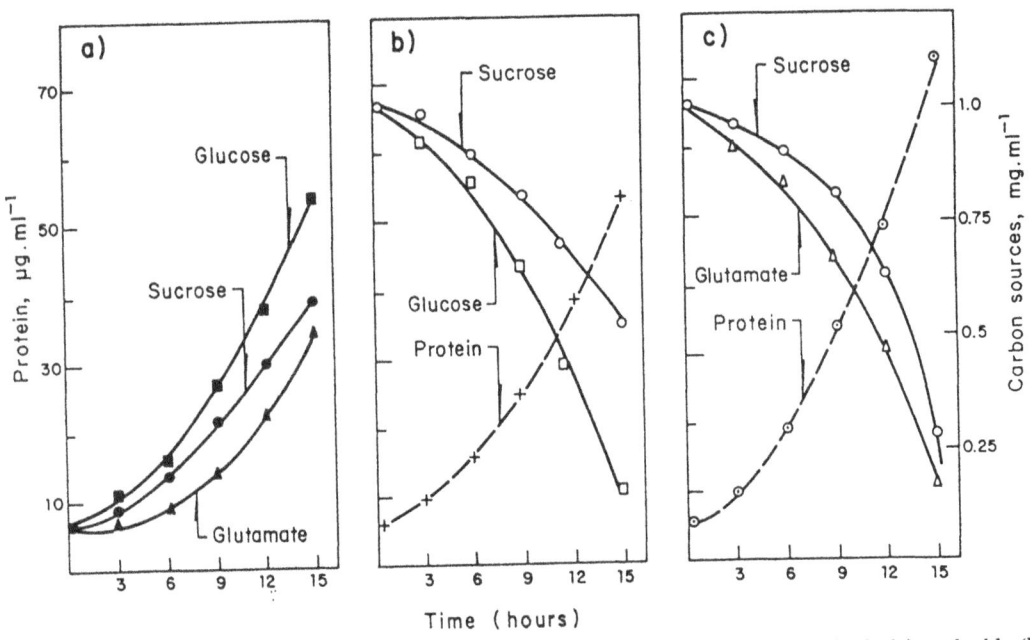

Fig. 2. Growth of *R. tropici* and carbon source concentration in minimal medium, containing single (**a**) or double (**b** and **c**) carbon sources. Growth was determined by protein measurements of cultures and plotted as function of time. a - growth on MM plus glucose (■), sucrose (●) or glutamate (▲). b - growth on MM plus glucose+sucrose (+); (O) sucrose and (□) glucose concentration in culture. c - growth on MM plus sucrose+glutamate (⊙); (O) - sucrose and (△) - glutamate concentration in culture. This figure represents the data from one of the two experiments which shows very similar results.

Both an uncoupler and inhibitors of terminal oxidation severely reduced the uptake of 0.1 m*M* sucrose in sucrose-grown cells (Table 2). The sulfhydryl reagent N-ethylmaleimide inhibited sucrose uptake by more than 96%. EDTA and arsenate however, had little effect. These data

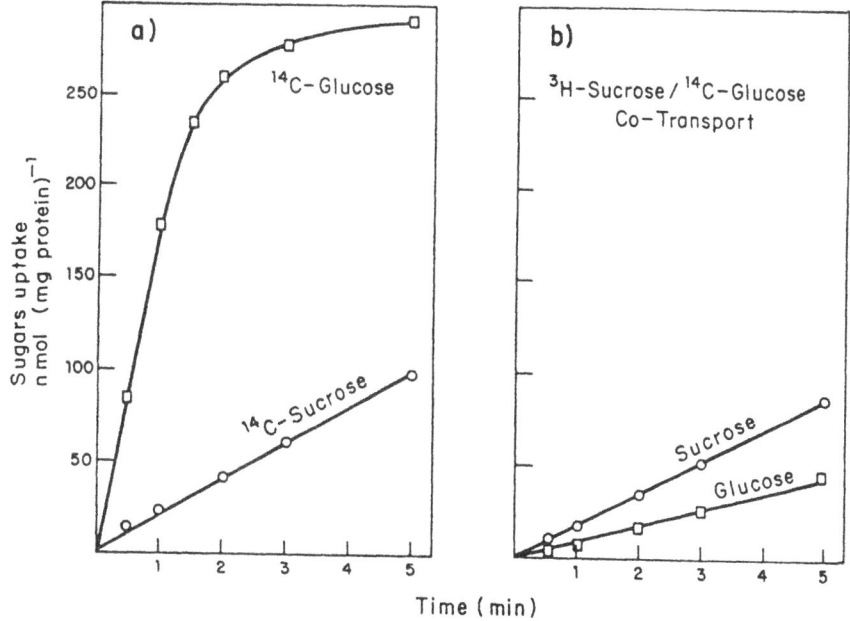

Fig. 3. Uptake of sucrose and glucose by *R. tropici* grown on minimal medium containing sucrose+glucose.**a**- independent measurements of (^{14}C)-sucrose and (^{14}C)- glucose uptake. **b** - simultaneous uptake of (^{3}H)-sucrose and (^{14}C)-glucose.

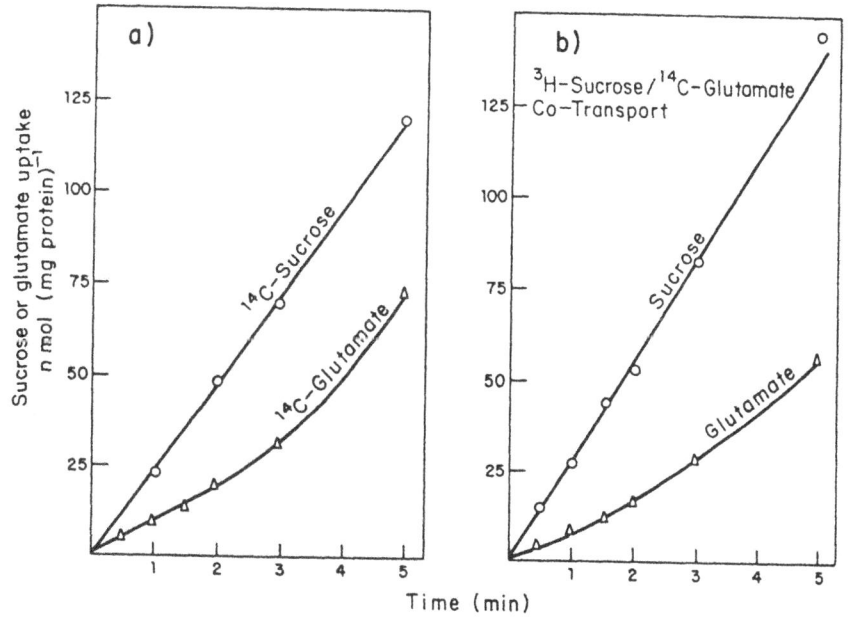

Fig. 4. Uptake of sucrose and glutamate by *R. tropici* grown on minimal medium containing sucrose+glutamate. **a**- independent measurements of (^{14}C)-sucrose and (^{14}C)-glutamate uptake. **b** - simultaneous uptake of (^{3}H)-sucrose and (^{14}C)-glutamate.

suggested that the uptake of sucrose in *R. tropici* is via an active process.

As for organic acids, in particular C_4-dicarboxylates, it was shown that sucrose+malate-grown bacteria were able to take up both substrates (data not shown), but that the presence of

malate in MM caused the bacteria to aggregate. To address this question, other growth condition for *R. tropici* would be needed.

In summary, we stress two important and interesting points: (a) *R. tropici* strain CFN 299 is well adapted for the uptake of sucrose in a broad

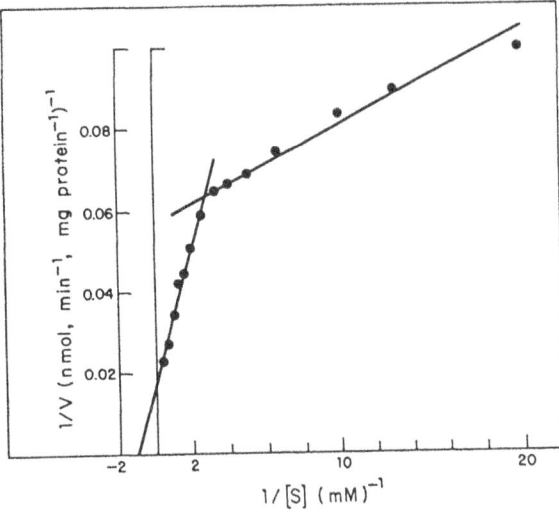

Fig. 5. Lineweaver-Burk plot of the rate of sucrose transport by sucrose-grown *R. tropici*. Specific radioactivity 0.1 mCi/mmol.; concentration range 0.05 to 3.0 m*M*. Uptake was determined at 2 min. This figure represents the data from one of the five experiments which shows similar results.

Table 2. Effect of metabolic inhibitors on sucrose transport by sucrose grown *Rhizobium tropici*[a]

Inhibitor	Inhibition %
Dinitrophenol (1 m*M*)	98.7±0.4
Potassium cyanide (1 m*M*)	95.8±0.2
Azide (2 m*M*)	88.8±0.2
N-Ethylmaleimide (1 m*M*)	96.7±1.5
EDTA (10 m*M*)	16.0±4.7
Arsenate (5 m*M*)	2.9±1.3

[a]Inhibitors were added 1 min before the addition of radioactivity and uptake was measured for 5 min. Results are average of three experiments and control rate was 42.6±5.1 nmol/min per mg protein.

range of sucrose concentration; (b) this organism lack catabolite repression control by glucose and, probably, could utilize different sugars and aminoacids (at least glutamate) simultaneously.

Acknowledgements

We acknowledge M A Rogel and L Candiani for technical help, DGAPA grant IN203691 and VLIR-ABOS grant from Belgium for financial support. V.I. Romanov was supported by CONACyT Cátedras Patrimoniales de Excelencia, Nivel II.

References

Day D A and Copeland L 1991 Carbon metabolism and compartmentation in nitrogen fixing legume nodules. Plant Physiol. Biochem. 29, 185–201.

De Hollander J A and Stouthamer A H 1979 Multicarbon-substrate growth of *Rhizobium trifolii*. FEMS Microbiol. Let. 6, 57–59.

Duncan M J 1981 Properties of Tn5-induced carbohydrate mutants in *Rhizobium meliloti*. J. Gen. Microbiol. 122, 61–67.

Glenn A R and Dilworth M J 1981 The uptake and hydrolysis of disaccharides by fast and slow-growing species of *Rhizobium*. Arch. Microbiol. 129, 233–239.

Herrada G, Puppo A and Rigaud J 1989 Uptake of metabolites by bacteroid-containing vesicles and by free bacteroids from french bean nodules. J. Gen. Microbiol. 135, 3165–3171.

Hoelzle I and Streeter J G 1990 Stimulation of α-glucosidases from fast-growing rhizobia and *Agrobacterium tumefaciens* by K$^+$, NH$_4^+$, and Rb$^+$. Can. J. Microbiol. 36, 223–227.

Martínez-Romero E, Segovia L, Mercante F M, Franco A A, Graham P and Pardo, M A 1991 *Rhizobium tropici*, a novel species nodulating *Phaseolus vulgaris* L. beans and *Leucaena* spp. trees. Int. J. Syst. Bacteriol. 41, 417–426.

Mackay I A, Dilworth M J and Glenn A R 1989 Carbon catabolism in continuous culture and bacteroids of Rhizobium Ieguminosarum MNF 3841. Arch. Microbiol. 152, 606–610.

Mieyal J J 1972 Sucrose phosphorylase from *Pseudomonas saccharophila*. Methods in Enzymology 28, 935–943.

Noel K D, Sanchez A, Fernandez L, Leemans J and Cevallos M A 1984 *Rhizobium phaseoli* symbiotic mutants with transposon Tn5 insertions. J. Bacteriol. 158, 148–155.

Romanov V I, Hajy-sadeh B R, Ivanov B F, Shaposhnikov G L and Kretowich W L 1985 Labelling of lupine nodule metabolites with ^{14}CO$_2$ assimilated from the leaves. Phytochemistry 24, 2157–2160.

Streeter J G 1991 Transport and metabolism of carbon and nitrogen in legume nodules. Adv. Bot. Res. 18, 129–187.

Thummer F and Verma D P S 1987 Nodulin-100 of soybean is the subunit of sucrose synthase regulated by the availability of free heme in nodules. J. Biol. Chem. 262, 14730–14736.

Vance C P and Hiechel G H 1991 Carbon in N$_2$ fixation:Limitation and exquisite adaptation. Annu. Rev. Plant Physiol. 42, 373–392.

Wong T Y 1990 Possible mechanism of mannose inhibition of sucrose-supported growth in N$_2$-fixing *Azotobacter vinelandii*. Appl. Environ. Microbiol. 56, 93–97.

Plant and Soil **161**: 97–101, 1994.
© 1994 *Kluwer Academic Publishers.*

Shoot/root assimilate allocation and nodulation of *Vigna unguiculata* seedlings as influenced by shoot light environment

M. J. Kasperbauer and P. G. Hunt
Coastal Plains Soil and Water Conservation Research Center, Florence, SC 29502–3039, USA

Key words: nodulation, photomorphogenesis, photosynthate partitioning, reflected light

Abstract

Spectral balance of light received by southern pea [*Vigna unguiculata* (L.) Walp.] seedling shoots affected photoassimilate allocation among leaves, stems and roots. A higher ratio of far-red (FR) relative to red (R) light resulted in longer stems, higher shoot/root biomass ratio, less massive roots and fewer nodules. The same response pattern to FR/R ratio was obtained in a controlled environment with artificial light sources, or in sunlight where the FR/R ratio was modified by reflection from different colored soil surfaces or by FR reflected from competing plants. The importance of early shoot/root photoassimilate allocation and nodulation may differ according to soil nitrogen availability and moisture content.

Introduction

Early nodulation and N_2 fixation are important in legume crops, especially when grown on N-deficient soils. Allocation of photoassimilates to roots is essential for nodule formation, and light environment of the shoot can influence allocation among growing plant parts. Downs et al. (1957) showed that brief exposures to red (R) or far-red (FR) light at the end of the day could influence stem length. Hendricks and Borthwick (1967) attributed regulation of seedling shoot morphogenesis to the action of phytochrome. Kasperbauer et al. (1964) noted that seedling shoots were morphologically responsive to prolonged exposures to FR wavelengths beyond the absorption peak for the FR-absorbing form of phytochrome. Kasperbauer (1971) used hydroponic grown seedlings to demonstrate the effect of FR/R ratio on biomass allocation among leaves, stems and roots. That work was extended to controlled-environment grown soybean [*Glycine max* (L.) Merr.] seedlings to demonstrate that the FR/R ratio at the end of the photosynthetic period could affect nodulation as well as shoot/root ratios (Kasperbauer et al., 1984). Greater photoassimilate allocation to roots was associated with formation of more nodules, and the affect of FR/R ratio on nodulation was greater on seedlings grown in potting medium with low available-N content. The research reported in this paper was conducted to determine whether spectral balance of light received by southern pea seedling shoots could affect photoassimilate allocation and nodulation during early seedling growth in sunlight.

Materials and methods

Southern pea [*Vigna unguiculata* (L.) Walp. cv. Mississippi Silver] seedlings were grown in individual containers in each of the three experiments so that roots as well as shoots could be evaluated after the treatment periods. At the end of each experiment, stems were measured, roots were washed, nodules were counted, and plants were separated into component parts. The parts were freeze-dried and weighed. Data are presented as means per plant ± S.E.

Controlled environment.

Seedlings were started and grown in 3-L pots of a 1:1 mixture of potting soil (Pro-Mix-BX, Premier Brands Inc., Red Hill, PA) and vermiculite. The seeds were inoculated with the S strain of *Bradyrhizobium japonicum* (Nitragin Co., Milwaukee, WI). All seedlings were grown in the same controlled environment chambers where they received 12 h photosynthetic periods from cool-white fluorescent lamps at 520 μmol m^{-2} s^{-1} and day/night temperatures of 27±2°C. At the end of the daily photosynthetic period, plants were exposed to either 5 min of R (3.6 Wm^{-2} in the 600–700 nm waveband), 5 min FR (3.6 Wm^{-2} in the 700–770 waveband), or 5 min FR followed immediately by 5 min R. The R and FR radiation units were as described by Kasperbauer (1987). Plants remained in uninterrupted darkness for 12 h after the R or FR treatments. Thus, the seedlings received the same treatment except for R and FR at the end of each day to put phytochrome predominantly in the FR-absorbing or the R-absorbing form, respectively, at the beginning of each night. The R and FR treatments began when the seedlings were in the unifoliate stage, and the same procedure was followed each day for 28 days until plants were evaluated.

Soil surface color effects.

Seeds were sown in 3-L pots of Norfolk loamy sand and placed 60 cm apart in groups of four on greenhouse benches. The loamy sand was taken from field plots previously used for southern pea. The soil was screened through a 2 mm sieve to remove old roots, and it was thoroughly mixed before equal amounts were added to each of the pots. A styrofoam insulation panel (122 × 122 × 2 cm) with four 2.5 cm (diam) holes was placed over each group of four pots so that an emerging

seedling was centered in each hole. The insulation panels were covered with about 1 cm of white or brick-red soil. Spectral distributions of light reflected from the soils were measured 10 cm above these surfaces. Spectral measurements were taken at 5 nm intervals from 400 to 800 nm with a LiCor-1800 Spectroradiometer (LiCor, Lincoln, NE.) equipped with a remote, hemispherical, cosine-corrected light collector on a 1.5 m fiber optic probe. Incoming sunlight was measured and the reflected light values were then calculated as percentages of the incoming sunlight at each measured wavelength. Spectral irradiances at 735 and 645 nm were used to calculate the FR/R ratios. These values were used because they approach the peaks for phytochrome action spectra in green plants; 645 nm was used instead of 660 nm because chlorophyll competition for light at 660 nm (the approximate phytochrome absorbance maximum in vitro) shifts the action peak to about 645 nm in green plants (Kasperbauer et al., 1964). The FR/R ratios shown in this report are relative to the ratio in direct sunlight, which was arbitrarily assigned a value of 1.00. The rationale for this approach is that plants are adapted to sunlight, and a FR/R ratio that deviates from that in direct sunlight might signal the plant to activate or repress genes that regulate adaptation to the altered light environment (Kasperbauer, 1988). The reflected light measurements were taken at solar noon ± 30 min on a clear day. Soil temperatures below the white and red surfaced insulation panels were monitored at depths of 2.5 cm at 5 min intervals and averaged each hour for 8 days. Temperatures were determined with copper-constantin thermocouples attached to a Campbell CR-7 Datalogger (Campbell Scientific, Logan, UT) as described by Hunt et al. (1989). Mean root temperature difference in the pots (root zone) below the insulation panels with the two different soil colors was less than 0.5°C. This approach allowed comparison of seedling morphological development over different colored soils when effects of soil surface color on rhizosphere temperature were minimized below the soil colors. There were eight plants per soil color. Seedlings were evaluated after 18 days.

Row orientation effects

Seeds were sown in 10 cm (diam.) × 45 cm polyvinyl chloride cylinders of Norfolk loamy sand that were embedded in field plots in north-south (N-S) versus east-west (E-W) rows that were 50 cm apart. This procedure was used in order to recover roots and nodules from seedlings grown in NS and EW rows. There were eight rows in each direction. Two cylinders were embedded in each of four rows in each direction. At emergence, seedlings were thinned to 5 cm apart within rows. Only one seedling was kept per cylinder. The plant spacing in rows between and within the cylinders provided similar amounts of leaf surface to reflect FR. Light measurements were recorded at about 1700 h on a cloudless day to determine row orientation effects on heliotropic influenced directional reflection of FR and on FR/R ratios received by seedlings that were growing in the cylinders. Seedlings were removed from the cylinders and evaluated after 30 days.

Results and discussion

Photoreversible effects of R and FR on assimilate allocation.

When photosynthetic light was held constant in a controlled environment, southern pea seedlings responded morphologically to R or FR received for 5 min at the end of the photosynthetic period (Table 1). The fact that plants responded to brief exposures to R (low FR/R ratio) and FR (high FR/R ratio) and that the effects of FR were reversed by R is consistent with the hypothesis that phytochrome functions in an environment-sensing mechanism that regulates photosynthate allocation among developing plant parts. In addition to having longer internodes, seedlings that received FR last each day (high FR/R ratio) developed less massive roots and a higher shoot/root biomass ratio. Seedlings whose shoots received the high FR/R ratio developed fewer nodules during the treatment period. Results of the controlled-environment study suggest that field management systems which modify the FR/R ratio received by growing southern pea seedlings should also result in modified photosynthate allocation and the amount of nodulation. To test this theory, southern pea seedlings were grown in soil with low available-N; and the FR/R ratio reaching the seedlings was modified by growing them over soil surfaces that reflected different FR/R ratios, and by growing seedlings in N-S versus E-W rows which differed in FR/R ratio because heliotropic leaf movement caused them to be directional FR reflectors (see Kasperbauer, 1987) near the end of each day.

Soil surface color effects.

Seedlings grown in sunlight over the soil surface color that reflected the higher FR/R ratio developed longer stems, a higher shoot/root biomass ratio, less massive roots, and fewer nodules (Table 2). The phytochrome system within the young seedlings apparently sensed the FR/R ratio reflected from the soil surface and initiated physiological processes that resulted in allocation of photoassimilate among growing plant parts, just as would occur if the FR/R ratio was altered in a controlled environment (Table 1) or by FR reflected from leaves of competing plants, as discussed below. Under field conditions the FR/R ratio reflected from the soil surface could also be influenced by plant residues from previous crops in conservation tillage systems, low growing weeds, or artificially colored mulches (Hunt et al., 1989; Kasperbauer and Hunt, 1987, 1992).

Row orientation effects.

In field plots, row orientation influenced the amount of reflected FR and the FR/R ratio received by nearby seedlings (Table 3). Green leaves absorbed most of the incoming R, but they reflected or transmitted most of the FR. Because of heliotropic (sun tracking) movement of southern pea leaves, they became directional FR reflectors. Seedlings in N-S rows received more reflected FR and higher FR/R ratios near the end of each day. The patterns were very similar to those pre-

Table 1. Shoot and root size and nodulation of southern pea seedlings in a controlled environment with 5 min end-of-day treatment with red (R, low FR/R ratio), far-red (FR, high FR/R ratio), or FR followed immediately by R each day for 28 consecutive days

Character	End-of-day-light		
	R	FR	FR,R
Stem length (mm)	164 ± 7[a]	277 ± 13	155 ± 8
Shoot/root (dry wt. ratio)	3.03 ± .27	4.01 ± .43	3.34 ± .20
Root dry wt. (mg)	997 ± 179	645 ± 71	906 ± 91
Nodules (no)	29 ± 10	8 ± 4	36 ± 13

[a] Values are means for 10 plants ± S.E.

Table 2. Shoot and root size and nodulation of southern pea seedlings grown on a greenhouse bench in sunlight for 18 days over different colored soils covering insulation panels

	Soil color	
	Gray-white	Brick-red
Upwardly reflected light (relative to incoming sunlight)[a]		
PAR(%)	24	10
FR/R (ratio)	1.00	1.18
Plant[b]		
Stem length (mm)	145 ± 3	156 ± 7
Shoot/root (dry wt. ratio)	2.47 ± .16	2.88 ± .13
Root dry wt. (mg)	1753 ± 157	1374 ± 119
Nodule dry wt. (mg)	93 ± 19	45 ± 14
Nodules (no)	176 ± 26	108 ± 29

[a] PAR, photosynthetically active radiation; FR/R, photon ratio at 735 nm relative to 645 nm. The FR/R ratios in upwardly reflected light are expressed relative to ratio in incoming sunlight, which was assigned a value of 1.00.
[b] Values are means for eight plants ± S.E.

viously reported for *Phaseolus vulgaris* L. (Kaul and Kasperbauer, 1988). Southern pea seedlings grown in cylinders that were embedded in the N-S rows developed longer stems, higher shoot/root biomass ratios, less massive roots, and fewer nodules (Table 3).

It is evident that the phytochrome system within the seedlings responded to the FR/R ratio as a "signal" of potential competition from other plants (FR reflectors). The adaptive response to a higher FR/R ratio was to allocate more pho-

tosynthate to developing stems, leaving less for new root growth. Less photoassimilate allocation to roots also resulted in less nodulation. In nature, the adaptive advantage of longer stems is that they increase the probability of plant survival because some leaves would be in photosynthetic sunlight above competing plants. Awareness of natural bioregulatory systems should be useful in developing field crop management systems. The importance of early shoot/root photoassimilate allocation and nodulation may differ

Table 3. Shoot and root size and nodulation of southern pea seedlings grown in field plots for 30 days in cylinders of loamy sand embedded in east-west (E-W) versus north-south (N-S) rows that were 50 cm apart

Character	Row orientation	
	E-W	N-S
Light[a]		
FR/R (ratio)	1.05	1.15
Plant[b]		
Stem length (mm)	140 ± 8	175 ± 6
Shoot/root (dry wt. ratio)	2.60 ± 0.23	3.06 ± 0.29
Root dry wt. (mg)	902 ± 81	806 ± 74
Nodule dry wt. (mg)	35 ± 9	19 ± 6
Nodules (no)	78 ± 10	55 ± 8

[a] FR/R, photon ratio at 735 nm relative to 645 nm. Spectra of light coming to shoot tips were measured parallel to the ground from N, S, E and W at 1700 h on a cloudless day, and the ratios are means for 8 measurements (2 each from N,S, E and W).
[b] Values are means for eight plants ± S.E.

according to soil N availability and soil moisture content. We hypothesize that E-W row orientation of southern pea and other legume seedlings on low-N, droughty soils might have the advantage of more massive roots, less moisture stress, more early nodulation and N_2 fixation, and greater productivity; whereas N-S row orientation could favor greater shoot growth and increased productivity on soils with high available-N and no water stress.

Acknowledgements

We thank W Sanders for technical assistance. Mention of a trademark, proprietary product, or vendor anywhere in this paper does not constitute a guarantee or warranty of the product by the U.S. Department of Agriculture and does not imply its approval to the exclusion of other products or vendors that may also be suitable.

References

Downs R J, Hendricks S B and Borthwick H A 1957 Photoreversible control of elongation of pinto beans and other plants. Bot. Gaz. 188,199–208.

Hendricks S B and Borthwick H A 1967 Function of phytochrome in regulation of plant growth. Proc. Nat. Acad. Sci. U.S.A. 58,2125-2130.

Hunt P G, Kasperbauer M J and Matheny T A 1989 Soybean seedling growth responses to light reflected from different colored soil surfaces. Crop Sci. 29, 130–133.

Kasperbauer M J 1971 Spectral distribution of light in a tobacco canopy and effects of end-of-day light quality on growth and development. Plant Physiol. 47,775–778.

Kasperbauer M J 1987 Far-red light reflection from green leaves and effects of phytochrome-mediated partitioning under field conditions. Plant Physiol. 85,350–354

Kasperbauer M J 1988 Phytochrome involvement in regulation of the photosynthetic apparatus and plant adaptation. Plant Physiol. Biochem. 26, 519–524.

Kasperbauer M J, Borthwick H A and Hendricks S B 1964 Reversion of phytochrome 730 (Pfr) to P660 (Pr) in *Chenopodium rubrum*. Bot. Gaz. 125,85-90.

Kasperbauer M J and Hunt P G 1987 Soil color and surface residue effects on seedling light environment. Plant and Soil 97,295–298.

Kasperbauer M J and Hunt P G 1992 Cotton seedling morphogenic responses to FR/R ratio reflected from different colored soils and soil covers. Photochem. Photobiol. 56,576–584.

Kasperbauer M J, Hunt P G and Sojka R E 1984 Photosynthate partitioning and nodule formation in soybean plants that received red or far-red light at the end of the photosynthetic period. Physiol. Plant. 61,549–554.

Kaul K and Kasperbauer M J 1988 Row orientation effects on FR/R light ratio, growth and development of field-grown bush bean. Physiol. Plant. 74,415–417.

Plant and Soil **161**: 103–108, 1994.

Mechanism of osmotically regulated N-acetylglutaminylglutamine amide production in *Rhizobium meliloti*

Linda Tombras Smith[1], Abdul Ameer Allaith[1] and Gary M. Smith[2]
[1]*Department of Agronomy and Range Science and* [2]*Department of Food Science and Technology, University of California, Davis, CA 95616, USA*

Key words: N-acetylglutaminylglutamine amide, osmoregulation, *Rhizobium meliloti*

Abstract

Rhizobium meliloti adapts to environments of high osmolarity by accumulating glutamate, trehalose, and the dipeptide N-acetylglutaminylglutamine amide (NAGGN) intracellularly. In this study, the mechanism of NAGGN production and accumulation was examined. NAGGN was produced in osmotically shocked cultures after a lag period of more than one hour, and NAGGN was undetectable in cultures treated with chloramphenicol, indicating that genetic induction is required for NAGGN accumulation. *In vitro* radiolabeling experiments demonstrated that the peptide synthesis step in NAGGN production did not occur ribosomally. Rather, it was catalyzed by an ATP-dependent enzyme that appeared to be both induced by high osmolarity and activated by K^+. Also, a mutant analysis suggested that NAGGN may be partly responsible for the osmotic tolerance observed in *R. meliloti*. 36% of mutants that were characterized as osmotically sensitive compared to the parent strain, were also found to contain reduced levels of NAGGN. The phenomenon of osmolyte accumulation as it relates to adaptation to other environmental stresses is discussed.

Abbreviations: MCAA – malate-casamino acids medium, NAGG – N-acetylglutaminylglutamine, NAG-GN – N-acetylglutaminylglutamine amide

Introduction

Many species of bacteria respond and adapt to hyperosmotic conditions in the environment by the intracellular accumulation of low molecular weight organic solutes called osmolytes (Csonka and Hanson, 1991). The accumulation of osmolytes is thought to counteract the dehydrating effect of low water activity in the medium, but not interfere with macromolecular structure or function. Several recent reports have shown that Rhizobia utilize this mechanism of osmotic adaptation, accumulating either glutamate (Botsford, 1984; Smith et al., 1994; Yap and Lim 1983; Yelton et al., 1983), trehalose (Elsheikh and Wood, 1990; Smith et al., 1994), or both (Smith and Smith, 1989). In *Rhizobium meliloti*, the root nodule symbiont of alfalfa, three osmolytes are accumulated: glutamate, trehalose, and an unusual dipeptide, N-acetylglutaminylglutamine amide (NAGGN) (Smith and Smith, 1989) which has thus far been found only in a few bacterial species (Severin et al., 1992; Smith et al., 1994). The NAGGN level in *R. meliloti* cells increases as the osmolarity of the growth medium is increased, so that it becomes the major osmolyte in highly stressed cells (i.e., with greater than 0.5 M NaCl in the growth medium). However, the relative concentrations of NAGGN, trehalose and glutamate are dependent on other environmental conditions, as well (Smith et al., 1994).

In this report the mechanism by which NAGGN production is osmotically regulated is described. Also, using both biochemical and genetic approaches the contribution of NAGGN to the osmotic tolerance of *R. meliloti* was examined.

Materials and methods

R. meliloti 102F34 was maintained on solid mannitol-yeast extract medium and cultures were grown in malate-casamino acids medium (MCAA) or malate-glutamate medium (Smith and Smith, 1989) with additions as noted. Mutations were chemically induced by N-methyl-N'-nitro-N-nitrosoguanidine using the procedure of Miller (Miller, 1972). Mutants were screened for growth on malateglutamate, MCAA, and MCAA plus 0.5 M NaCl. Salt sensitive mutants were assayed for NAGGN accumulation and N-acetylglutaminylglutamine (NAGG) synthetase activity. The intracellular concentration of NAGGN was quantitated by high pressure liquid chromatography using an Alltech Econosil NH$_2$5U (4.8 mm id × 250 mm) column with acetonitrile:H$_2$O, 80:20 v/v as mobile phase. A flow rate of 1 mL/min was used. Peaks were detected by absorbance at 202 nm using a Gilson Holochrome UV monitor, and peaks were integrated using a SpectraPhysics SP4270 integrator. Thin layer chromatography was carried out with Silica gel G60 plates and the solvent used was phenol:H$_2$O, 4:1. The NAGG synthetase activity assay was based on N-acetylglutamine- and ATP-dependent incorporation of [^{14}C] glutamine into NAGG. The reaction mixture contained 0.5 mM [^{14}C] glutamine (0.05 μCi), 10 mM N-acetylglutamine, 10 mM Mg/ATP, 50 mM Tris HCl, pH 8.5, 10 mM NH$_4$Cl, 1 mM dithiothreitol, and KCl as indicated. The product was separated from the radioactive substrate by pas-

sage over a Dowex 50 column as described previously (Powers-Lee, 1985)

Results

Osmoregulation of NAGG synthetase

Previously, it was reported that NAGGN accumulated in *R. meliloti* only in osmotically stressed cultures and that the intracellular concentration of NAGGN was proportional to the NaCl in the medium (Smith and Smith, 1989). Here, the kinetics of NAGGN production are probed in more detail. The possibility of genetic induction of NAGGN biosynthesis was investigated by using cultures osmotically upshocked with 0.4 M NaCl. No NAGGN was detected in cultures until two hours after upshock, and the maximum level of NAGGN (about 320 nmol/mg protein) was observed 10 hours after upshock. In cultures osmotically shocked in the presence of chloramphenicol the level of NAGGN after 10 hours was less than 11 nmol/mg protein. Hence, the addition of chloramphenicol reduced the NAGGN level over 28 fold, indicating that genetic induction likely occurred.

Although the upshock experiments provided evidence for the genetic induction of the NAGGN pathway, it did not address the issue of whether the synthesis of the dipeptide itself is ribosomal or nonribosomal. This step in NAGGN production was investigated using ^{14}C labeled glutamine, N-acetylglutamine, and ATP (Table 1). In the presence of all substrates the production of NAGG was observed; however, if either ATP or N-acetylglutamine were omitted from the reaction mixture, enzyme activity was either not observed or greatly reduced. To demonstrate that the product of the reaction was in fact NAGG, it was authenticated by mass spectral analysis (data not shown) and thin layer chromatography. Although much of the radiolabel co-migrated with glutamate, because of a high level of glutaminase activity in the crude extract, 80% of the remaining ^{14}C label co-migrated with authentic NAGG (R$_f$ = 0.21), while 20% migrated with the final product,

Table 1. NAGG synthetase activity in *R. meliloti* 102F34

Extract	Reaction mixture[a]	KCl (mM)	N-acetylgln-gln produced (nmol/30 min)
Crude extract of stressed cultures	complete	50	1.6
	complete	300	3.4
	minus N-acetylgln	50	0
	minus Mg/ATP	50	0.02
Crude extract of unstressed cultures	complete	50	0.15
	complete	300	0.72

[a]The complete reaction mixture contained [^{14}C] glutamine, N-acetylglutamine, Mg/ATP, Tris buffer, NH$_4$Cl, dithiothreitol and KCl as shown. Details are given in Materials and Methods.

Table 2. Activation of NAGG synthetase by salt

Addition to reaction mixture	Activity[a] (%)
No addition	49
0.03 M KCl	55
0.3 M KCl	100
0.6 M KCl	117
0.3 M NaCl	64
0.3 M NH$_4$Cl	61
0.3 M LiCl	20

[a]100% NAGG synthetase activity was that obtained with 0.3 M KCl in the reaction mixture.

NAGGN. Hence, dipeptide synthesis appears to be via an ATP-dependent synthetase.

NAGG synthetase activity was also measured in crude extracts of unstressed cultures. These extracts contained much less activity than those of stressed cultures (Table 1), consistent with the upshock experiment which indicated that genetic induction is required for a high level of activity.

In addition to genetic induction, NAGG synthetase activity is increased by allosteric activation of the synthetase enzyme. K$^+$ ions stimulated the in vitro production of NAGG at least two-fold, while Na$^+$, Li$^+$, or NH$_4^+$ could not substitute for K$^+$ effectively (Table 2). It should be noted that K$^+$ also stimulated the low level of enzyme activity observed in extracts from unstressed cultures (Table 1). However, when 0.5 M KCl was added to the growth medium as osmotic stress, it did not stimulate the accumulation of intracellular NAGGN more than NaCl did.

Mutant analysis of osmoregulation

Of approximately 40,000 colonies that were screened subsequent to chemical mutagenesis, 39 were osmotically sensitive. These mutants grew well in low salt media, but unlike the parent *R. meliloti* 102F34, was incapable of growth at 0.5 M NaCl. To determine the role that NAGGN plays in the osmotic tolerance of *R. meliloti*, the mutants were tested for osmoregulated NAGGN production and NAGG synthetase activity. Cultures grown on MCAA were incubated for at least 12 hours in MCAA plus 0.4 M NaCl to allow NAGGN production, if any, to occur and then assayed. Compared to the wild type, 14 out of the 39 mutants were reduced in NAGGN concentration by at least 2-fold, and 7 of these 14 mutants were reduced by at least 10-fold. Also,

10 of the 14 NAGGN mutants were also reduced in NAGG synthetase activity by 2-fold or more. In only one mutant was NAGG synthetase undetectable, and this mutant also contained about 10-fold less NAGGN than stressed wild type cultures. These results support the concept that NAGGN is at least partly responsible for the osmotic tolerance of *R. meliloti*.

Discussion

Results presented in this report are consistent with the view that NAGGN accumulation is osmotically controlled at the genetic level. Also, the mutant analysis indicated that NAGGN accumulation contributes to the overall osmotic tolerance of the species. Previously, only indirect evidence for this suggestion was available. For example, a natural isolate of *R. meliloti* that was considerably less salt tolerant than the laboratory strains was also deficient in NAGGN accumulation (Smith and Smith, 1989).

Although NAGGN appears to be important in osmoregulation in *R. meliloti*, it is not the sole participant. This point is illustrated by the fact that only 14 out of 39 osmotically sensitive mutants contained at least 2-fold less NAGGN than the wild type. This result is not surprising considering that *R. meliloti* also accumulates K^+, glutamate, and trehalose in response to osmotic stress. Hence, blocks in the pathway leading to the accumulation of these osmolytes may also result in the salt sensitive phenotype. Furthermore, mutations leading to salt sensitivity of any essential enzyme might also produce this phenotype. These possibilities are currently being investigated.

An overview of the current picture of osmoregulation in *R. meliloti* is given in Fig 1. In the absence of osmotic stress protectants in the growth medium, such as choline or glycine betaine, the cell transports K^+ and produces glutamate immediately after osmotic upshock (Botsford and Lewis, 1990). Genetic induction does not appear to be required for the accumulation of glutamate. Later, the dipeptide NAGGN is produced via nonribosomal biosynthesis and reach-

es a maximum concentration in the cell after 10 hours. The reaction by which the intermediate in the biosynthetic pathway, N-acetylglutamine, is produced is unknown at this time, but it is presumably via acetylation of glutamine. Also the donor of the nitrogen in the amidation of NAGG has not been unequivocally identified because of the high level of glutaminase in crude extracts. Likely candidates for this step are glutamine, glutamate, or NH_4^+. Experiments are under way to resolve these issues. Trehalose, the least abundant osmolyte in *R. meliloti*, seems to be important under the following environmental conditions: in media of exceedingly high osmolarity, when cultures are grown in poor carbon sources, during stationary phase, or when trehalose is supplied in the medium (Smith et al., 1994). To our knowledge, trehalose is the only osmolyte reported to accumulate by both uptake and biosynthesis in the cell.

Thus far, osmolyte accumulation in bacteria has been viewed almost exclusively as a mechanism of adaptation to stress caused by media of elevated salinity. But, hypothetically any stress that decreases the water activity of the environment could be alleviated by the accumulation of osmolytes. Already eucaryotes have been shown to accumulate osmolytes during desiccation (Crowe et al., 1992) and freeze (Storey and Storey, 1983) stress, in addition to salt stress (Yancey et al., 1982). Although no direct evidence for this hypothesis has been offered for procaryotes, it has been shown that incubation of *R. meliloti* in liquid media of reduced water activity improved its tolerance to desiccation (Mary et al., 1986).

Finally, the phenomenon of intracellular accumulation of osmolytes should be considered in an even broader context (Fig. 2). Beyond their role in protection against water stress, osmolytes may be key participants in other stress responses. For example, in *Escherichia coli*, the intracellular concentration of the osmolyte trehalose is controlled by high osmolarity, by elevated temperature, and by a stationary phase sigma factor (Hengge-Aronis et al., 1991). In fact, Hengge-Aronis et al. suggested that trehalose accumula-

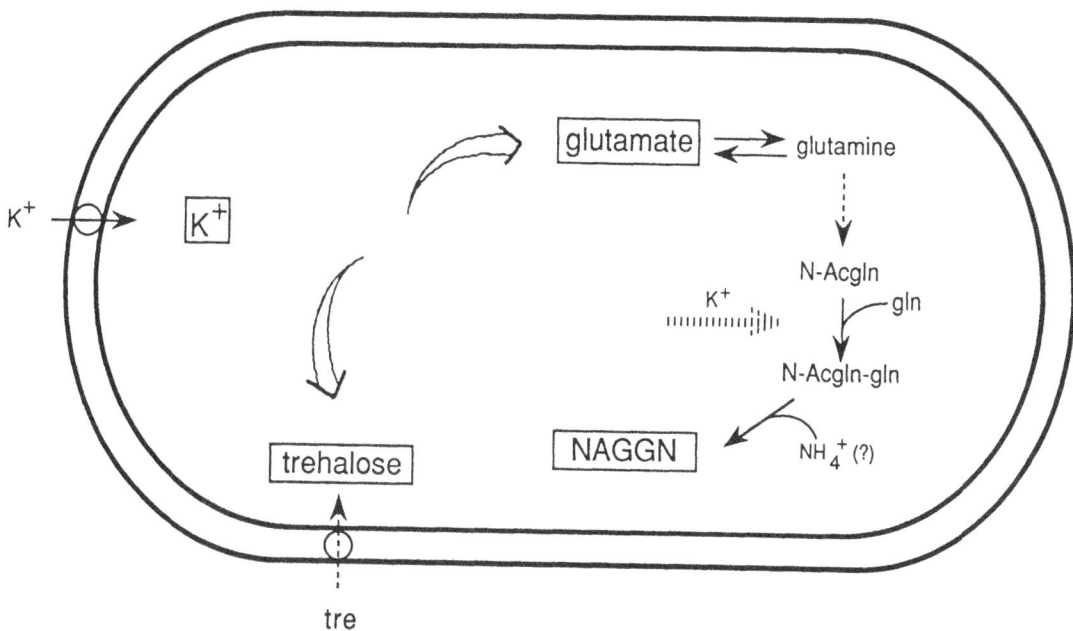

Fig. 1. Current view of osmoregulation in *R. meliloti*. In the absence of osmotic stress protectants, (e.g., glycine betaine) added to the growth medium, the osmotically stressed cell accumulates K^+, glutamate, NAGGN and low levels of trehalose. The absolute and relative concentrations of these osmolytes depend significantly on the environmental conditions.

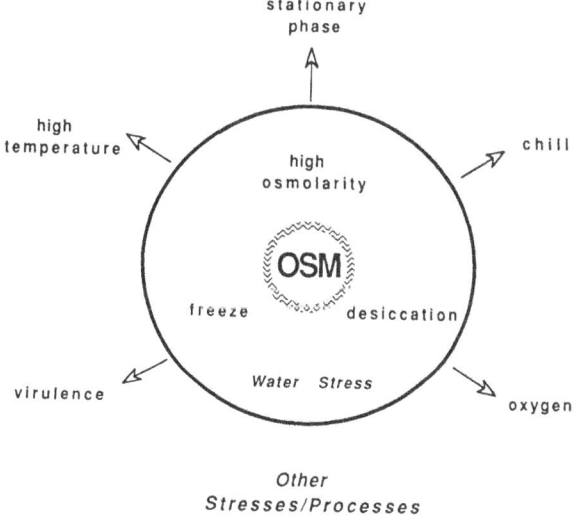

Fig. 2. Schematic representation of the relationships of osmolyte accumulation to other stresses and processes.

tion may be a general stress response in *E. coli*. Other osmotically regulated genes are controlled by the environmental oxygen concentration and reduced temperature, as well (Eshoo, 1988). Still others may be involved in the process of virulence in bacteria (Shortridge et al., 1992). Undoubtedly the elucidation of these interrelationships will be the focus of the next phase of research in this field.

Acknowledgements

This work was supported by National Science Foundation grant DCB-8903923 (L.T.S.)

References

Botsford J L 1984 Osmoregulation in *Rhizobium meliloti*: inhibition of growth by salts. Arch. Microbiol. 137, 124–127.

Botsford J L and Lewis T A 1990 Osmoregulation in *Rhizobium meliloti*: production of glutamic acid in response to osmotic stress. Appl. Environ. Microbiol. 56, 488–494.

Crowe J H, Hoekstra F A and Crowe L M 1992 Anhydrobiosis. Annu. Rev. Physiol. 54, 579–599.

Csonka L N and Hanson A D 1991 Prokaryotic osmoregulation: genetics and physiology. Annu. Rev. Microbiol. 45, 569–606.

Elsheikh E A E and Wood M 1990 Rhizobia and bradyrhizobia under salt stress: possible role of trehalose in osmoregulation. Lett. Appl. Microbiol. 10, 127–129.

Eshoo M W 1988 *lac* Fusion analysis of the *bet* genes of *Escherichia coli*: regulation by osmolarity, temperature, oxygen, choline and glycine betaine. J. Bacteriol. 170, 5208–5215.

Hengge-Aronis R, Klein W, Lange R, Rimmele M and Boos W 1991 Trehalose synthesis genes are controlled by the putative sigma factor encoded by *rpoS* and are involved in stationary-phase thermotolerance in *Escherichia coli*. J. Bacteriol. 173, 7918–7924.

Hua S-S T, Tsai V Y, Lichens G M and Noma A T 1982 Accumulation of amino acids in *Rhizobium* sp. strain WR1001 in response to sodium chloride salinity. Appl. Environ. Microbiol. 44, 135–140.

Mary P, Ochin D and Tailliez R 1986 Growth status of Rhizobia in relation to their tolerance to low water activities and desiccation stresses. Soil Biol. Biochem. 18, 179–184.

Miller J H 1972 Experiments in Molecular Genetics. Cold Spring Harbor Laboratory.

Powers-Lee S G 1985 N-Acetylglutamate synthase. Meth. Enzymol. 113, 27–35.

Shortridge V D, Lazdunski A and Vasil M L 1992 Osmoprotectants and phosphate regulate expression of phospholipase C in *Pseudomonas aeruginosa*. Molec. Microbiol. 6, 863–871.

Severin J, Wohlfarth A and Galinski E A 1992 The predominant role of recently discovered tetrahydropyrimidines for the osmoadaptation of halophilic eubacteria. J. Gen. Microbiol. 138, 1629–1638.

Smith L T and Smith G M 1989 An osmoregulated dipeptide in stressed *Rhizobium meliloti*. J. Bacteriol. 171, 4714–4717.

Smith L T, Smith G M, D'Souza M R, Pocard J-A, Le Rudulier D and Madkour M A 1994 Osmoregulation in *Rhizobium meliloti*: mechanism and control by other environmental signals. J. Exper. Zool. 268, 162–165.

Storey K B and Storey J M 1983 Biochemistry of freeze toleranace in terrestrial insects. TIBS 8, 242–245.

Yancey P H, Clark M E, Hand S C, Bowlus R D and Somero G N 1982 Living with water stress: evolution of osmolyte systems. Science 217, 1214–1222.

Yap S F and Lim S T 1983 Response of *Rhizobium* sp. UMKL 20 to sodium chloride stress. Arch. Microbiol. 135, 224–228.

Yelton M M, Yang S S, Edie S A and Lim S T 1983 Characterization of an effective salt-tolerant, fast-growing strain of *Rhizobium japonicum*. J. Gen. Microbiol. 129, 1537–1547.

Plant and Soil **161**: 109–114, 1994.
© 1994 *Kluwer Academic Publishers.*

What triggers the regulation of nitrogenase activity in forage legume nodules after defoliation?

U. A. Hartwig and J. Nösberger
Institute of Plant Sciences, Federal Institute of Technology , 8092 Zürich, Switzerland

Key words: defoliation, nitrogen fixation, nitrogen sink-strength, nitrogenase activity, oxygen diffusion, white clover

Abstract

The dramatic decrease in nitrogenase activity after the defoliation of forage legumes has been recognized for a long time; however, the underlying mechanisms are not understood yet. The impact of current photosynthesis can be excluded. The precise role of carbohydrate availability is still unclear and remains to be established. From current knowledge we can conclude that, after defoliation, nitrogenase activity in legume nodules is down-regulated by a variable oxygen diffusion resistance. The triggering elements are not known; there is, however, increasing evidence that the plant's demand for symbiotically fixed nitrogen plays an important role. The possibility is here discussed that, after defoliation, a nitrogen feedback mechanism regulates nitrogenase activity through a variable oxygen diffusion resistance in the nodules.

Introduction

During a growing season, forage legumes around the world are inevitably exposed to repeated defoliation either through grazing or cutting. The persistence of these legumes is highly dependent on their ability to withstand the physiological stress induced by leaf removal. One particular effect of defoliation on forage legumes is the rapid and severe decrease in N_2 fixing activity in the root nodules; hence, it is assumed that the post defoliation growth rate and persistence of these plants are affected by the sensitivity of N_2 fixation to defoliation. The decrease in nitrogenase activity due to defoliation has been reported many times with respect to white clover (*Trifolium repens* L.) (Hartwig et al., 1987; Moustafa et al., 1969; Ryle et al., 1985), alfalfa (*Medicago sativa* L.) (Cralle and Heichel, 1981; Denison et al., 1992; Fishbeck and Phillips, 1982; Vance et al., 1979), and many other legume species. A first important observa-

tion was that no bacteroid degradation occurs in the nodules within several hours after defoliation (Gordon et al., 1986; Vance et al., 1980).

Since symbiotic N_2 fixation is a highly energy demanding process, very active nodule respiration is needed in order to provide the nitrogenase enzyme with the required amounts of energy and reducing power (Day and Copeland, 1991). Therefore, numerous investigators have directed their efforts towards exploring the response of nodule respiration to defoliation in an attempt to explain the dramatic decrease in nitrogenase activity after defoliation. Carbohydrates and oxygen are the substrates which support respiration and hence nitrogenase activity. It can be assumed that any limitation of the carbohydrate or oxygen supply to the infected nodule cells decreases nitrogenase activity. A number of attempts to explain the response of nitrogenase activity to defoliation with varying carbohydrate or oxygen supply are briefly reviewed here. We also discuss the hypothesis that the demand for symbiotically fixed nitrogen is involved in the regulation of nitrogenase activity.

Does current photosynthesis provide carbon for nitrogenase activity?

The observation that darkening during day time or over a period of several days results in a decrease in nitrogenase activity led to the conclusion that products from current photosynthesis are the primary carbon source driving nitrogenase activity. However, this interpretation was very controversial. Hartwig et al. (1990) demonstrated that both illumination by night and darkening during the day decreased nitrogenase activity significantly. It was concluded that nitrogenase activity responded negatively to the disturbance of the regular diurnal light cycle and not to the interruption of photosynthesis itself. If current photosynthesis determined nitrogenase activity, then an increase or decrease in atmospheric CO_2 would also cause immediately an increase or decrease in nitrogenase activity. However, Finn and Brun (1982) and Williams et al. (1982) could not show a short term increase in nitrogenase activity due to elevated CO_2. Likewise, Heim (1993) did not find a significant and fast decrease in nitrogenase activity after the exposure of plant leaves to less than $1O\ \mu l\ l^{-1}\ CO_2$, whereas in only two hours after defoliation, nitrogenase activity was reduced by 60%. Culvenor and Simpson (1990) also were not able to restore nitrogenase activity after the partial defoliation of *Trifolium subterraneum* by increasing net photosynthesis to the pre-defoliation value with elevated CO_2 and supplemental light. From these results, it follows that current photosynthesis does not directly regulate nitrogenase activity.

Do stored carbohydrates play a role in maintaining nitrogenase activity after defoliation?

Because stored carbohydrates act transiently as a carbon source to support regrowth and metabolism after defoliation, it was generally assumed that availability of stored carbohydrates largely determines nitrogenase activity after defo-

liation. In an attempt to test this hypothesis, Hartwig et al. (1990) modified carbohydrate reserves by exposing clover plants to different atmospheric CO_2 concentrations for three weeks prior to defoliation. However, the resulting distinct differences in reserve carbohydrate contents did not influence the response of nitrogenase activity to defoliation. Moreover, the time course of the re-establishment of nitrogenase activity during regrowth did not correspond to concentrations of the carbohydrate reserves. Therefore we can conclude that stored carbohydrates are usually in excess and hence do not account for differences in nitrogenase activity, unless concentrations fall below a certain threshold value.

The role of oxygen in nitrogenase activity

Although oxygen is an irreversible inhibitor of nitrogenase, it is required by the bacteroids to drive respiration and hence nitrogenase activity. As a result, a high oxygen flux at a low concentration has to be maintained. Investigating the nodules' regulatory potential at various oxygen partial pressures in the soil, Criswell et al. (1976) and Weisz and Sinclair (1987) demonstrated that the N_2 fixing system in soybean nodules is able to adapt very quickly to a wide range of pO_2. Minchin et al. (1983, 1986) and Witty et al. (1984) reported several cases in which environmental effects obviously limited nitrogenase activity increasing the oxygen diffusion barrier in the nodules. In spite of the knowledge that oxygen possibly plays a key role in regulating nitrogenase activity, the hypothesis that oxygen supply could limit nitrogenase activity after defoliation was not tested until Hartwig et al. (1987) demonstrated that oxygen supply limits nitrogenase activity in white clover nodules after defoliation. This observation was confirmed with soybean (*Glycine max* [L.] Merr.) (Denison et al., 1991), alfalfa (*Medicago sativa* L.) and birdsfoot trefoil (*Lotus corniculatus* L.) (Denison et al., 1992) and pea (*Pisum sativum* L.) and lupin (*Lupinus albus* L.) (Diaz del Castillo et al., 1992). Applying several degrees of defoliation, Heim (1993) showed that the decrease in nitrogenase activity was closely accompanied by an appropriate increase in the oxygen limitation of nitrogenase activity. He found no response if up to 50% of the leaves were removed and a gradual increase in oxygen limitation if more than 50% of the leaf area was removed.

What triggers the regulation of O_2 diffusion to the infected zone of the nodules?

"Long term mechanisms" that affect nodule oxygen diffusion were reported to be associated with morphological features (Dakora and Atkins, 1991; Parsons and Day, 1990) or glycoprotein deposition (James et al., 1991). However, in order to explain the extremely rapid change in the nodule oxygen diffusion resistance, e.g. after defoliation, alternative mechanisms must be tested. Such mechanisms could involve changes in nodule cell shape or in the intercellular air spaces in the nodules. Such effects could be related to water movement into and out of intercellular spaces, induced by osmotically active compounds (Layzell and Hunt, 1990). Nevertheless, the question about the plant elicitor that initiates such responses would still not be answered.

We suggest that the decrease in nodule carbohydrate concentrations after defoliation (Gordon et al., 1986; Hartwig et al., 1990; Heim, 1993) is the result rather than the cause of the decrease in nitrogenase activity after defoliation. This hypothesis is consistent with a recent finding by Denison et al. (1992) demonstrating that, after the defoliation of alfalfa and birdsfoot trefoil, V_{max} of nodule respiration (oxygen saturated respiration) decreases only after the nodule permeability decreased.

The possible impact of the plant's demand for symbiotically fixed nitrogen on nitrogenase activity

Using nitrate reductase-minus pea mutants, Jacobsen (1984) and Shelp et al. (1991) observed that nitrogenase activity did not respond significantly to nitrate supply, whereas activity was halved within two days in the wild-type pea. It can be assumed that in nitrate reductase-minus plants, nitrate does not provide an additional assimilable nitrogen source and consequently the plants have to keep fixing N_2 in order to satisfy their nitrogen requirements. From this result, one can hypothesize that the demand for symbiotically fixed nitrogen governs nitrogenase activity. In fact, several environmental effects and manipulation of either the plant nitrogen sink or source (N_2 fixing system) are reported to affect nitrogenase activity (Table 1).

After defoliation, growth and likewise plant nitrogen demand are temporarily reduced (Boller and Nösberger, 1988). As a result, it would seem feasible that N compounds from current N_2 fixation (e.g. amino acids) are not exported from the nodules and that hence a nitrogen feedback mechanism could decrease nitrogenase activity. The concept of feedback regulation is a very common feature in biochemical pathways. In the case of the regulation of nitrogenase activity after defoliation, such a hypothesis is strongly supported in findings by Heim et al. (1993), who succeeded in significantly diminishing the negative response of nitrogenase activity to defoliation by preventing the nodules from fixing N_2 after defoliation. This was achieved by continuously exposing the nodulated root system to an $Ar:O_2$ (80:20, v:v) atmosphere. It was suggested that under conditions that prevented N_2 fixation ($Ar:O_2$), a possible nitrogen feedback mechanism could not become effective because no N-containing products were formed from current N_2 fixation.

The concept that the demand for symbiotically fixed nitrogen (plant N sink-strength) is a key element in regulating nitrogenase activity might be generally applicable: For instance, Zhu et al. (1991) reported enhanced nitrogenase activity and increased nodule dry weights after growing soybean seedlings under $Ar:O_2$ for 21 days. Monitoring the transport of products from current N_2 fixation after pod removal and reciprocal grafting on soybean, Fujita et al. (1991a; b) postulated a feedback control of nitrogen fixation. The same hypothesis was proposed by Parsons et al. (1993) after they demonstrated increased nodulation under $Ar:N_2:O_2$. How such a feedback regulation could indeed operate was demonstrated by Oti-Boateng and Silsbury (1993). They could sup-

Table 1. Environmental effects or plant manipulation that either reduce or increase the demand for symbiotically fixed nitrogen (N sink) or that reduce or increase nitrogenase activity (N source). Only short term effects (<2 days) are summarized in this table

Effect	Plant species[a]	Response	Reference
defoliation, reducing N sink strength	alfalfa pea and lupin white clover	decrease in nitrogenase activity within few hrs; decrease in nodule permeability	Denison et al., 1992 Diaz del Castillo et al., 1992 Hartwig et al., 1987
nitrate supply, reducing demand for fixed N	soybean white clover soybean	decrease in nitrogenase activity within 48 hrs; decrease in nodule permeability	Carroll et al., 1987 Minchin et al., 1986 Vessey et al., 1988
removing 50% of nodules, reducing N source capacity	faba bean	increase in specific nitrogenase activity (maintaining total nitrogenase activity) within 24 hrs	Herdina and Silsbury, 1990
increasing rhizosphere pO_2, stimulating N source activity	subclover soybean	transient increase in nitrogenase activity, then unchanged activity; decrease in nodule permeability	Davey and Simpson, 1989 Weisz and Sinclair, 1987
decreasing rhizosphere pO_2, depressing N source activity	subclover soybean	transient decrease in nitrogenase activity, then unchanged activity; increase in nodule permeability	Davey and Simpson, 1989 Weisz and Sinclair, 1987
dehydration, depressing nodule outflow	soybean	decrease in nitrogenase activity; decrease in nodule permeability	Durand et al., 1987

[a]alfalfa (*Medicago sativa* L.), pea (*Pisum sativum* L.), lupin (*Lupinus albus* L.), soybean (*Glycine max* [L.] Merr.), white clover (*Trifolium repens* L.), faba bean (*Vicia faba* L.), subclover (*Trifolium subterraneum* L.).

press nitrogenase activity by injecting asparagine or glutamine into stems.

Model

We propose that the plant nitrogen sink-strength - through a nitrogen feedback mechanism - is involved in controlling the quick response of nitrogenase activity after defoliation. The following response chain is suggested: Due to defoliation, the plant's nitrogen sink-strength is reduced and ongoing N_2 fixation leads to an accumulation of nitrogen compounds in the nodules. As a result, through an unknown mechanism, O_2 diffusion resistance in the nodules increases, resulting in an inhibition of respiration, and thereby decreased nitrogenase activity. This concept would provide a simple explanation for the strong response of nitrogenase activity to defoliation.

References

Boller B C and Nösberger J 1988 Influence of dissimilarities in temporal and spatial N-uptake patterns on ^{15}N-based estimates of fixation and transfer of N in ryegrass-clover mixtures. Plant and Soil 112, 167–175.

Carroll B J, Hansen A P, McNeil D L and Gresshoff P M 1987 Effect of oxygen supply on nitrogenase activity of nitrate- and dark-stressed soybean (*Glycine max* (L.) Merr.) plants. Aust. J. Plant Physiol. 14, 679–687.

Cralle H T and Heichel G H 1981 Nitrogen fixation and vegetative regrowth of alfalfa and birdsfoot trefoil after successive harvests or floral debudding. Plant Physiol. 67, 898–905.

Criswell J G, Havelka U D, Quebedeau B and Hardy R W F 1976 Effect of rhizosphere pO_2 on nitrogen fixation by excised and intact nodulated soybean roots. Crop Sci. 17, 39–44.

Culvenor R A and Simpson R J 1990 Studies on the relation between net photosynthesis and nitrogenase-linked respiration in subterranean clover. J. Exp. Bot. 41, 933–939.

Dakora F D and Atkins C A 1991 Adaptation of nodulated soybean (*Glycine max* L. Merr.) to growth in rhizospheres containing nonambient pO_2. Plant Physiol. 96, 728–736.

Davey A G and Simpson R J 1989 Changes in nitrogenase activity and nodule diffusion resistance of subterranean clover in response to pO_2. J. Exp. Bot. 40, 149–158.

Day D A and Copeland L 1991 Carbon metabolism and compartmentation in nitrogen-fixing legume nodules. Plant Physiol. Biochem. 29, 185–201.

Denison R F, Hunt S and Layzell D B 1992 Nitrogenase activity, nodule respiration, and O_2 permeability following detopping of alfalfa and birdsfoot trefoil. Plant Physiol. 98, 894–900.

Denison R F, Smith D L, Legros T and Layzell D B 1991 Non-invasive measurement of internal oxygen concentration of field-grown soybean nodules Agron. J. 83, 166–169.

Diaz del Castillo L, Hunt S and Layzell D A 1992 O_2 regulation and O_2-limitation of nitrogenase activity in root nodules of pea and lupin. Physiol. Plant. 86, 269–278.

Durand J L, Sheehy J E and Minchin F R 1987 Nitrogenase activity, photosynthesis and nodule water potential in soybean plants experiencing water deprivation. J. Exp. Bot. 38, 311–21.

Finn G A and Brun W A 1982 Effect of atmospheric CO_2 enrichment on growth, nonstructural carbohydrate content, and root nodule activity in soybean. Plant Physiol. 69, 327–331.

Fishbeck K A and Phillips D A 1982 Host plant and *Rhizobium* effects on acetylene reduction in alfalfa during regrowth. Crop Sci. 22, 251–254.

Fujita K, Masuda T and Ogata S 1991a Analysis of factors controlling dinitrogen fixation in wild and cultivated soybean (*Glycine max*) plants by reciprocal grafting. Soil Sci. Plant Nutr. 37, 233–240

Fujita K, Masuda T and Ogata S 1991b Effect of pod removal on fixed-N (^{15}N_2) export from soybean (*Glycine max* L) nodules. Soil Sci. Plant Nutr. 37, 463–469

Gordon A J, Pyle G J A, Mitchell D F, Lowry K H and Powell C E 1986 The effect of defoliation on carbohydrate, protein and leghaemoglobin content of white clover nodules. Ann. Bot. 58, 141–156.

Hartwig U A, Boller B C, Baur-Höch B and Nösberger J 1990 The influence of carbohydrate reserves on the response of nodulated white clover to defoliation. Ann. Bot. 65, 97–105.

Hartwig U A, Boller B C and Nösberger J 1987 Oxygen supply limits nitrogenase activity of clover nodules after defoliation. Ann. Bot. 59, 285–291.

Heim I 1993 Regulation of nitrogenase activity after defoliation of white clover: The involvement of current photosynthesis, current N_2 fixation and nodular carbohydrate content. PhD-thesis No 10007, Federal Institute of Technology, Switzerland.

Heim I, Hartwig U A and Nösberger J 1993 Current nitrogen fixation is involved in the regulation of nitrogenase activity in white clover (*Trifolium repens* L.). Plant Physiol. 103, 1009–1014.

Herdina and Silsbury J H 1990 The effect of reduction in the number of nodules on nodule activity of faba bean (*Vicia faba* cv. Fiord). Ann. Bot. 65, 473–481.

Jacobsen E 1984 Modification of symbiotic interaction of pea (*Pisum sativum* L.) and *Rhizobium leguminosarum* by induced mutations. Plant and Soil 82, 427–438.

James E K, Sprent J I, Minchin F R and Brewin N J 1991 Intercellular localization of glycoprotein in soybean nodules: effect of altered oxygen concentration. Plant Cell Environ. 14, 467–476.

Layzell D B and Hunt S 1990 Oxygen and the regulation of nitrogen fixation in legume nodules. Physiol. Plant. 80, 322–327.

Minchin F R, Minguez M I, Sheehy J E Witty J F and Skot L 1986 Relationship between nitrate and oxygen supply in symbiotic nitrogen fixation by white clover. J. Exp. Bot. 37, 1103–1113.

Minchin F R, Witty J F, Sheehy J E and Müller M 1983 A major error in the acetylene assay: decreases in nodular nitrogenase activity under assay conditions. J. Exp. Bot. 34, 641–649.

Moustafa E, Ball R and Field R R O 1969 The use of acetylene reduction to study the effect of nitrogen fertilizer and defoliation on nitrogen fixation by field-grown white clover. N. Z. J. Agric. Res. 12, 691–696.

Oti-Boateng C and Silsbury J H 1993 The effects of exogenous amino acid on acetylene reduction activity of *Vicia faba* L. cv. Fiord. Ann. Bot. 71, 71–74.

Parsons R and Day D A 1990 Mechanism of soybean nodule adaptation to different oxygen pressures. Plant Cell Environ. 13, 501–512.

Parsons R, Stanforth A, Raven A J and Sprent J I 1993 Nodule growth and activity may be regulated by a feedback mechanism involving phloem nitrogen. Plant Cell Environ. 16, 125–136.

Ryle G J A, Powell C E and Gordon A J 1985 Short-term changes in CO_2 evolution associated with nitrogenase activity in white clover in response to defoliation and photosynthesis. J. Exp. Bot. 165, 634–643.

Shelp B J, Taylor D C and Nelson L M 1991 Carbon and nitrogen partitioning in young nodulated pea (wild type and nitrate reductase-deficient mutant) plants exposed to NO_3^- or NH_4^+. Can. J. Bot. 69, 1780–1786.

Vance C P, Heichel G H, Barnes D K, Bryan J W and Johnson L E 1979 Nitrogen fixation, nodule development, and vegetative regrowth of alfalfa (*Medicago sativa* L.) following harvest. Plant Physiol. 64, 1–8.

Vance C P, Johnson J E B, Halvorsen A M, Heichel G H and Barnes D k 1980 Histological and ultrastructural observations of *Medicago sativa* root nodule senescence after foliage removal. Can. J. Bot. 58, 295–309.

Vessey J K, Walsh K B and Layzell D B 1988 Oxygen limitation of N_2 fixation in stem-girdled and nitrate-treated soybean. Physiol. Plant. 73, 113–121.

Weisz P R and Sinclair T R 1987 Regulation of soybean nitrogen fixation in response to rhizosphere oxygen. II. Quantification of nodule gas permeability. Plant Physiol. 84, 906–910.

Williams L E, DeJong T M and Phillips D A 1982 Effect of change in shoot carbon-exchange rate on soybean nodule activity. Plant Physiol. 69, 432–436.

Witty J F, Minchin F R, Sheehy J E and Minguez M 11984 Acetylene induced changes in the oxygen diffusion resistance and nitrogenase activity of legume root nodules. Ann. Bot. 53, 13–20.

Zhu Y, Schubert K R and Kohl D H 1991 Physiological responses of soybean plants grown in a nitrogen-free or energy limited environment. Plant Physiol. 96, 305–309.

Plant and Soil **161**: 115–125, 1994.
© 1994 *Kluwer Academic Publishers.*

Nodulation and nitrogen fixation in extreme environments

L.M. Bordeleau and D. Prévost
Research Station, Agriculture Canada, 2560 Hochelaga Blvd., Sainte-Foy, Québec, Canada G1V 2J3

Key words: nodulation, nitrogenase, adaptation, extreme environments, arctic rhizobia

Abstract

Biological nitrogen fixation is a phenomenon occurring in all known ecosystems. Symbiotic nitrogen fixation is dependent on the host plant genotype, the *Rhizobium* strain, and the interaction of these symbionts with the pedoclimatic factors and the environmental conditions. Extremes of pH affect nodulation by reducing the colonization of soil and the legume rhizosphere by rhizobia. Highly acidic soils (pH <4.0) frequently have low levels of phosphorus, calcium, and molybdenum and high concentrations of aluminium and manganese which are often toxic for both partners; nodulation is more affected than host-plant growth and nitrogen fixation. Highly alkaline soils (pH >8.0) tend to be high in sodium chloride, bicarbonate, and borate, and are often associated with high salinity which reduce nitrogen fixation. Nodulation and N-fixation are observed under a wide range of temperatures with optima between 20-30°C. Elevated temperatures may delay nodule initiation and development, and interfere with nodule structure and functioning in temperate Iegumes, whereas in tropical legumes nitrogen fixation efficiency is mainly affected. Furthermore, temperature changes affect the competitive ability of *Rhizobium* strains. Low temperatures reduce nodule formation and nitrogen fixation in temperate legumes; however, in the extreme environment of the high arctic, native legumes can nodulate and fix nitrogen at rates comparable to those observed with legumes in temperate climates, indicating that both the plants and their rhizobia have successfully adapted to arctic conditions. In addition to low temperatures, arctic legumes are exposed to a short growing season, a long photoperiod, low precipitation and low soil nitrogen levels. In this review, we present results on a number of structural and physiological characteristics which allow arctic legumes to function in extreme environments.

Introduction

Biological nitrogen fixation is a phenomenon occurring in all known ecosystems and is undoubtedly of greatest agricultural importance. Symbiotic N_2 fixation involves different hosts and microsymbionts between legumes and bacteria belonging to the genera *Rhizobium, Bradyrhizobium* and *Azorhizobium*. In this paper, the term "rhizobia" embraces strains of all three genera.

Symbiotic N_2 fixation is dependant on host cultivar and rhizobia, but as well may be limited by pedoclimatic factors especially those associated with the acid soil complex of high aluminium and manganese, low calcium and phosphorus. Some legumes are especially sensitive to these stresses when dependent on symbiotic nitrogen fixation which requires vigorous plant growth. The establishment of an effective symbiosis requires: (a) colonization and survival in soil by rhizobia as saprophytes in competition with other endogenous microbes; there is evidence that this phase is also limited by soil infertility and extremes of temperature, (b) a rapid colonization of the rhizosphere prior to root infection and genetic compatibility between host and root nodule bacteria to establish an effective nodule, and (c) a favourable environment to allow maximum fixation. Microorganisms live in most of the environments of the earth, from frigid arctic soils to near boiling hot springs. From the microbial point of view, their ubiquity raises the question of what is normal and what is extreme, it is perhaps best to think of gradients of temperature, pH, salinity, nutrients and moisture, and to define extreme conditions as the ends of these gradients.

The purpose of this review is to consider two aspects of the environment on the symbiosis. At first, we highlight the limits of the symbiosis as imposed by some environmental factors, particularly soil pH and soil fertility, water stress and salinity, and temperature. These have been reviewed in recent papers of (cf. Alexander, 1985; Brockwell et al., 1991; Eaglesham and Ayanaba, 1984; Graham, 1992; Munns, 1986; Zahran, 1991). We identify stresses that affect the microsymbiont, those that affect the host plant and the functioning of the symbiotic system mainly on the nodulation and nitrogen fixation. Factors important in the survival and function of rhizobia in soil differ with the organisms and environment. The choice of factors is selective rather than exhaustive and it mainly deals with the symbiotic associations in the temperate environments. In the second part, we examine the adaptation of legume nitrogen fixation to the extreme environment of the high arctic and we highlight some unusual structural and physiological characteristics of this symbiosis encountered in an environment of low temperatures, short growing seasons, long photoperiods, low precipitation and low soil fertility.

Soil pH and soil fertility factors

Most leguminous plants require a neutral or only a slightly acid soil pH for growth with nodulation problems to be expected once the pH falls much below 5.5. Acid soils usually have some inherent adverse concentrations of elements coupled with related nutrient deficiencies. The principal effects of soil acidity may be resolved into hydrogen ion concentration, deficiencies of calcium, phosphorus and molybdenum, and excessive quantities of aluminium and manganese. Soil acidity adversely affects the survival, growth and nitrogen fixation of micro-organisms, while nutritional disorder affects legume-*Rhizobium* symbiosis (Lie, 1981). These problems have been reviewed in detail by Munns (1986) and Graham (1992). The following comments highlight some aspects of the situation. Each phase of the legume-*Rhizobium* symbiosis may differently be affected: i) rhizobial survival in the soil and growth in the rhizosphere, ii) infection and nodule establishment, iii) nodule function and iv) growth of the host plant. Species of *Rhizobium* are known to differ in their tolerance of soil acidity; the slow-growing *Bradyrhizobium* strains being generally more acid tolerant than the fast-growing species, especially *R. meliloti* (Munns and Keyser, 1981). Hydrogen ion activity is a major factor restricting the survival and growth of rhizobia in soil. Differences have been observed between *R. meliloti* strains in their ability to grow and to nodulate alfalfa (*Medicago sativa*) under acid conditions (Bordeleau et al., 1977; Rice et al., 1977). Brockwell et al. (1991) suggested that acid pH was the major determinant limiting the number of naturally-occurring *R. meliloti* in non-cultivated soils of New South Wales. Some strains of *R. meliloti* are more effective than others in nodulating alfalfa grown on moderately acid soils (Rice, 1982). Low pH tolerant strains have been selected and were shown to effectively nodulate alfalfa cultivars in moderately acid soil (Rice and Olsen, 1983, 1988). Differences in acid tolerance among common bean (*Phaseolus vulgaris*) cultivars has also been reported (Vargas and Graham, 1988). Use of an acid-tolerant *R. meliloti* strain with *Medicago polymorpha*, adapted to acid conditions, has extended the area of Australia sown to pasture by some 350,000 ha since 1985 (Howieson et al., 1988; Loi et al., 1993). Few cultivated legume species are adapted to low pH levels; the primary protective mechanisms of acid tolerance in certain cultivars of lentil (*Lens culinaris*) is excess production of citric, malic, aspartic, gluconic and succinic acids in root exudate under acidic condition (Rai, 1992). Excess production of citric acid was also reported on white lupin (*Lupinus albus*) (Gardner et al., 1983).

In acid soils with pH <5.0 where heavy metal activity is really only relevant, the presence of available aluminium inhibits nodulation directly and indirectly by stunting root growth, and also tends to compound the effect of low levels of calcium by inhibiting its uptake (Bell and Edwards, 1987). The nodule number decreases with decreasing calcium availability and with increasing aluminium level in the soil. Aluminium is a potent stress to the growth of free-living rhizobia; fast-growing rhizobia appear to be less tolerant to aluminium than slow-growing rhizobia. Aluminium not only prevents some plants from nodulating, but also delays and depresses nodulation. However, high aluminium content does not have effect on the functioning of nodules in N_2-fixation. Therefore, for acid soils with high aluminium content, improvement is achieved by manipulating the plant rather than the *Rhizobium* (Taylor et al., 1990). Foy and Lee (1987) suggested that aluminium-tolerant plants contain and exude more organic acid and other ligands that form stable chelates with aluminium and thereby reduce its chemical activity and toxicity.

Unsuccessful symbiosis in the presence of excess manganese may be explained by the failure in the infection process rather than to the malfunctioning of nitrogenase activity. Manganese toxicity mainly affects legume growth; tolerance to manganese varies considerably between and within legume species (Keyser and Munns, 1979; Munns and Keyser, 1981). The adverse effects of manganese are alleviated by liming which reduce hydrogen ion concentration and increase calcium cation concentration (Bell and Edwards, 1987).

Phosphorous deficiency is common in acid soil where this element may be immobilized as iron and aluminum phosphates. Phosphorus deficiency limits nodulation indirectly by reducing legume growth rather by direct action on the infection process, although evidence implies some rhizo-

bia able to nodulate at lower phosphate levels than others (Eaglesham and Ayanaba, 1984). Nodule development requires adequate phosphorus and nodules accumulate a higher phosphorus content than roots. A *Rhizobium*-mycorrhiza association may succeed in establishing symbiosis in low phosphorus acid soil. An exception found was lupin group which does not establish successful plant-mycorrhiza association, and still effectively extracts fixed soil phosphorus to fulfil the plant demand; lupins have adapted to this situation by producing organic acid in root exudates (Gardner et al., 1983).

Generally, lime application in acid soils corrects the observed toxicity and element deficiencies (Rice, 1975). However, liming in many areas is not always economically feasible because of the high application rates needed and high transportation cost. Therefore, the development of adapted systems by acid and high aluminium tolerant strains of *Rhizobium*, as well as tolerant plants to acid pH and adapted to low soil fertility is the strategy to favour.

Alkalinity can develop from saline soils with low calcium reserve. In lowering the water table, soluble salts are washed down the profile and exchangeable calcium is replaced by sodium. Soil carbon dioxide forms carbonate and bicarbonate ions and these react with sodium to raise the pH (Hayward and Wadleigh, 1949). This situation is frequently encountered in the northern Great Plains of North America with the Mollisols and the Chernozems (Rieger, 1983) and these soils are often considered undesirable for legumes (Alexander, 1985). Very few data have been reported on the effects of high pH on rhizobial growth, nodulation or legume growth. Yadav and Vyas (1971) in two surveys of 23 rhizobial isolates from eight diverse legume species (not including *Glycine max*) reported that all grew well in non-saline conditions at pH values up to 10. By contrast, none of the 17 strains of *Bradyrhizobium japonicum* tested showed significant growth in liquid media at pH 8.5 (Diatloff, 1970). However, beneficial effects on root hair infection and nodulation on alfalfa were reported on high extremes of soil pH (Lakshmi-Kumari et al., 1974), whereas pH above 6.0 reduces nodulation in lupins (Tang and Robson, 1993). Eaglesham and Ayanaba (1984) are of the opinion that constraints to rhizobial survival, nodulation and legume growth pertaining to saline-alkaline soils also apply to saline soils where similar effects are observed.

Water stress and salinity

Legumes are intolerant to shortage and excess of water and this is primarily due to the ultrasensitivity of the symbiosis to water stress (Sprent, 1984). In the field it is common to have waterlogged conditions after a heavy rainfall, but a few days of sunny weather, combined with wind, will dry the soil sufficiently to induce temporary wilting of the plants. In dry soils infection is restricted because of the absence of normal root hairs; instead, short, stubby root hairs appear, which are inadequate for infection by rhizobia (Lie, 1981). Upon watering, these may resume normal growth, resulting in a slender outgrowth which eventually may become infected. Nodules initiated under conditions of adequate water are retarded in growth if exposed to dry conditions, resulting in partially developed organs, embedded in the root cortex. Many of the crop legumes studied have root hairs which are an outgrowth of epidermal cells, and through which rhizobial entry is effected. For these plants two variances of the general infection pattern are observed: 1) in plants such as soybean (*Glycine max*) which has only one infectible period, close to the beginning of hair growth; and 2) in others such as clovers (*Trifolium* spp.) and alfalfa which have a second infectible period later in hair development. These two types of legume may also differ in the response of their root hairs to growth and infection by rhizobia at low water potentials. Subsequent nodule growth is sometimes enhanced under stress, in an apparent attempt by the plant to compensate for reduced nodule efficiency. This has been found in fababean (*Vicia faba*) grown at low temperature and under low water potentials (Sprent and Zahran, 1988).

Some legumes, such as alfalfa, fababean and clovers, produce indeterminate meristematic nodules which are more salt and drought tolerant than the determinate (non-meristematic) nodules formed by soybean and common bean (Sprent and Zahran, 1988). Only indeterminate nodules have the potential to regenerate activity in structures affected by stress treatment. Levels of salinity that inhibit the symbiosis between legumes and rhizobia differ from those that inhibit the growth of the individual symbiont. NaCl concentrations that affect the symbiosis between *R. meliloti* and alfalfa are lower than those that affect the growth and survival of individual alfalfa genotypes or *Rhizobium* species (Sprent, 1984). Legumes are generally more sensitive to osmotic stress than their microsymbionts are. In contrast to their host legumes, some rhizobia can survive in the presence of extremely high levels of salt both in culture and in soil. Rhizobia isolated from arid lands are better able to nodulate the legume host under saline and drought conditions.

Legumes have long been recognized as either sensitive or only moderately resistant to salinity (for comprehensive review, see Zahran, 1991). Most legumes response to moderate salinity with a decrease in growth. This growth depression can be attributed to the accumulation of toxic ions such as sodium and chlorine in plant tissues, where they can disturb enzyme activities. In some legumes, distribution pattern of carbohydrate and organic solutes, such as proline, are changed under salt stress. Most legumes respond to these saline conditions by exclusion of sodium or chlorine or both ions from the leaves. Particularly salt-tolerant legumes include alfalfa and narrow-leafed lupins (*L. angustifolius*), which can grow in salinities equivalent to sea water. Other mechanism of adaptation is infection by mycorrhizal fungi; the rate of *Glycine max* mycorrhizal infection has been shown to increase under drought stress, and mycorrhizal inoculation has alleviated drought stress under arid conditions with *Acacia* and *Leucaena* (Robson and Bottomley, 1991) It has been argued that organic solutes such as proline and glycine enhance the salt tolerance of legumes and rhizobia. Intracellular accumulations of these solutes have been correlated with salt stress tolerance in *R. meliloti* (Botsford and Lewis, 1990).

Instead of osmotic adjustment by organic solutes, a more energy efficient way of achieving salt tolerance is to select for subcellular compartmentation of inorganic ions coupled with salt sequestration in the cell vacuole or plant tissue (Kramer, 1984). Nodules with vacuolate cells seem to be more efficient in compartmentalizing and distribution of ions, hence avoid salt toxicity (Zahran, 1991). Improved ion compartmentation within the cells of nodules and their organelles could promote growth under saline conditions and appears to be a basic criterium for the selection and breeding of salt tolerant legumes (Shannon, 1984).

Water stress reduces both N_2 fixation and respiration of nodules, and within certain limits this reduction is proportional to the degree of water loss of the nodules. During periods of drought, osmotic damage to N_2 fixation may occur because of the high salt concentrations near or on the nodules. In soybean, the decrease in nitrogenase activity is closely related to the decrease in energy charge of the nodules and to the changes in photosynthate pool sizes (Guerin et al., 1990). However, the decline in N_2 fixation activity of nodules during stress does not correlate with the availability of carbohydrates, since a large accumulation of sucrose is reported in nodules of *G. max* under stress; water stress induces a 5% decline in photosynthesis in this plant, while nodule acetylene reduction activity shows a 70% decrease (Durand et al., 1987), indicating the involvement of other mechanisms in the drop of N_2 fixation. Thus, the possibility for an increase in the resistance to oxygen diffusion through the nodule cortex to the central tissue was proposed (Durand et al., 1987). The dense packing of the cortical cells of nodules may be responsible for the limitation of oxygen diffusion to the central tissue. Bacteroids from stressed nodules from fababean appear more sensitive to oxygen, and their optimal activity declines with increasing nodule water deprivation. This defect could be partly due to decrease bacteroid respiration capacity with water stress. Water stress is also responsible for a decrease of the cytosolic protein content of the nodule and more specifically of leghemoglobin.

Excess water is particularly detrimental to N_2 fixation. A thin layer of water on the nodule surface reduces N_2 fixation considerably presumably due to the low diffusion of oxygen. In the rhizosphere under water logged conditions, the build-up of carbon dioxide may occur, and at high concentrations inhibit nodule formation. Another gas known to be produced in anaerobic soils is ethylene which, at low concentrations, can also restrict nodulation (Eaglesham and Ayanaba, 1984). Many products derived from microbial fermentation under anaerobic conditions are phytotoxic, and contribute to the waterlogging syndrome. With prolonged exposure to waterlogging, the plant may undergo structural change on the nodule surface (ruptures, lenticels, protuberances) to acquire more oxygen for the functioning of the nodules by offering a greater surface : volume ratio. There is a variation in susceptibility amongst plants, pea (*Pisum sativum*) being susceptible to excess water and fababean rather more tolerant. A rapid adaptation of the nodulated roots to low oxygen tensions was noted in soybean plants, indicating the flexibility of the intact system (Durand et al., 1987).

Temperature

Temperature affects legumes non-specifically and through plant metabolic processes such as respiration, photosynthesis, transport, and transpiration. The temperature range for the symbiotic system is narrower than that of the plant supplied with fertilizer nitrogen, and symbiosis ceases when it is exposed to extreme temperatures. All leguminous plants so far investigated have a normal Calvin photosynthetic cycle, with an optimum temperature of 15–25°C (Lie, 1981). At higher temperatures, photosynthesis is severely reduced. No legumes have been found with C_4 photosynthesis, which functions optimally at 30–40°C.

The root system can be regarded as a heterotrophic organ, depending on the shoot for the supply of carbohydrates. Root respiration may use up to 50% of the day-acquired photosynthates. It is well-known that respiration is increased with higher temperatures, and this may imply that less carbon (C) will be available for the symbiosis. Low temperature at night contributes to higher amounts of N_2 fixed, presumably by saving C from respiration (Lie, 1981). At high temperatures, the amount of root produced is low, and the roots produced are thin, unbranched, and with few lateral root and hairs. Elevated temperatures may delay nodule initiation and development, and interfere with nodule structure and functioning in temperate legumes, whereas in tropical legumes, N_2 fixation efficiency is mainly affected. Moderate (30–20°C) day-night temperatures give good early nodulation and biological N_2fixation in common bean however, duration of active N_2-fixation is shortened because of rapid degeneration of nodules (Graham, 1979).

Low temperatures delay root hair infection, and decrease nodulation and nitrogenase activity (Waughman, 1977). Initiation and establishment of N_2 fixation in alfalfa is subject to severe constraints at low temperatures; in controlled environment studies with this legume, growth is reduced by 24% and 75% at root temperatures of 13 and 8°C, respectively when compared to growth at 21°C (Cralle and Heichel, 1982). Effective nodule development and nitrogenase activity is also completely inhibited at 8°C. A specific example of extreme low temperature stress is the high arctic, where native legumes nodulate and fix N_2 at rates comparable to those observed with legumes in temperate climates. This is an indication that both the plant and their rhizobia have successfully adapted to arctic conditions (Schulman et al., 1988). Arctic legumes are exposed to a short growing season, a long photoperiod, low precipitation and low soil nitrogen levels. However, the greatest environmental stress northern organisms must deal with is the enormous difference between the temperature of the coldest month and that of the warmest month, which can be as great as 50°C in the high arctic. Mean daily air temperatures rise above 0°C for only three months of the year, and daily temperatures may drop to 0°C or below at any time. We have worked extensively in this environment and, the remainder of this paper will consign some unusual aspects of this extreme symbiosis.

Arctic legumes

Plant habitat

Our field of observation was in the area around Sarcpa Lake (68°32'N, 83°19'W) on the Melville Peninsula, approximately 90 km WSW of Hall Beach, NWT, Canada, and roughly 430 km north of the tree line. The underlying Precambrian shield rocks form a landscape of high relief with deep valleys containing numerous lakes and rounded highlands rising 100 to 150 m above valley floors. The soils are classified as Cryic Regosol with a slight moisture deficit and are saturated for moderately long periods. Regosolic soils are generally well to imperfectly drained, with weak horizon development of 0.5 to 1 m depth, and moderate to good oxidizing conditions. A pH range of 6.5 to 7.1 has been measured at our experimental sites. Three species of legumes were studied in the Sarcpa Lake area. The most common, *Oxytropis maydelliana* Trautv, is widely distributed across the North American Arctic and reaches into East Asia. It grows in a variety of habitats, varying from dry to moist, with an approximate density of 13 plants m^{-2}. It is a low growing, tufted perennial herb with a stout taproot. The second plant is *Oxytropis arctobia* Bunge which is endemic to the eastern arctic coastal areas, and it is relatively rare at Sarcpa Lake with less than one plant m^{-2}, being restricted to dry exposed gravely fel-field sites. The third species, *Astragalus alpinus* L., has a circumpolar distribution, sometimes extending to more southern alpine areas. It forms low, spreading mats on moist solifluction lobes and benches where it occurs at similar densities to *O. maydelliana*.

Nitrogen fixation in arctic area

Arctic legumes growing under the described conditions exhibit significant rates of nitrogenase activity (Karagatzides et al., 1985, Schulman et al., 1988). Rates are correlated with the number of nodules per plant, with field soil temperatures and are limited by water shortage. In natural habitats, the phenological stage of *O. maydelliana* affects nitrogenase activity throughout the growing season, with a peak of specific activity at full bloom, followed by a gradual decrease during pod fill. The rapid increase in nitrogenase activity during pre-bloom and bloom is likely to be due to a higher metabolic rate in existing tissue, rather than to the production of a more active one, assuming that meristematic tissue was produced at the end of the previous season. Throughout the growing season, fluctuation in nitrogenase activity followed soil temperature changes closely. However, maximum activity does not occur at the highest temperature, but rather follows growth stage development (Tipping, 1984).

Arctic legume adaptation

While N_2 fixation requires a high level of plant respiration, glucose supply is unlikely to be a major limitation to N_2 fixation in arctic plants. Such plants exhibit high levels of non-structural carbohydrate accumulation through most of the growing season, but growth is limited by factors such as nutrient availability, low temperature and drought (Tieszen et al., 1981). In our observations, irradiance during the assay period was above the 700 $\mu E.m^{-2}.s^{-1}$ necessary for maximum photosynthesis. During the peak growing season (mid-June to mid-July) daily pattern of irradiance shows a large diurnal fluctuation with a maximum at 14 h 00 reaching 1500 to 1700 $\mu E.m^{-2}.s^{-1}$ on clear days, and dropping to 50 to 100 $\mu E.m^{-2}.s^{-1}$ at night, to create a period of 16 hours of irradiance exceeding 300 $\mu E.m^{-2}.s^{-1}$ (Tipping, 1984). It has been shown that arctic plants have higher rates of alternate respiration (cyanide insensitive) (McNulty and Cummins 1987) which may contribute to the ability to grow and translocate at low temperatures (Tieszen et al., 1981). Since this pathway is less temperature sensitive than the normal cytochrome pathway, and functions effectively at low temperatures, it is possible that alternate respiration provides a significant por-

tion of the energy needed for N_2 fixation at soil temperatures around freezing. This may explain why the nitrogenase activity with excised nodules extends to below O°C. In controlled temperature assays with *O. maydelliana*, nitrogenase activity was maximum between 15°C and 25°C and dropped abruptly at 30°C. Detectable activity was measurable to -4°C (Tipping, 1984). Clearly, temperature affects N_2 fixation in this species in the field. The pattern of relative activity is similar to that of alfalfa (Cralle and Heichel, 1982), although the maximum appears to occur at lower temperatures in the arctic species. Lower nitrogenase activity at night is due mostly to lower soil temperature, and not to reduced irradiance. This is in agreement with other findings that whole plants maintain full activity after 10–12 hours in the dark (Halliday and Pate, 1976). Arctic plants have adapted to diurnal changes by functioning with stored energy during the night and activity is more affected by temperature.

Nodulation pattern

The arctic legumes studied have perennial nodules in which the apical meristem persists and resumes growth each summer, with most growth being made in the late summer when soil temperatures are higher. In the two *Oxytropis* species studied, nodules occurred on both lateral and tap roots, while in *A. alpinus*, nodules were restricted to lateral roots. This pattern of annual growth results in segmented (articulated, beaded) nodules, in which each segment represents one year's growth. Nodules also branch by subdivision of the apical meristem; older nodules, particularly those attached to the tap root of *O. maydelliana* may assume a coralloid appearance. Nodules are not uniformly distributed over the root system, but tend to be concentrated between 2 and 10 cm depth in soil (Schulman et al., 1988). Each season's growth appears to produce a distinctive bulge on nodule development, with some nodules still functional after 10 or more years (Tipping, 1984). This may be of adaptive significance since conditions favourable for infection and growth of new nodules may not occur every year (Pate, 1977). It has been shown that low temperatures can have a considerable effect on nodule development, causing, for example, a greatly prolonged pre-infection period in fababean (Fyson and Sprent, 1982) and clovers (Kumarasinghe and Nutman, 1979). The perennial nodules of these arctic legumes are in marked contrast to at least a majority of temperate legumes in which nodules are annual with nodulation occurring each spring. The presence of lipid bodies in the whole cells of *Oxytropis* is the most interesting feature observed either from mature nodules collected in the arctic environment (Newcomb and Wood, 1986) or from those developed under controlled conditions (Prévost et al., 1988). These lipid bodies are more numerous in the uninfected cells of the cortex at the distal part of the nodule than in the infected cells. This may indicate that the lipids are used up in the infected zone for growth and multiplication of the bacteroid in the formation of membranes where energy demand is very high.

Rhizobium *isolates*

Temperature is one of the factors which limits microbial growth in cold environments, but most of the bacteria isolated from arctic regions are able to grow at low temperatures (Nelson and Parkinson, 1978). In general, rhizobia are tolerant to 4°C, but very little growth occurs (Trinick, 1982). Clover rhizobia isolated from the sub-arctic region in Scandinavia show a better adaptation to low temperature and grew faster at 10°C than isolate from southern areas (Ek-Jander and Fahraëus, 1971). In a detailed studies conducted with 48 strains isolated from three arctic legumes (Prévost et al., 1987c), arctic rhizobia could grow from 5 to 30°C, with growth, as estimated by protein synthesis, at its maximum from 5 to 25°C, and very poor at 30°C. Optimum growth temperatures were 25°C for one selected arctic strain as compared to 30°C for most of the temperate strains; however, at 10°C the arctic strain showed a shorter doubling time than the temperate strains and, at 5°C, the temperate strains were unable to grow. Growth responses of arctic strains are similar to those found with psychrophilic and mesophilic bacteria (Innis and Ingraham, 1978). The fact that arctic rhizobia grow faster at 10°C than temperate rhizobia demonstrates a clear adaptation to low temperature stress. Furthermore, arctic rhizobia can synthesize proteins at temperatures not permissive for growth (Cloutier et al., 1992); they produce more cold shock proteins under freezing conditions (-10°C) and can synthesize heat shock proteins at temperature higher (46.4 versus 41.4°C) than temperate rhizobia. This adaptation to low temperature confer to arctic strains some properties which are advantageous over the other rhizobia, especially when temperature is limiting for bacterial growth during cold phases of the growing season.

The ability of bacteria to survive and grow under low temperature might be related to their energy of maintenance (Dawes, 1985). The amount of chemical energy represented by the intracellular ATP pool is only a very small fraction of the potential energy flux throughout the cells. In arctic rhizobia, the potential energy flux through the bacterial cells, as reflected by the total amount of nucleotides, increases with lowering temperature from 25°C to 15°C (Bordeleau, unpublished). Furthermore, when arctic rhizobia were grown at 15°C on glycerol, succinate or malate, nucleotide production was greater on glycerol medium. Glycerol is metabolized via the pentose phosphate pathway which is known to be functional at low temperature (Dawes, 1985). Arctic rhizobia possess the key enzymes of the pentose phosphate pathway, as well as those for the Entner-Doudoroff and the Embden-Meyerhof-Parnas pathways (Diaz Del Castillo, 1987). The operation of pentose phosphate pathway in arctic rhizobia would provide a source of ribose-5-phosphate, which could, in turn, be converted to phosphoribosyl pyrophosphate for the biosynthesis of nucleotides (Hong and Copeland, 1990). This pathway may play an important role for low temperature adaptation because nucleotide biosynthesis is very efficient by this metabolic route (Karl, 1980).

Rhizobium *effects on the legume*

It is well established that the strain of *Rhizobium* plays an important role in determining the efficiency of nitrogen fixation at low temperature (Layzell et al., 1983, Pankurst and Layzell, 1984). For example, strains of *R. leguminosarum* bv. *trifolii* isolated from sub-arctic regions of Scandinavia show earlier nodulation and higher nitrogenase activity with clover at 10°C than those isolated from southern regions, while no significant differences were observed between isolates at 20°C (Ek-Jander and Fahraëus, 1971). Prévost et al. (1987a, b) examined the effects of low temperature on nitrogenase activity in sainfoin nodulated by arctic and temperate rhizobia. At 5 and 10°C, the arctic rhizobia expressed 12 and 33% of the N$_2$-fixing activity they showed at 20°C, while the temperate strains show an average of only 3.7 and 22.4% of the respective activity; the Q$_{10}$ value for the arctic strain was 3.3 between 5 and 15°C, while that of the temperate strain was 23.5. This indicates that the nitrogenase system formed by the arctic strain is affected by temperature like a simple enzymatic reaction which shows a Q$_{10}$ of about 2. Similar values have been reported for the nitrogenase activity of detached nodules of alfalfa evaluated between 5–25°C (Cralle and Heichel, 1982). Our results suggest that some of the metabolic activities involved in the symbiotic nitrogen fixation process were affected by arctic rhizobia. It is possible that arctic rhizobia are using specific photosynthates as energy sources for the nitrogenase reaction. These photosynthates might be present either with arctic or temperate strains, or synthesize only when arctic rhizobia are microsymbionts in the sainfoin symbiosis. Nevertheless, under controlled low temperatures and particular field conditions, arctic strains enhanced dry matter yield production with sainfoin by a factor of 1.5 to 3 (Prévost et al. 1993), indicating that cold-adapted rhizobia could be used with benefit to improve legume production in cool temperate zone. Arctic rhizobia were demonstrated to be more competitive than temperate rhizobia to form nodules on sainfoin at low temperature (Prévost and Bromfield 1991). Such

adaptation to grow and nodulate at low temperature were also demonstrated with *R. meliloti* on alfalfa (Rice et al. 1992). The present knowledge on the adaptational mechanisms which operate in the establishment of effective N$_2$-fixing systems in extreme environments suggests that both rhizobia and their host have coevolved to cope in a particular environment. Such evolution resulted in opportunistic extreme symbiosis which can be exploited for the mankind.

References

Alexander M 1985 Ecological constraints on nitrogen fixation in agricultural ecosystems. Adv. Microbiol. Ecol. 8, 163–183.

Bell L C and Edwards D G 1987 The role of aluminium in acid soil infertility. International Board for Soil Research and Management Inc. (IBSRAM), Soil Management under Humid Conditions in Asia (ASIALAND), Proceedings of the First Regional Seminar on Soil Management under Humid Conditions in Asia and the Pacific, Khon Kaen, Phitsanulok, Thailand, October 13–20, 1986, pp 201–223.

Bordeleau L M, Antoun H and Lachance R A 1977 Effets des souches de *Rhizobium meliloti* et des coupes successives de la luzerne (*Medicago sativa*) sur la fixation symbiotique d'azote. Can. J. Plant Sci. 57, 433–439.

Botsford J L and Lewis T A 1990 Osmoregulation in *Rhizobium meliloti*: Production of glutamic acid in response to osmotic stress. Appl. Environ. Microbiol. 56, 488–494.

Brockwell J, Pilka A and Holliday R A 1991 Soil pH is a major determinant of the numbers of naturally-occurring *Rhizobium meliloti* in non-cultivated soils of New South Wales. Aust. J. Exp. Agric. 31, 211–219.

Cloutier J, Prévost D, Nadeau P and Antoun H 1992 Heat and cold shock protein synthesis in arctic and temperate strains of rhizobia. Appl. Environ. Microbiol. 58, 2846–2853.

Cralle H T and Heichel G H 1982 Temperature and chilling sensitivity of nodule nitrogenase activity of unhardened alfalfa. Crop Sci. 21, 300–304.

Dawes E A 1985 Starvation, survival and energy reserves. *In* Bacteria in their natural environments, pp 43–79. Eds. M Fletcher and G D Floodgate, Special Publications of the Society for General Microbiology, Academic Press Inc (London).

Diaz Del Castillo L 1987 Catabolisme du glucose chez des *Rhizobium* de l'Arctique. M.Sc Thesis 8022, 60 p. Graduate School of Laval University, Québec, Canada.

Diatloff A 1970 Relationship of soil moisture, temperature and alkalinity to a soybean nodulation failure. Q. J. Agric. Anim. Sci. 27, 279–293.

Durand J L, Sheehy J E and Minchin F R 1987 Nitrogenase activity, photosynthesis and nodule water potential in soyabean plants experiencing water deprivation. J. Exp. Bot. 38, 311–321.

Eaglesham A R J and Ayanaba A 1984 Tropical stress ecology of rhizobia, root nodulation and legume fixation. *In* Current Developments in Biological Nitrogen Fixation. Ed. N S Subba Rao. pp 1–35. Edward Arnold Ltd, London, UK.

Ek-Jander J and Fahraëus G 1971 Adaptation of rhizobia to subarctic environment in Scandinavia. Plant and Soil, Spec. Vol., 129–137.

Foy C D and Lee R H 1987 Differential aluminium tolerance of two barley cultures related to organic acids in their roots. J. Plant Nutr. 10, 1089–1101.

Fyson A and Sprent J I 1982 The development of primary root nodules on *Vicia faba* L. grown at two temperatures. Ann. Bot. 50, 681–692.

Gardner W K, Barber D A and Parbery D G 1983 The acquisition of phosphorus by *Lupinus albus* L. III. The probable mechanism by which phosphorus movement in the soil/root interface is enhanced. Plant and Soil 70, 107–124.

Graham P H 1979 Influence of temperature on growth and nitrogen fixation in cultivars of *Phaseolus vulgaris* L., inoculated with *Rhizobium*. J. Agric. Sci. 93, 365–370.

Graham P H 1992 Stress tolerance in *Rhizobium* and *Bradyrhizobium*, and nodulation under adverse soil conditions. Can. J. Microbiol. 38, 475–484.

Guerin V, Trinchant J C and Rigaud J 1990 Nitrogen fixation (C_2H_2 reduction) by broad bean (*Vicia faba* L.) nodules and bacteroids under water-restricted conditions. Plant Physiol 92, 595–601.

Halliday J and Pate J S 1976 The acetylene reduction assay as a means of studying nitrogen fixation in white clover under sward and laboratory conditions. J. British Grassl. Soc. 31, 29–35.

Hayward H E and Wadleigh C H 1949 Plant growth on saline and alkali soils. Adv. Agron. 1, 1–38.

Hong Z Q and Copeland L 1990 Pentose phosphate pathway enzymes in nitrogen-fixing leguminous root nodules. Phytochemistry 29, 2437–2440.

Howieson J G, Ewing M A and D'Antuono M F 1988 Selection for acid tolerance in *Rhizobium meliloti*. Plant and Soil 105, 179–188.

Innis W E and Ingraham J E 1978 Microbial life at low temperatures: mechanisms and molecular aspects. *In* Microbial Life in Extreme Environments. pp 73–99. Ed D J Kushner. Academic Press, New York, NY.

Karagatzides J D, Lewis M C and Schulman H M 1985 Nitrogen fixation in the high arctic tundra at Sarcpa Lake, Northwest Territories. Can. J. Bot. 63, 974–979.

Karl D M 1980 Cellular nucleotide measurements and applications in microbial ecology. Microbial Rev 44, 739–796.

Keyser H H and Munns D N 1979 Effects of calcium, manganese, and aluminium on growth of rhizobia in acid media. Soil Sci. Soc. Am. J. 43, 500–503.

Kramer D 1984 Cytological aspects of salt tolerance in higher plants. *In* Salinity tolerance in plants: Strategies for crop improvements. Ed. R C Staples. John Wiley, New York, pp 3–15.

Kumarasinghe R M K and Nutman P S 1979 The influence of temperature on root hair infection of *Trifolium parvflorum* and *T. glomeratum* by root nodule bacteria. II. The effects of changes in root temperature. J. Exp. Bot. 30, 517–528.

Lakshmi-Kumari M, Singh C S and Subba Rao N S 1974 Root hair infection and nodulation in lucerne (*Medicago sativa* L.) as influenced by salinity and alkalinity. Plant and Soil 40, 261–268.

Layzell D B, Rochman P and Canvin D T 1983 Low root temperatures and nitrogenase activity in soybean. Can. J. Bot. 62, 965–971.

Lie T A 1981 Environmental physiology of the legume-*Rhizobium* symbiosis. *In* Nitrogen Fixation Vol. 1: Ecology, pp 104–134. Ed W J Broughton. Clarendon Press, Oxford.

Loi A, Howieson J G, Cocks P S and Caredda S 1993 The adaptation of *Medicago polymorpha* to a range of edaphic and environmental conditions - Effect of temperature on growth, and acidity stress on nodulation and nod gene induction. Aust. J. Exp. Agric. 33, 25–30.

McNulty A K and Cummins W R 1987 The relationship between respiration and temperature in leaves of the arctic plant *Saxifraga cernua*. Plant, Cell Environ. 10, 319–325.

Munns D N 1986 Acid soil tolerance in legumes and rhizobia. Adv. Plant Nutr. 2, 63–91.

Munns D N and Keyser H H 1981 Tolerance of rhizobia to acidity, aluminium and phosphate. Soil Sci. Soc. Am. J. 34, 519–523.

Nelson L M and Parkinson D 1978 Growth characteristics of three bacterial isolates from an arctic soil. Can. J. Microbiol. 24, 909–914.

Newcomb W and Wood S W 1986 Fine structure of nitrogen-fixing leguminous root nodules from the Canadian arctic. Nordic J. Bot. 6, 609–626.

Pankurst C E and Layzell D B 1984 The effect of bacterial strain and temperature changes on the nitrogenase activity of *Lotus pedunculatus* root nodules. Physiol. Plant. 62, 404–409.

Pate J S 1977 Functional biology of dinitrogen Fixation by legumes. *In* A Treatise on Dinitrogen Fixation. Section III: Biology, pp 473–517. Eds R W F Hardy and W S Silver. John Wiley, Toronto.

Prévost D and Bromfield E S P 1991 Effect of low root temperature on symbiotic nitrogen fixation and on competitive nodulation of *Onobrvchis viciifolia* (sainfoin) by strains of arctic and temperate rhizobia. Biol. Fertil. Soils 12, 161–164.

Prévost D, Antoun H and Bordeleau L M 1987a Effects of low temperatures on nitrogenase activity in sainfoin (*Onobrychis viciifolia*) nodulated by arctic rhizobia. FEMS (Fed.Eur. Microbiol. Soc.) Microbiol. Ecol. 45, 205–210.

Prévost D, Bordeleau L M and Antoun H 1987b Symbiotic effectiveness of arctic rhizobia on a temperate forage legume sainfoin (*Onobrychis viciifolia*). Plant and Soil 104, 63–69.

Prévost D, Bordeleau L M and Antoun H 1988 Effet des souches arctiques de Rhizobium sur la structure des nodules du Sainfoin (*Onobrychis viciifolia*) et de légumineuses arctiques (*Astragalus* et *Oxytropis* spp.). Can. J. Bot. 67, 3164–3168.

Prévost D, Bordeleau L M, Caudry-Reznick S, Schulman H M and Antoun H 1987c Characteristics of rhizobia isolated from three legume indigenous to the Canadian high arctic *Astragalus alpinus*, *Oxytropis maydelliana* and *Oxytropis arctobia*. Plant and Soil 98, 313–324.

Prévost D, Bordeleau L M, Michaud R, Lafreniére C, Waddington J and Biederbeck V O 1994 Nitrogen fixation efficiency of cold-adapted rhizobia dn sainfoin (*Onobrychis viciifolia*): in laboratory and field. Plant and Soil pages.

Rai R 1992 Effect of acidity factors on aspects of symbiotic N$_2$ fixation of *Lens culinaris* in acid soils. J. Gen. Appl. Microbiol. 38, 391–406.

Rice W A 1975 Effects of CaCO$_3$ and inoculum level on nodulation and growth of alfalfa in an acid soil. Can. J. Soil Sci. 55, 245–250.

Rice W A 1982 Performance of *Rhizobium meliloti* strains selected for low pH tolerance. Can. J. Plant Sci. 62, 941–948.

Rice W A and Olsen P E 1983 Inoculation of alfalfa seed for increased yield on moderately acid soil. Can. J. Soil Sci. 63, 541–545.

Rice W A and Olsen P E 1988 Soil inoculants for alfalfa grown on moderately acid soil. Communi. in Soil Sci. Plant Anal. 19, 947–956.

Rice W A, Penney D C and Nyborg M 1977 Effects of soil acidity on rhizobia numbers, nodulation and nitrogen fixation by alfalfa and red clover. Can. J. Soil Sci. 57, 197–203.

Rice W A, Olsen P E and Collins M M 1992 Nitrogen fixation by alfalfa at low root temperature. First Circumpolar Agricultural Conference, Whitehorse, YT. Sept 28 - Oct 2, 1992.

Rieger S 1983 The genesis and classification of cold soils. Academic Press, New York, 230 p.

Robson A D and Bottomley P J 1991 Limitations in the use of legumes in agriculture and forestry. *In* Biology and Biochemistry of Nitrogen Fixation. Ed M J Dilworth and A R Glenn. pp 320–349. Elsevier Publisher, Amsterdam, New York, Oxford, Tokyo.

Schulman H M, Lewis M C, Tipping E M and Bordeleau L M 1988 Nitrogen fixation by three species of *leguminosae* in the Canadian High Arctic Tundra. Plant, Cell Environ. 11, 721–728.

Shannon M C 1984 Breeding, selection and the genetics of salt. *In* Salinity tolerance in plants: Strategies for crop improvement. Ed. R C Staples. John Wiley, New York, pp 231–254.

Sprent J I 1984 Effects of drought and salinity on heterotrophic nitrogen fixing bacteria and on infection of legumes by rhizobia. *In* Advances in Nitrogen Fixation Research, pp 295–302. Ed C Veeger and W E Newton. Martinus Nijhoff, The Hague.

Sprent J I and Zahran H H 1988 Infection, development and functioning of nodules under drought and salinity. *In* Nitrogen fixation by legumes in Mediterranean agriculture, pp 145–151. Eds D P Beck and L A Materon. Martinus Nijhoff/Dr. W. Junk, Dordrecht.

Tang C and Robson A 1993 pH above 6.0 reduces nodulation in *Lupinus* species. Plant and Soil 152, 269–276.

Taylor R W, Sistani K R and Patel S 1990 Soybean-*Rhizobium* combination for tolerance to low P - high aluminium. J. Agron. Crop Sci 165, 54–60.

Tieszen L L, Lewis M C, Miller P C, Mayo J, Chapin F S and Oechel W 1981 An analysis of primary production in tundra growth forms. *In* Tundra ecosystems: a comparative analysis, pp 285–386. Eds L C Bliss, O W Heal and J J Moore, Cambridge University Press, Cambridge.

Tipping E M 1984 Symbiotic dinitrogen fixation in *Oxytropis maydelliana* Trautv., an arctic legume. M.Sc Thesis 6105, 92 p. Graduate School Laval University, Quebec, Canada.

Trinick M J 1982 Biology. *In* Nitrogen Fixation, Vol. 2: *Rhizobium*, pp 76–146. Ed W J Broughton, Clarence Press, Oxford.

Vargas A A T and Graham P H 1988 *Phaseolus vulgaris* cultivar and *Rhizobium* strain variation in acid-pH tolerance and nodulation under acid conditions. Field Crops Res. 19, 91–101.

Waughman G J 1977 The effect of temperature on nitrogenase activity. J. Exp. Bot. 28, 949–960.

Yadav N K and Vyas S R 1971 Response of root-nodule rhizobia to saline, alkaline and acid conditions. Indian J. Agric. Sci. 41, 875–881.

Zahran H H 1991 Conditions for successful *Rhizobium*-legume symbiosis in saline environments. Biol. Fertil. Soils 12, 73-80.

Plant and Soil **161**: 127–134, 1994.
© 1994 *Kluwer Academic Publishers.*

Analysis and regulation of legume inoculants in Canada: The need for an increase in standards

P.E. Olsen[1], W.A. Rice[1], L.M. Bordeleau[2] and V.O. Biederbeck[3]
[1]*Agriculture Canada, Beaverlodge, Alberta,* [2]*Agriculture Canada, Sainte-Foy, Quebec and* [3]*Agriculture Canada, Swift Current, Saskatchewan, Canada*

Key words: legume inoculant, quality control, *Rhizobium*, standards

Abstract

Rhizobial inoculants for use in Canada are regulated and have been evaluated in a formal testing program since 1975. This program is carried out by Agriculture Canada under authority of the Fertilizers Act and involves inoculant strain and formulation registration (with Food Production and Inspection Branch) as well as analysis (by Research Branch) of approximately 220 inoculants and pre-inoculated seed products yearly. Inoculant evaluation is based upon the calculated number of viable rhizobia which would be provided per seed if the inoculant was applied at the manufacturer's recommended rate. Current standards are 10^3, 10^4, and 10^5 viable rhizobia per seed, of the proper cross-inoculation group, for small, intermediate, and large seeded legumes, respectively. Application of these standards means that some inoculants are considered "satisfactory" even though they yield test results as low as 9.4×10^6 rhizobia per gram. No standards are currently applied relative to permissible levels of contaminants in inoculant products, despite the fact that some inoculants contain many more contaminating microorganisms than they do *Rhizobium* cells. The demands of modern sustainable agriculture, taken together with advances in inoculant formulation technology, warrant an increase by a factor of ten in the minimum acceptable Canadian standards for legume inoculants and pre-inoculated seed products.

Introduction

Agriculture Canada conducts the only program in North America which regulates inoculant products and pre-inoculated seed. This program evolved as a result of surveys conducted in 1973 which found that 80% of the inoculants collected from 36 retailers in Quebec contained fewer than 1×10^6 rhizobia per gram. These results were published in an in-house Agriculture Canada newsletter where they aroused the concern of officials and led to efforts to formalize a testing program for rhizobial inoculant products. The program came into effect, under the authority of the Fertiliz-

ers Act, in 1975. The first Canadian standard for inoculants required only that the product contain a minimum of 1×10^6 rhizobia per gram. A new set of standards was established in 1979 requiring that 10^3, 10^4, and 10^5 nodulating rhizobia be delivered per seed, when applied at the manufacturer's recommended rate, for small, intermediate, and large seed sizes, respectively (Bordeleau and Prevost, 1981). Granular inoculants are required to deliver a minimum of 10^{11} rhizobia per hectare. Analysis of inoculants was carried out throughout the 1980s at four Agriculture Canada research stations, but the testing program was consolidated in 1991 and analyses are now conducted only at the Ste-Foy,

Quebec and Beaverlodge, Alberta research stations. Each group annually tests approximately 110 officially collected samples of inoculants (or pre-inoculated seed products). Administrative, regulatory, and enforcement functions of the program are carried out by Fertilizers Section, Plant Products Division of Food Production and Inspection Branch in Ottawa, Ontario. The results of inoculant product testing are published annually (since 1992) as the "Canadian Legume Inoculant and Pre-Inoculated Seed Product Testing Report" (Anonymous, 1992). This report lists all samples analyzed by cross inoculation group, manufacturer, brand, sampling location, carrier type and test result.

The Canadian Legume Inoculant Testing Program

Registration

Rhizobial inoculants are classified as supplements under the Fertilizers Act and are subject to registration. Information required for registration includes a list of constituent materials, a guarantee of the minimum number of viable rhizobia, the product label, and efficacy data. A unique registration number is assigned to each product, and this number must appear on the product label. Registration remains in effect for two years and must be amended to reflect any changes in product formulation or label. Renewal of registration is contingent upon meeting prescribed standards and label claims.

Sampling

Inoculant and pre-inoculated seed products are sampled by inspectors of the Food Production and Inspection Branch of Agriculture Canada. Sampling normally takes place at wholesale warehouses and retail distributors, but may also occur at sites of inoculant manufacture or seed inoculation. Inspectors are provided with control inoculants of known cell count which accompany the test sample from the point of sampling until arrival

at the testing laboratory. Test samples, together with the corresponding control inoculant, are transported together by courier air express, stored under refrigeration upon arrival, and are tested as soon as possible following arrival.

Enforcement

Products failing to meet standards may have registration revoked or may be refused registration at the time of renewal. When an inoculant is tested and found unsatisfactory, the rest of the product of that type and lot number present at the sampling site is placed under detention. Unsatisfactory pre-inoculated seed products must have labelling attesting to the pre-inoculation removed and cannot be sold as pre-inoculated seed. Unregistered or improperly labelled products are also detained. Detained products may not be moved or sold. Products are released from detention when the conditions that caused the detention are corrected, or at the end of six months. Violation of provisions of the Fertilizers Act may result in fines or imprisonment.

Analytical methods

General
Official analysts within Research Branch closely supervise or perform the laboratory analyses. The analytical procedures applied are detailed in "The Methods of Testing Legume Inoculant and Pre-inoculated Seed Products" (Anonymous, 1991). This "Official Methods" describes two acceptable methods of analysis, leaving the choice of method up to the analyst. The two methods include the most-probable-number (MPN) plant nodule grow-out test and an immuno-identification of plate count colonies known as the colony immunoblot test. The Official Methods also describes the standards applied, the statistics used, and the procedures for rating test samples. Official analysts must strictly adhere to the Official Methods and must issue a Certificate of Analysis for all samples submitted by government inspectors. Control inoculants are pre-distributed to sampling inspectors and must accompany all

inoculant samples collected from the point of sampling to the testing laboratory. These controls, prepared with sterile peat carrier, are evaluated by plate count at the time that the product sample is tested.

MPN test

The MPN test used in the Canadian inoculant testing program is similar in principle to that described by Vincent (1970). Growth units are prepared by germinating surface sterilized seed in troughs of sanitary growth pouches containing sterile, nitrogen-free plant nutrient solution. Tenfold dilutions of the test inoculant sample are prepared in buffer. One mL of the last 10-fold dilution is used to prepare the first of six consecutive 5-fold dilutions. Each 5-fold dilution is applied to four growth units, and an uninoculated (control) growth unit is included after each group of four inoculated units. The growth units are then placed in growth chambers providing appropriate temperature, lighting, and humidity conditions. Three weeks following inoculation the plants are examined for the presence or absence of nodules and each unit is scored "+" or "−". The nodulation score is a six digit "code" which is analyzed in the MPNES computer program (Woomer et al., 1990) to yield the most-probable-number of rhizobia in the sample. The MPNES program also reports the corresponding 95% confidence limits. Presence of nodules on any plants in uninoculated control units invalidates the test.

Colony immunoblot analysis

The colony immunoblot test used in the Canadian inoculant testing program was derived from methods described by Olsen and Rice (1989, 1991). Appropriate tenfold dilutions of test or control inoculant samples are made and 100 μL portions of the appropriate dilution are plated on yeast-extract mannitol agar (Vincent, 1970). The plates are incubated to permit colony growth and a nitrocellulose membrane is imprinted with surface colonies by gentle application of the membrane to the agar surface. The membrane is incubated in a solution of strain or species specific anti-*Rhizobium* monoclonal antibody mixed with com-

mercial biotin-labelled anti-mouse immunoglobulin, and subsequently incubated with alkaline phosphatase-labelled streptavidin. The membrane is then reacted with a precipitating substrate for alkaline phosphatase and sites on the membrane which had touched colonies reactive with the antibody turn purple. The spots are then counted and multiplied by the proper dilution factor to provide an estimate of the number of viable rhizobia present in the original sample.

Sample rating

Standards

Current regulatory standards require a minimum of 10^3, 10^4, and 10^5 viable rhizobia, of the proper cross-inoculation group, per seed for small, intermediate, and large size legume seeds, respectively. These standards are used in the calculation of the required value (see below). Granular inoculants require rhizobial delivery at a rate of 10^{11} cells per hectare.

Required value

The required value is a calculated number of rhizobia per g of inoculant product needed to meet the standards, on the basis of seed size, when the inoculant is applied to the seed at the manufacturer's recommended rate as printed on the package label.

Confidence limits

Confidence limits for MPN tests are based on a confidence factor which varies with change in base dilution ratio and the number of units inoculated per dilution, but remains constant for a particular test performed in a particular way. The 95% confidence factor associated with the Canadian official methods MPN test and based on probability estimates generated by the MPNES computer program is 2.88. In practice, the actual MPN test result is simply multiplied by 2.88 to obtain the upper 95% confidence limit and divided by 2.88 to obtain the lower 95% confidence limit. The 95% confidence limits for the colony immunoblot procedure can be determined from tables of confidence limits of a single count, or a close approximation can be calculated as plus and minus 1.96 times the square root of the colony count (Wardlaw, 1985).

Rating

Samples are rated satisfactory if the upper 95% confidence limit of the test result (number of viable rhizobia per g of inoculant product) is equal to or larger than the calculated required value. Samples are rated unsatisfactory if the upper 95% confidence limit value is less than the required value and the accompanying control sample is rated satisfactory. Pre-inoculated seed samples are rated in a similar manner using the specified rhizobia per seed standard in place of the required value. The following example may help to clarify rating calculations: an alfalfa peat-base inoculant, with a package label recommending application at the rate of 227 g per 25 kg seed or 9.08 g per kg seed, is submitted for testing. The rhizobial standard for small seeded legumes is 1000 per seed and the analyst uses (from the Official Methods) a figure of 441,000 alfalfa seeds per kg. Therefore, 9.08 g inoculant must provide $441,000 \times 1000 = 4.41 \times 10^8$ rhizobia, or 4.86×10^7 rhizobia per g of inoculant (the required number). The MPN value for which 4.86×10^7 is the upper 95% confidence limit is 1.69×10^7 rhizobium per g of inoculant (the minimum acceptance number). Therefore, an inoculant with an MPN test result of 1.69×10^7 rhizobia per g, when recommended for application at the rate of 9.08 g per kg of seed, would result in a "satisfactory" rating.

Problems with the current rating system

Since rating of inoculants in the Canadian regulatory system is accomplished by application of the upper 95% confidence limit of the actual MPN test result, the critical rhizobial number used in inoculant rating is not the prescribed standard value,

(e.g., 10^3 rhizobia per small size legume seed), but 1/2.88 of the standard value (e.g., 347 rhizobia per small seed). An inoculant yielding an MPN test result of only 347 rhizobia per small legume seed will therefore be rated "satisfactory". The reason for application of the upper 95% confidence limit rule is that it provides the analyst with 97.5% confidence (95% plus the lower 2.5% "tail") that a failed product did not meet the standard. The obvious problem associated with this level of confidence in a "unsatisfactory" result is that an inoculant with a test result as low as 347 rhizobia per seed is termed "satisfactory" even though there is only 2.5% probability that such a product met the 10^3 rhizobia per seed standard.

Implementation of the Canadian inoculant testing program in 1975 resulted initially in a rapid and significant increase in the quality of inoculant products offered to the Canadian farmer. Day (1991) wrote "It is now well established and documented that the introduction of standards and quality testing services has improved the situation regarding inoculants in at least Australia, Canada and the UK." Despite the general improvement in quality of inoculant products tested in Canada during the late 1970s and early 1980s, pre-inoculated seed products have since become problematic and traditional inoculant products have remained unchanged or have even decreased in quality, particularly in the 1990s. The presence of large numbers of non-rhizobial microorganisms (contaminants) in inoculant products is a continuing concern. It is widely understood that the products made with non-sterile peat carrier carry contaminants, but it is not well known that they often contain many times more contaminants than they do rhizobia. No Canadian standards are currently enforced in relation to levels of contamination found in inoculant products. There is, however, a legal requirement in the Fertilizers Act that any contaminants present in inoculants must not have any detrimental effect on the rhizobia present, and regulations also exist relating to the import or dissemination of plant pathogens.

As new technologies and improved formulations utilizing gamma-irradiated carriers have arrived on the marketplace, manufacturers have used the high rhizobial numbers common to these products to increase the amount of seed that the product is recommended to treat. It has become common in North America for high quality inoculants (in terms of rhizobia per g) to be recommended for application to seed at such a low rate that the per seed dose of rhizobial cells is no higher than traditional inoculants applied at traditional rates. Such practices would be acceptable if the present regulatory standards were optimal and ensured satisfactory nodulation and nitrogen fixation. Otherwise, the standards should be reevaluated on the basis of agronomic and biological need.

Relationship between seed inoculum dosage and nodulation

Evidence accumulated over the years indicates that increases in rhizobial cell numbers applied per seed results in increased nodulation response and nitrogen fixation, especially under stress conditions. Burton (1976) wrote "It is generally conceded that large numbers of rhizobia on seed favour survival before planting and that large numbers of rhizobia on seed at planting favour rhizobia multiplication in the rhizosphere and early nodulation." In field and greenhouse studies, Weaver and Frederick (1974a, 1974b) showed that the percentage of nodules formed by an inoculant strain in a soil containing low levels of *Bradyrhizobium japonicum* varied proportionally with the logarithm of the number of bacteria in the inoculum. They concluded that commercial inoculants in the United States were probably not supplying adequate numbers of rhizobia for successful competitive nodulation of soybean. Amarger and Lobreau (1982) cited from previous work (Amarger, 1974) an identical relationship between inoculum dosage and competition for nodulation of faba bean. Their strain competition studies with faba bean and alfalfa also showed that increases in rhizobial inoculum dosage resulted in increases in nodule occupancy by the inoculant strain. They demonstrated a logarithmic function between inoculant rhizobia numbers and nodule occupancy and stated that the logarithmic dose-

response effect also applied to the majority of quantitative strain competition results reported in the literature. Similarly, Hume and Blair (1992) found that increasing rhizobia from 10^5 to 10^6 per seed improved yields in first-time soybean fields by an average of 24% and suggested that the existing Canadian standard of 10^5 per seed should be increased. Brockwell et al. (1987) reported that indigenous soil *Bradyrhizobium* dominated nodule formation in soybean, but that the magnitude of the domination could be reduced by increased rates of inoculation. Working with alfalfa, Rice (1982) showed that increases in yield, nitrogenase activity, and nodules per plant were related to inoculum dosage up to 10^7 rhizobia per seed. Bordeleau (1988) varied *R. meliloti* rates from 10^3 to 10^8 per seed in soils with high and low indigenous rhizobial populations and found that a minimum of 10^5 per alfalfa seed was required for good stand establishment and crop yield in both situations. High seed inoculation dosage had a beneficial effect for three subsequent years at one site and two years at another. Bhuvaneswari et al. (1988) showed that formation of early nodules on both soybean and cowpea was linearly dependent on the log of the inoculum dosage up to 10^6 rhizobia per seed. De Oliveira and Graham (1990) showed that inoculation dosage level has an effect on speed of nodulation and on the apparent competitiveness of strains of *R. leguminosarum* bv. *phaseoli*. Smith (1992) cited studies showing that soybean nodule numbers in a tropical soil increased with increasing rates of rhizobia applied in a liquid formulation from log 4.59 up to the highest rate of log 9.59 viable cells per cm of seed row. Thiess et al. (1992), working with a variety of legumes and environments, concluded that competition between applied inoculant and indigenous rhizobia was most strongly affected by the size of the indigenous rhizobial population. Thus, the basic aim of legume inoculation is to provide the maximum numbers of suitable rhizobia at the time of nodule initiation as excess numbers cannot adversely affect nodulation (Thompson, 1991a). Maintenance of rhizobial populations, i. e., inoculant quality, at levels much above the Canadian regulatory minimum is not only desirable for inoculant distribution and marketing, but is of critical importance when forage legumes are grown under adverse soil conditions (Biederbeck and Geissler, 1993). As stated by Burton (1976), "Large inocula are not always essential but there is safety in numbers regardless of whether inoculated seeds are stored or planted immediately. An abundance of rhizobia is particularly important when seeds are planted under adverse conditions or in soils which harbour a large population of ineffective rhizobia."

World perspective of rhizobial inoculant standards

Regulations on inoculum quality vary from country to country and no set of international standards exists. Nevertheless, standards for levels of viable rhizobia in inoculants have been established in several countries and it is generally conceded that the introduction of governmental standards and testing services has led to improvements in inoculant quality in the countries involved. Furthermore, this positive effect often extends to other markets where the regulated products are sold. France currently has the highest standards worldwide for inoculant quality, and these standards include field testing and expiry date testing as well as a requirement of 10^6 *Bradyrhizobium japonicum* per seed for soybean. France also has a strict requirement that no contaminants be present in rhizobial inoculant products (Catroux, 1991). Australian standards are also high, requiring 1×10^9 rhizobia per g moist inoculant and fewer than 10^6 contaminants per g (Thompson, 1991b). England has no governmental standard, but Agricultural Genetics Company's self imposed minimum is 2×10^9 rhizobia per g with a maximum of 10^6 contaminants per g (Day, 1991). The Thai standard for its government program which produces inoculant is 5×10^7 rhizobia per g, but this minimum is sensibly tied to a requirement for a high inoculant application rate with the net result that 10^5-10^6 *Bradyrhizobium japonicum* per soybean seed are applied (Boonkerd, 1991). The Netherlands require from 4 to 25×10^9 rhizobia per g,

and South Africa and New Zealand both require a minimum of 1×10^8 rhizobia per g (Smith, 1992). Rwanda allows no inoculant containing less than 1×10^9 rhizobia per g to be sold and permits contaminants at a level of no more than 0.001% of the rhizobial count (Scaglia, 1991). The United States applies no standards and inoculant quality is left to the manufacturer's discretion (Smith, 1992), but most US inoculants exported to Canada contain at least 1×10^8 rhizobia per g (Anonymous, 1992). Canadian standards are specifically tied to the manufacturer's recommended rate of application so as to require delivery of a specific number of rhizobia per seed. The Canadian standard for minimum numbers of rhizobia in inoculants therefore varies with manufacturer, legume host, and application rate. For inoculants tested in Canada in 1992, the required number varied from 2.7×10^7 to 1.6×10^9 rhizobia per g (Anonymous, 1992), but application of Canada's upper 95% confidence limit rule in effect lowers the actual MPN test result values required for product acceptance in these cases to 9.4×10^6 and 5.6×10^8 rhizobia per g. By this comparison, current Canadian standards are not high on the world scale.

Conclusions

The existing Canadian Inoculant Testing Program standards are 10^3, 10^4, and 10^5 cells per seed for small, intermediate and large sized legume seed. As previously discussed, the current actual MPN test values required for product acceptance are 347, 3,470, and 34,700 rhizobia per seed. Based on the ability of the inoculant manufacturing industry to provide the necessary combination of quality and dose rate to achieve high levels of rhizobia on seed, and the proven agronomic need for higher levels of rhizobia on the seed, the Canadian inoculant standards should be raised by a factor of ten to 10^4, 10^5, and 10^6 per seed for the three respective seed sizes. The upper 95% confidence limit should continue to be applied and the minimum MPN acceptance values would therefore become 3470, 34,700 and 347,000 rhizobia per seed. The Canadian regulatory agency should also consider routine monitoring of contamination levels in inoculant products with a view towards the establishment of reasonable limits.

We agree with Thompson (1991a) when he wrote "There is no point in setting standards which are unobtainable ..." but that "...an undesirable or unsuccessful producer may justifiably be excluded from the industry by standards which others can consistently meet." The new standards suggested for the Canadian Legume Inoculant Testing Program are readily attainable. Some manufacturers may find it necessary, however, to increase their dosage recommendations and others may find it necessary to improve internal quality control procedures. The underlying purpose of an increase in standards is to ensure that farmers are provided with inoculants that are effective, thereby more fully utilizing the potential of biological nitrogen fixation. In the long run, it seems to us that this is inextricably tied to the success of the inoculant industry as a whole because poor products and field results damage the reputation of microbial aids to agriculture in general. The demands of modern sustainable agriculture, taken together with advances in inoculant formulation technology, warrant the increases in the minimum acceptable Canadian standards for legume inoculants which are suggested in this paper.

References

Amarger N 1974 Competition pour la formation des nodosites sur la feverole entre souches de *Rhizobium leguminosarum* et souches du sol. C. R. Acad. Sci. (Paris) 279, 527–530.

Amarger N and Lobreau J P 1982 Quantitative study of nodulation competitiveness in *Rhizobium* strains. Appl. Environ. Microbiol. 44, 583–588.

Anonymous 1991 The methods of testing legume inoculant and pre-inoculated seed products. Fertilizers Act, Section 23, Regulations. Feed and Fertilizer Division, Government of Canada, Ottawa. 15 p.

Anonymous 1992 Canadian legume inoculant and pre-inoculated seed product testing report. Food Production and Inspection Branch, Feed and Fertilizers Division, Government of Canada, Ottawa. 30 p.

Bhuvaneswari T V, Lesniak A P and Bauer W D 1988 Efficiency of nodule initiation in cowpea and soybean. Plant Physiol. 86, 1210–1215.

Biederbeck V O and Geissler H J 1993 Effect of storage temperatures on *Rhizobium meliloti* survival in peat and clay-based inoculants. Can. J. Plant Sci. 73, 101–110.

Boonkerd N 1991 Inoculant quality control and standards in Thailand. *In* Report of the Expert Consultation on Legume Inoculant Production and Quality Control. Ed. J A Thompson. pp 121–129. Food and Agriculture Organization of the United Nations, Rome.

Bordeleau L M and Prevost D 1981 Quality of commercial legume inoculants in Canada. *In* Proceedings of the 8th North American *Rhizobium* conference. Eds. K W Clark and J H G Stephens. pp 562–565. University of Manitoba, Winnipeg.

Bordeleau L M 1988 Effects of inoculation rate on the establishment and performance of alfalfa in Quebec. Proceedings of the Inoculant Product Technology and Seed Inoculation Research Workshop Sept. 27–29, 1988. Agriculture Canada Research Station, Swift Current, Saskatchewan.

Brockwell J, Roughley R J and Herridge D F 1987 Population dynamics of *Rhizobium japonicum* strains used to inoculate three successive crops of soybean. Aust. J. Agric. Res. 38, 61–74.

Burton J C 1976 Methods of inoculating seeds and their effect on survival of rhizobia. *In* Symbiotic Nitrogen Fixation in Plants. Ed. P S Nutman. pp 175–189. International Biological Programme 7. Cambridge University Press.

Catroux G 1991 Inoculant quality standards and controls in France. *In* Report of the Expert Consultation on Legume Inoculant Production and Quality Control. Ed. J A Thompson. pp 113–120. Food and Agriculture Organization of the United Nations, Rome.

Day J M 1991 Inoculant production in the UK. *In* Report of the Expert Consultation on Legume Inoculant Production and Quality Control. Ed. J A Thompson. pp 75–85. Food and Agriculture Organization of the United Nations, Rome.

De Oliveira L A and Graham P H 1990 Speed of nodulation and competitive ability among strains of *Rhizobium leguminosarum* bv *phaseoli*. Arch. Microbiol. 153, 311–315.

Hume D J and Blair D H 1992 Effect of numbers of *Bradyrhizobium japonicum* applied in commercial inoculants on soybean yield in Ontario. Can. J. Microbiol. 38, 588–593.

Olsen P E and Rice W A 1989 *Rhizobium* strain identification and quantification in commercial inoculants by immunoblot analysis. Appl. Environ. Microbiol. 55, 520–522.

Olsen P E and Rice W A 1991 Use of monoclonal antibodies in a colony immunoblot analysis of viable *Rhizobium* cell numbers in legume inoculants and on preinoculated seed. Can. J. Microbiol. 37, 430–432.

Rice W A 1982 Performance of *Rhizobium meliloti* strains selected for low-pH tolerance. Can. J. Plant Sci. 62, 941–948.

Scaglia J A 1991 Production and quality control of inoculants in Rwanda. *In* Report of the Expert Consultation on Legume Inoculant Production and Quality Control. Ed. J A Thompson. pp 61–69. Food and Agriculture Organization of the United Nations, Rome.

Smith R S 1992 Legume inoculant formulation and application. Can. J. Microbiol. 38, 485–492.

Thiess J E, Bohlool B B and Singleton P W 1992 Environmental effects on competition for nodule occupancy between introduced and indigenous rhizobia and among introduced strains. Can. J. Microbiol. 38, 493–500.

Thompson J A 1991a Legume inoculant production and quality control. *In* Report of the Expert Consultation on Legume Inoculant Production and Quality Control. Ed. J A Thompson. pp 15–32. Food and Agriculture Organization of the United Nations, Rome.

Thompson J A 1991b Australian quality control and standards. *In* Report of the Expert Consultation on Legume Inoculant Production and Quality Control. Ed. J A Thompson. pp 107–111. Food and Agriculture Organization of the United Nations, Rome.

Vincent J M 1970 A Manual for the Practical Study of the Root-Nodule Bacteria. International Biological Programme Handbook No. 15. Blackwell Scientific Publications, Ltd., Oxford.

Wardlaw A C 1985 Practical Statistics For Experimental Biologists. John Wiley and Sons, New York.

Weaver R W and Frederick L R 1974a Effect of inoculum rate on competitive nodulation of *Glycine max* L. Merrill. I. Greenhouse studies. Agron. J. 66, 229–232.

Weaver R W and Frederick L R 1974b Effect of inoculum rate on competitive nodulation of *Glycine max* L. Merill. II. Field Studies. Agron. J. 66, 233–236.

Woomer P, Bennett J and Yost R 1990 Overcoming the inflexibility of most-probable-number procedures. Agron. J. 82, 349–353.

Plant and Soil **161**: 135–145, 1994.

Recent developments in the actinorhizal symbioses

Alison M. Berry
Department of Environmental Horticulture, University of California, Davis, CA 95616, USA

Key words: actinorhizal, *Frankia*, nitrogen fixation, root nodule, symbioses

Abstract

Over 200 species of angiosperms in eight different families are capable of forming root nodule symbioses with the actinomycetal genus *Frankia* as endosymbiont. Several thorough reviews of the biology of these actinorhizal associations have appeared in recent years (Benson and Silvester, 1993; Schwintzer and Tjepkema, 1990; Tjepkema et al., 1986). The purpose of the present discussion is to provide a summary overview of the actinorhizal symbioses, with an emphasis on recent research activities. A few areas of comparative interest with other symbiotic diazotrophs will be highlighted, especially regarding the question of oxygen protection and nitrogen fixation.

The actinorhizal host plants

The actinorhizal hosts are perennial plants in eight angiosperm families, which are diverse in habit and habitat (Table 1). None of the host species are crop plants, with the exception of certain arborescent species of *Casuarina* and *Alnus* (alders) which are harvested for timber and fuelwood (Diem and Dommergues, 1990; Hibbs and Cromack, 1990). The significance of the actinorhizal symbioses in the global nitrogen budget is primarily ecological. The plants are mostly found in nutrient poor soils, disturbed sites, sand dunes and gravels. Many of the actinorhizal plants are of further ecological value because they are adapted to extreme environments such as saline sands, semi-arid plains, bogs or polar environments. They are usually considered pioneer species. While there is no doubt that the actinorhizal plants enhance site productivity in many plant communities (Conard et al. 1985; Dawson, 1990; Hibbs and Cromack, 1990), their influence on primary succession needs to be carefully assessed. Recent findings

of Kohls et al. (in press), based on ^{15}N isotopic natural abundance patterns, suggest that for *Dryas* spp., rosaceous plants that form dense recumbent mats following deglaciation in the northern hemisphere, nodulation was delayed by up to 70 years after colonization in some sites, where elements of later seres had already become established. In other sites, *Dryas* was well nodulated and contributed fixed nitrogen early in the chronosequence (Kohls, pers. comm.). Seasonal environmental parameters such as thermal regime or moisture availability that limit actinorhizal nitrogen fixation may determine whether these plants influence early successional patterns, or rather function more in enhancing site productivity over time.

The rates of nitrogen fixation measurable in nodules of actinorhizal genera such as *Alnus* and *Elaeagnus* (10–90 μmol g^{-1}(f.w.) h^{-1} ethylene as reduced acetylene) are higher than or comparable to rates attained in legumes such as peas, soybean, and sweet clover using similar assays (Torrey, 1978). Estimates of annual rates of N accretion in actinorhizal genera have likewise been within

Table 1. Representative genera of woody angiosperm plants nodulated *Frankia*

Family	Genus	Habitat
Betulaceae	*Alnus*	bogs, riparian
Casuarinaceae	*Allocasuarina*	sand dunes, saline, desert
	Casuarina	
	Gymnostoma	
Coriariaceae	*Coriaria*	gravel, poor soils
Datiscaceae	*Datisca*	gravel streams
Elaeagnaceae	*Elaeagnus*	poor soils, disturbed sites
	Hippophae	
	Shepherdia	
Myricaceae	*Comptonia*	bog, ocean dunes
	Myrica	
Rhamnaceae	*Ceanothus*	chaparral, upland
	Colletia	
	Trevoa	
Rosaceae	*Cercocarpus*	semiarid soils, sand
	Dryas	gravelly soil
	Purshia	

the same range as those of many legumes (40–350 kg·ha^{-1}·y^{-1}; Torrey, 1978).

The phylogenetic distribution of the actinorhizal host genera within the angiosperms has raised interesting questions concerning the evolution of nitrogen-fixing symbioses. By no means all of the genera in each family are capable of forming root nodules: Within the Betulaceae, for example, *Alnus* is actinorhizal, while *Betula* is not; in the Rhamnaceae, *Ceanothus* is actinorhizal while *Rhamnus* is not. In generally-accepted classification schemes (Cronquist, 1988), three of the actinorhizal families have been considered to cluster closely within the sub-class Hamamelidae (Betulaceae, Casuarinaceae, and Myricaceae). Three other families were grouped within the Rosidae (Rosaceae, Elaeagnaceae, and Rhamnaceae), along with the Leguminosae, and were considered to be of more recent origin. The remaining two families, Datiscaceae and Coriariaceae, were placed in two other sub-classes (Dilleniidae and Magnoliidae, respectively), although their precise assignments have been regarded as somewhat uncertain. Because of this apparently scattered taxonomic distribution, the possibility exists of polyphyletic origins within the angiosperms for the capacity to form actinorhizal symbioses.

Other lines of evidence suggest that differences among the actinorhizal taxa are not profound. In a significant recent advance, Swensen et al. (in press) have used nucleotide sequence analysis to show much stronger affinities among several actinorhizal genera than have been reflected in prior classifications. Such molecular phylogenetic analyses are likely to reshape our thinking on the evolution of root nodule symbioses. From what we know of early infection patterns in these symbioses, root hair infections occur among members of the Betulaceae, Casuarinaceae and Myricaceae, while intercellular infection mechanisms have been observed in genera of the Elaeagnaceae, Rhamnaceae, and Rosaceae (reviewed in Berry and Sunell, 1990; Benson and Silvester, 1993). The pattern of nodule initiation is apparently under host control: single *Frankia* strains have been shown to nodulate successfully hosts from both infection types (Miller and Baker, 1986; Racette and Torrey, 1989). Thus even though the capacity to nodulate has a disjunct distribution, the underlying host physiological or developmental basis for nodulation appears to be broadly conserved.

Frankia

Frankia is a gram-positive, septate filamentous prokaryote. The genus *Frankia* is classified as an actinomycete, based on morphological and biochemical characteristics, high DNA G+C content (approximately 70%), and 16s rRNA-DNA sequence analysis. *Frankia* is closely related to certain high G+C genera within the Actinomycetales, especially *Geodermatophilus* (Hahn et al., 1989). The first stable *Frankia* strain was isolated from nodules in pure culture in 1978 by Callaham et al, and since that time several hundred strains have been isolated from many, although by no means all, of the actinorhizal host genera. Although *Frankia* strains were categorized primarily based on host compatibility, increasing use

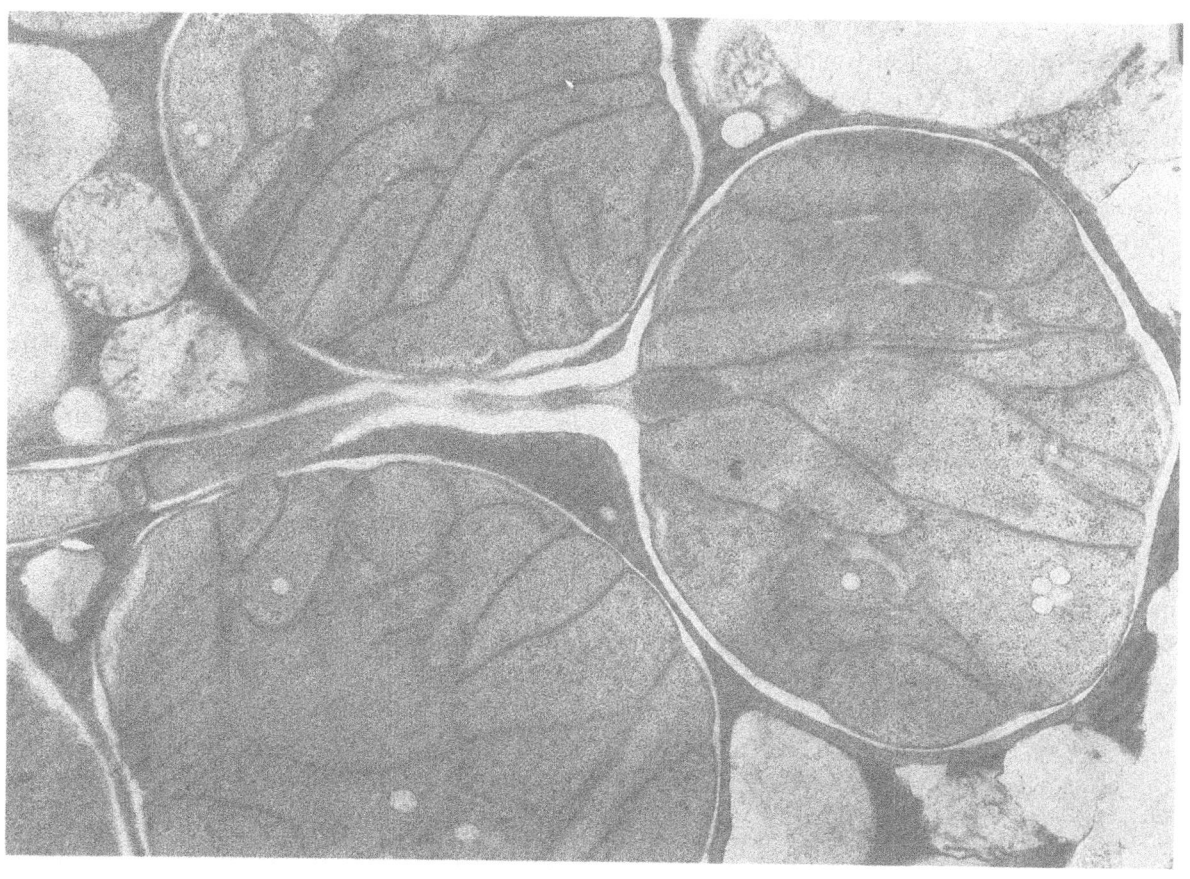

Fig. 1. Mature *Frankia* vesicles in root nodule of *Alnus serrulata*. Note extensive internal septations. In this high-pressure frozen, freeze-substituted preparation, the vesicle envelope is preserved as a thin, unstained, apparently void, space surrounding the septate vesicle cytoplasm (Berg 1993). The envelope is particularly thick at the junction between the stalk and the vesicle. The stalk cytoplasm appears constricted in this region. Vesicle diameter ± 3 μm. Transmission electron micrograph and vesicle preparation by R.H. Berg.

of DNA hybridization and restriction fragment analysis has led to the development of genomic species of *Frankia*. The conceptual framework for classifying *Frankia* has expanded significantly in the past few years, and the contributing research is reviewed by Benson and Silvester (1993).

Morphologically, *Frankia* is a complex organism. Vegetative hyphae proliferate in culture and in early stages of nodule development. Sporangia with thick-walled resting spores can form, but little is yet understood about what factors might govern sporangium formation or spore germination. The *Frankia* vesicle is a unique structure specialized for nitrogen fixation (Fig 1). Vesicles develop as terminal swellings of hyphal side branches, in conditions of ammonium limitation. The role of

the vesicle in nitrogen fixation is discussed in section 4, below.

Nif *genes in* Frankia

Several structural *nif* genes have been identified by heterologous probing of *Frankia* DNA and sequenced. These include *nif*HDK (Normand et al., 1988, Twigg et al., 1990), and a region that includes *nif*WZ(B) and *nif*X, separated by two open reading frames with sequence homology to *orf1* and *orf3* of *Azotobacter vinelandii* Benson et al., 1993). No *nif*A gene has been identified (Benson et al., 1993). Normand et al. (1988, 1992) observed that *Frankia nif*H and *nif*D genes and gene products have a higher degree of sequence homology to *Azotobacter* and cyanobacterial *nifH* and *nif*D (NifH, NifD) than would be expected based on 16S rRNA-DNA sequence comparisons, and suggested the possibility of lateral transfer of *nif* genes. The comparatively high sequence homology between frankial and cyanobacterial DNA has now been confirmed for *nif*K (A Hirsch, pers. comm.).

Symbiotic interactions

Actinorhizal root nodules are modified roots that branch repeatedly, forming lobed, coralloid structures. The nodules are perennial. A single nodule may persist for up to 5–10 years, while its active, nitrogen-fixing nodule lobes are formed annually (Schwintzer et al., 1982). What we know of early nodulation signals, and the events of nodule establishment, indicates that there are points of similarity between the *Frankia* symbioses and rhizobiallegume interactions, and areas of difference.

Nodulation signals

It is of considerable interest whether the molecular chain of events shown to occur in rhizobial symbioses involving plant flavonoid signalling, *nodD* gene expression, **nod** factor secretion, and subsequent nodule initiation, has a broad evolutionary basis. Searches have been made of *Frankia* DNA for sequence homology to the common *nod* genes of rhizobia, and for functional complementation of rhizobial nod mutants. To date, no definitive demonstration of *nod* genes in *Frankia* exists. Complementation of a rhizobial *nodD* mutant by *Frankia* DNA has recently been reported (Chen et al., 1991): however, unusual results associated with complementation of a *nod*A::Tn5 rhizobial mutant by *Frankia* DNA have also been described (Reddy et al. 1992).

Molecular approaches to understanding regulatory events in actinorhizal nodulation are currently limited by several factors, including slow growth rates of symbionts and hosts, high G+C content of *Frankia* DNA, and the lack of a genetic transformation system for *Frankia*. Mur-

ry (1993) has reported preliminary evidence that stable transformation of some *Frankia* strains can be achieved by electroporation, if non-methylated DNA is used.

Host-derived flavonoids may influence actinorhizal nodulation. In one study (Benoit and Berry, 1993), several flavonoidlike compounds were purified from *Alnus* seed eluates, and added exogenously to root systems at the time of *Frankia* inoculation. Differential effects on nodulation were seen: some compounds enhanced the number of nodules formed, while other compounds reduced nodulation. The compounds were not identified chemically, but had spectral characteristics typical of known flavonoids. The effect of these flavonoidlike compounds may be on *Frankia* gene expression, or alternatively on host process (as discussed by Hirsch, 1992, for rhizobial symbioses).

Lipo-oligosaccharidic **nod** factors by *Rhizobium* have powerful effects on root hair curling (RHC) and early hoist nodulation events in legumes (Denarié et al., 1992). A positive RHC response was recently reported to occur following application of concentrated filtrates of *Frankia* growth medium to axenic host root systems (Van Ghelue et al., 1993). The active RHC factor has yet to be identified. Some of the purified rhizobial **nod** factors have been tested for their effects on root hair curling in *Alnus* seedlings (Van Ghelue et al., 1993). The RHC response was variable, but always negative in the presence of *R. meliloti* **nod** factors. Incubation of *Frankia* cells in host exudates prior to inoculation of roots with *Frankia* culture filtrate has been reported both to enhance root hair curling in *Alnus* (Prin and Rougier, 1987), and to have no additive effect (Van Ghelue et al., 1993).

Nodule morphogenesis

Nodules are initiated in actinorhizal plants either via root hair infection by *Frankia,* or by intercellular invasion. As is the case in legume symbioses, the mode of infection is host-determined, since single strains of *Frankia* have been shown to nodulate different hosts by different infection mechanisms (Miller and Baker, 1986). Early stages in the infection process include root hair wall deformation, or in intercellular infections, the secretion of host extracellular matrix containing pectic polysaccharides (Liu and Berry, 1991). Host cortical cell proliferation is followed by expansion and even hypertrophy in groups of cells infected by *Frankia*. The endosymbiont penetrates postmeristematic cortical cells probably during cell expansion (Berry and Sunell, 1990). Once it is within the host cell lumen, the endosymbiont is always surrounded by a plant-derived encapsulation layer enriched in polygalacturonans, in addition to host plasmalemma and cytoplasm. Modification of the organization of cell wall synthesis and/or the composition of host extracellular matrix appears to be an important part of the establishment of an endosymbiotic association. Within the host cells, *Frankia* proliferates and, in all effective symbioses except those within the Casuarinaceae, vesicles are formed where nitrogen fixation takes place. The functional physiology of actinorhizal symbiotic nitrogen fixation has been reviewed by Huss-Danell (1990).

Controlling factors for nodule differentiation in actinorhizal plants have not been identified. Phytohormones are secreted by *Frankia* (Berry et al. 1989; Stevens and Berry 1988), and early active nodule tissue contains high levels of auxin conjugates and cytokinin derivatives (Wheeler et al., 1979), but their origin and functional significance is unknown. Nodule-specific gene expression is currently investigated by differential hybridization of a nodule mRNA/cDNA library with root and nodule cDNA probes. Several nodule specific genes have been isolated. One of these, a sequence present at relatively high abundance, shows homology to cysteine proteases (Goetting-Minesky and Mullin, 1993).

Oxygen protection and nitrogen fixation

Because the nitrogenase enzyme is oxygen-labile, the ability to regulate molecular oxygen is one of the key determinants of nitrogen fixation in aerobic microorganisms. Nitrogen-fixing organisms have evolved a number of mechanisms to regulate nitrogenase activity and energy production in relation to oxygen. Of the major groups of symbiotic nitrogen-fixers, Frankia spp. and certain of the cyanobacteria are capable of carrying out nitrogen fixation at atmospheric oxygen concentration. They accomplish this by localizing nitrogenase activity within specialized compartments, bounded by a gas diffusion barrier. In the case of *Frankia,* nitrogen depletion results in the development of vesicles-stalked, spheroidal structures that differentiate from vegetative hyphae (Fig. 1).

In symbiosis, host plant tissue adaptations are also evidently involved in oxygen diffusion regulation. Such host specialization can include cell packing, the occurrence of hemoglobin, the formation of nodule roots, and apparent cell wall modification (Tjepkema et al., 1986). In contrast to the legume symbioses, diffusion kinetics characteristic of a variable oxygen diffusion barrier within the nodule do not occur in most actinorhizal plants (Benson and Silvester, 1993), with one exception, *Coriaria* (Silvester and Harris, 1989).

Vesicle development and nitrogen fixation

Several lines of evidence — physiological, cytological and genetic — indicate that the function of the vesicle is to provide oxygen protection to nitrogenase. Provesicles begin to differentiate in free-living *Frankia* within 6 h following removal of external combined nitrogen. The onset of nitrogen fixation coincides with the appearance of mature vesicles detectable by about 24 h after nitrogen removal of external combined nitrogen. The onset of nitrogen fixation coincides with the appearance of mature vesicles, detectable by about 24 h after nitrogen removal (Fontaine et al., 1984). Mature vesicles are larger in diameter than provesicles, and are characterized by internal septations and by the formation of an external lamellate lipid envelope layer, which is not formed around vegetative hyphae (Fig. 2). Vesicle envelopes consist of multiple lamellae of uniform (3–5 nm) thickness. The lipidic nature

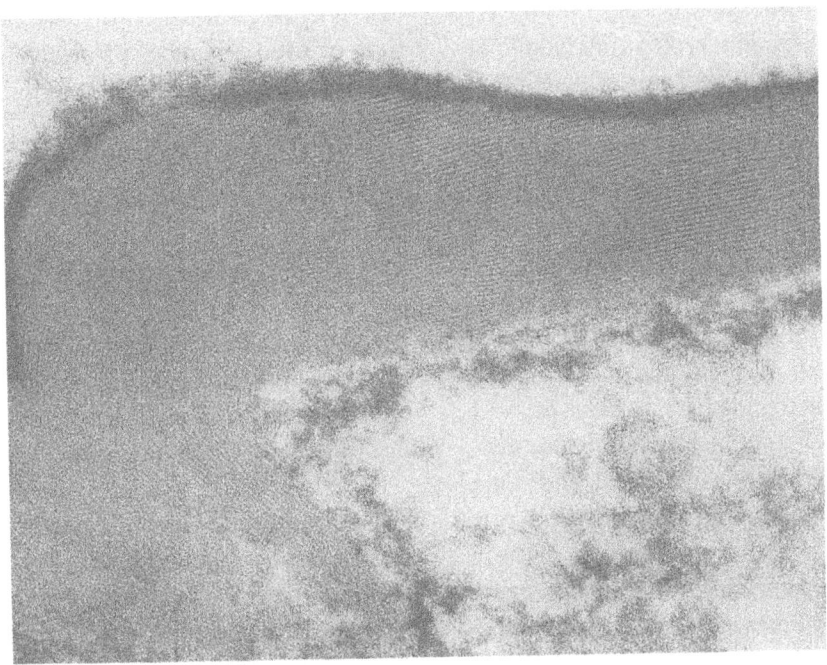

Fig. 2. Frankia vesicle envelope *Alnus serrulata* nodule stained with chromic acid to show individual laminae. Each lamnia is approximately 3 nm in width. Transmission electron micrograph and preparation by R.H. Berg.

of the envelope can be inferred ultrastructurally from patterns of heavy-metal staining (Harriott et al., 1992, Torrey and Callaham, 1982) and demonstrated with solvent extraction, lipid-specific fluorochromes and cytological observations (Lamont et al., 1988).

Nitrogen fixation in vesicles is thus dependent on conditions established during the later stages of vesicle development. Not only nitrogenase activity, but apparently the synthesis of nitrogenase proteins, also depends on vesicle maturation: Huss-Dannell and Bergman (1990) used immunogold labelling of antisera to Fe and MoFe proteins to show the absence of these proteins in provesicles forming within nodules of *Alnus incana,* but their specific localization in mature and maturing (at least partially septate) vesicles. The vesicle envelope was not observed in these preparations. The timing of envelope formation in relation to septation appears to be variable, but approximately coincident: while septae were still incomplete, vesicle envelopes thicker than adjacent hyphal walls were observed in *A. serrulata* nodule tissue, using transmission electron

microscopy (Berg, pers. comm.). These cytological observations suggest that either septa formation, or vesicle envelope formation, or both, are necessary preconditions for nitrogenase synthesis and activity.[1]

Vesicle physiology

Studies of gas diffusion kinetics in *Frankia* cells induced to fix nitrogen in culture have provided a physiological perspective on the interactions between oxygen and nitrogen fixation in the vesicle. When *Frankia* cells were grown under nitrogen-limited conditions and at very low external oxygen concentration, the saturation of respiration exhibited Michaelis-Menton kinetics in response to increasing external oxygen partial pressures. When otherwise similar cultures were grown at atmospheric pO_2, however, respiration saturation approximated diffusion-limited kinetics, implying the existence of a diffusion barrier (Murry et al., 1985). Murry et al. (1985) also showed for the first time that such a diffusion barrier was adapted to the oxygen condi-

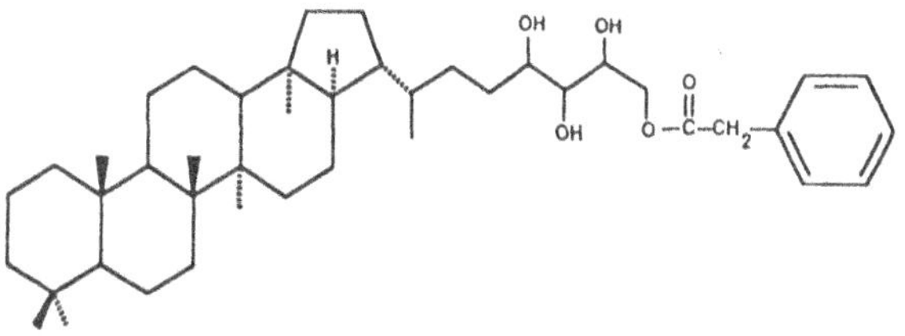

Fig. 3. Putative structure of the vesicle-specified hopanoid lipid, bacteriohopanetetrol phenylacetate monoester.

tions under which the vesicles differentiated: cells grown at different oxygen partial pressures (pO_2) under nitrogen limitation were assayed for nitrogen fixation (acetylene reduction) over a range of external oxygen conditions. The optimum oxygen concentration for nitrogenase activity either coincided with, or was somewhat lower than, the oxygen concentration at which the cells were grown. Acetylene reduction rates dropped off abruptly from this optimum with increasing pO_2, or with lower oxygen tension. Parsons et al. (1987) showed that oxygen adaptation for nitrogen fixation was observable in hyperbaric oxygen conditions(40kPa and even 70kPa oxygen partial pressure).

The vesicle envelope

What is the nature of the oxygen-adapted oxygen protection mechanism? The adaptability of the vesicle diffusion barrier is apparently primarily a function of the vesicle envelope. Parsons et al. (1987) found that upon increasing oxygen tension during vesicle differentiation, the thickness of the vesicle envelope increased dramatically. Not only did the optical birefringence of the envelope increase with increasing pO_2, but the number of lipid layers in the envelope also increased, from about 15 layers at 4kPa oxygen to about 50 layers at 40kPa O_2. Such a variation in envelope thickness suggests a very specific level of oxygen-dependent regulation of oxygen in the vesicle, i.e. by regulating synthesis and/or deposition of the envelope lipid layers.

Recent progress has been made in identifying the lipid composition of the vesicle envelope (Berry et al., 1993). Previously it was postulated that the vesicle envelope was composed of glycolipids, by analogy to the well-characterized cyanobacterial heterocyst envelope (Lambein and Wolk, 1973); but no major vesicle-specific glycolipids were detected in *Frankia* cultures (Lopez et al., 1983). When purified vesicle envelope preparations were recently analyzed using high performance liquid chromatography and mass spectrometry, the envelope was found to be virtually entirely composed of hopanoid lipids (Berry et al., 1993). Hopanoid lipids are a class of pentacyclic triterpenoids with sterol-like properties which are known to alter membrane permeability properties in a wide range of prokaryotes. The *Frankia* vesicle envelope is a unique structure in that the hopanoids are evidently secreted in quantity and assembled extracellularly rather than integrating solely in the cell membrane. Hopanoids are abundant in *Frankia* even in vegetative cells, where bacteriohopanetetrol ($C_{35}H_{62}O_4$) is the predominant lipid present (Berry et al., 1991). The vesicle envelope of *Frankia* HFPCpI1 is comprised predominantly of bacteriohopanetetrol and a phenylacetate derivative (Fig. 3), in roughly equal amounts. The latter molecule, bacteriohopanetetrol phenylacetate monoester, is either vesicle specific, or greatly vesicle-enhanced, since it is not detected in hyphal preparations.

The mechanism controlling oxygen concentration in the *Frankia* vesicle probably requires two elements– consumption of oxygen internally, as well as diffusion limitation. Especially high rates of respiration are characteristic of mature vesicles (Vikman, 1992). The possibility of a vesicle-specific electron transport chain remains to be studies in *Frankia*, and mutants defective in vesicle respiration have not yet been reported. Recently, a series of putative mutants unable to grow diazotrophically and lacking measurable acetylene reducing activity under aerobic conditions, have been described (Murry 1993). All such mutants with altered vesicle wall/envelope properties reduced acetylene under microaerophilic (0.5kPa O_2) conditions.

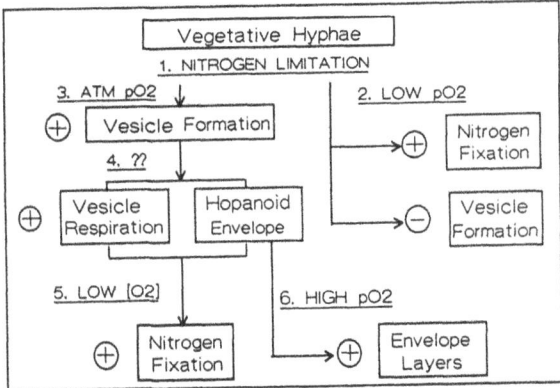

Fig. 4. The two pathways to nitrogen fixation in *Frankia*. Nitrogen limitation (1.) is required for nitrogenase expression, for either pathway. At very low oxygen tension (2.), vesicles do not differentiate but nitrogen fixation is induced. At atmospheric pO₂ (3.), vesicles differentiate. Nitrogen fixation is not detected until later stages of vesicle maturation, which are characterized by envelope formation and high rates of respiration (5.). The factors that trigger vesicle maturation (4.) are not known. Changes in internal oxygen concentration (5.) resulting from vesicle maturation may influence expression of nitrogenase. The thickness of the hopanoid vesicle envelope is determined by the oxygen concentration under which the vesicle differentiates (6.).

Taken together, these studies indicate a) that the vesicle envelope functions as an oxygen-adaptive gas diffusion barrier with a specific function in the oxygen protection of nitrogenase; and b) that while vesicle formation is triggered by ammonium limitation, nitrogenase synthesis appears to be dependent at least in part on envelope formation, perhaps at the level of gene expression, and most likely as a consequence of low internal oxygen concentrations. Oxygen modulation of nitrogenase gene expression has been described in several systems, including *R. meliloti* (Batut et al., 1993, De Bruijn et al., 1990). Interestingly, nitrogenase activity can be induced independently of vesicle formation, if *Frankia* cells are nitrogen-starved at very low external oxygen concentrations (Murry et al 1985). The two pathways to nitrogen fixation in *Frankia* described above are summarized in Figure 4.

The Frankia *vesicle and the cyanobacterial heterocyst*

The parallels between vesicle differentiation in *Frankia* and heterocyst differentiation in cyanobacteria such as *Anabaena* are striking. In both organisms, N-starvation is the initial condition that triggers what appears to be an alternate development pattern. Oxygen diffusion limitation is provided by a lipid envelope which is deposited at a later stage in structural differentiation (see Fay 1992). Nitrogenase activity is also detected during the later stages of development. The oxygen-adaptive nature of the diffusion barrier appears to be very similar in the two organisms: a direct correlation exists between oxygen tension during development and oxygen tension for optimum nitrogenase activity (Murry et al. 1984, 1985; Parsons et al. 1987). Nitrogenase activity can also be expressed under very low pO₂ or anaerobiosis in *Frankia* without any apparent cellular specialization, suggesting that regulation of expression of nitrogen fixation can be explained by a combination of nitrogen depletion and oxygen modulation. The same appears to be the case in heterocystous cyanobacteria, although there is debate whether some developmental factor is also required for nitrogenase expression in *Anabaena* (Buikema and Haselkorn, 1991; Ernst et al., 1992; Fay, 1992).

The spatial and temporal specificity of envelope deposition, as distinct from nitrogenase synthesis, suggests that some vesicle-specific developmental factor is required in this process. The genetic/cellular basis of developmental patterning for either vesicles or heterocysts is not known. Finally, it is interesting to note from an evolutionary standpoint that the lipid envelopes of *Frankia* and *Anabaena,* while analagous in function and development, are composed of molecules derived from completely different biosynthetic pathways. In *Frankia*, an understanding of hopanoid biosynthesis and its regulation during vesicle development should provide a new basis for interpreting oxygen interactions with nitrogen fixation, and perhaps vesicle development as well.

Concluding remarks

Advances in molecular genetic studies will provide increasingly solid frames of reference for comparison among the three classes of nitrogen-fixing plant symbioses– the actinorhizal, the cyanobacterial, and the legume-rhizobial. As the molecular phylogenetic data now indicate (Swensen et al., in press), plant taxa capable of root nodulation share a closer genetic relationship than had previously been perceived. Yet it must be noted that many other closely-related taxa are not nodulated. There is no clear information yet on the molecular nature of barriers or common factors in actinorhizal nodulation. On the endosymbiont side, structural *nif* gene sequences in *Frankia* show a greater degree of homology with *Anabaena* DNA (and with sequences from *Azotobacter* sp.) than would be predicted from 16S rRNA/DNS comparisons (Benson and Silvester, 1993; Normand et al., 1988). It is possible that similarities in regulatory and/or developmental aspects of nitrogen fixation in these organisms have a genetic basis as well.

Acknowledgements

The willingness of colleagues to communicate research findings, and the electron micrographs provided by R H Berg, are gratefully acknowledged. R. Edberg assisted technically in manuscript preparation. Research on hopanoid lipids was supported in part by USDA-NRICGP 91-37305-6704 and NSF DCB-91-05841.

Note

In some vesicles formed in symbiosis, and in certain free-living cultures, septa are not formed, but envelope formation and nitrogen fixation still take place. For a thorough review of *Frankia* morphogenesis, see Newcomb W and Wood S M 1987 Morphogenesis and fine structure of *Frankia* (Actinomycetales): the microsymbiont of nitrogen-fixing actinorhizal root nodules. Int. Rev. Cytol. 109, 1–88.

References

Batut J, dePhilip P, Reyrat J M, Waelkens F and Boistard P 1993 Oxygen regulation of nitrogen fixation gene expression in *Rhizobium meliloti In* Advances in Molecular Genetics of Plant-Microbe Interactions. Eds. E W Nester and D P S Verma. pp 183–191. Kluwer Academic Publishers, Dordrecht.

Benoit L F and Berry A M 1993 Flavonoidlike compounds from seeds of *Alnus rubra* Bong. influence host nodulation by *Frankia* (Actinomycetales). Physiol. Plant. (In press).

Benson D R, Harriott O T, Josted T J and Zhang X 1993 Analysis of nitrogen metabolism genes in *Frankia alni* strain CpI1. 9th International Conference on *Frankia* and Actinorhizal Plants Proceedings, Ruapehu NZ (Abstract).

Benson D R and Silvester W B 1993 Biology of *Frankia* strains, actinomycete symbionts of actinorhizal plants. Microbiol. Rev. 57, 2, 293–319.

Berg R H 1993 Imaging the symbiotic vesicle envelope in thin sections. 9th International Conference on *Frankia* and Actinorhizal Plants Proceedings, Ruapehu NZ (Abstract).

Berry A M, Harriott O T, Moreau R A, Osman S F, Benson D R and Jones A D 1993 Hopanoid lipids compose the *Frankia* vesicle envelope, presumptive barrier of oxygen diffusion to nitrogenase. Proc. Natl. Acad. Sci. (USA) 90, 6091–6094.

Berry A M, Moreau R A and Jones A D 1991 Bacteriohopanetetrol: abundant lipid in *Frankia* cells and in nitrogen-fixing nodule tissue. Plant Physiol. 95, 11–115.

Berry A M, Kahn R K S and Booth M C 1989 Identification of indole compounds secreted by *Frankia* HFPArI3 in defined culture medium. Plant and Soil 118, 205–209.

Berry A M and Sunell L A 1990 The infection process and nodule development. *In* The Biology of *Frankia* and Actinorhizal Plants. Eds. C R Schwintzer and J D Tjepkema. pp 61–81. Academic Press, San Diego, CA.

Buikema W J and Haselkorn R 1991 Isolation and complementation of nitrogen fixation mutants of the cyanobacterium *Anabaena* sp. strain PCC 7120. J. Bacteriol. 173, 6, 1879–1885.

Callaham D, DelTredici P and Torrey J G 1978 Isolation and cultivation in vitro of the actinomycete causing root nodulation in *Comptonia*. Science 199, 899–902.

Chen L, Cui Y, Qin M, Wang Y, Bai X and Ma Q 1991 Identification of a *nodD*-like gene in *Frankia* by direct complementation of a *Rhizobium nodD*-mutant. Mol. Gen. Genet. 233, 311–314.

Conard S G, Jaramillo A E, Cromack K and Rose S 1985 The Role of the Genus *Ceanothus* in Western Forest Ecosystems. USDA Forest Service, Pacific Northwest Forest and Range Esperiment Station, General Technical Report PNW-182, Portland OR.

Cronquist A 1988 The Evolution and Classification of Flowering Plants. The New York Botanical Garden, Bronx, NY.

Dawson J O 1990 Interaction among Actinorhizal and associated plant species. *In* The Biology of *Frankia* and Actinorhizal Plants. Eds. C R Schwintzer and J D Tjepkema. pp 299–316. Academic Press, San Diego, CA.

DeBruijn F J, Hilgert U, Stigter J, Schneider M, Meyer H, Klosse U and Pawlowski K 1990 Regulation of nitrogen fixation and assimilation genes in the free-living versus symbiotic state. *In* Nitrogen Fixation: Achievements and Objectives. Eds. P M Gresshoff, L E Roth, G Stacey, W E Newton. pp 33–44. Chapman and Hall, New York-London.

Denarié J and Roche P 1991 *Rhizobium* nodulation signals. *In* Molecular Signals in Plant-Microbe Communication. Ed. D P S Verma, pp 259–324. CRC Press, Boca Raton, FL.

Diem H G and Dommergues Y R 1990 Current and potential uses and management of Casuarinaceae in the tropics and subtropics. *In* The Biology of *Frankia* and Actinorhizal Plants. Ed. C R Schwintzer and J D Tjepkema. pp 317–342. Academic Press, San Diego, CA

Ernst A, Black T, Cai Y, Panoff JM, Tiwari D N and Wolk C P 1992 Synthesis of nitrogenase in mutants of the cyanobacterium *Anabaena* strain PCC 7120 affected in heterocyst development of metabolism. J. Bacteriol. 174, 19, 6025–6032.

Fay P 1992 Oxygen relations of nitrogen fixation in cyanobacteria. Microbiol. Rev. 56, 2, 340–372.

Fontaine M F, Lancelle S A and Torrey J G 1984 Initiation and ontogeny of vesicles in cultured *Frankia* sp strain HFPArI3. J. Bacteriol. 160, 921–927.

Goetting-Minesky P and Mullin B C 1993 Differential host gene expression in the *Alnus glutinosa-Frankia* symbiosis. 9th International Conference on *Frankia* and Actinorhizal Plants Proceedings, Ruapehu NZ (Abstract).

Hahn D, Lechevalier M P, Fischer A and Stackerbrandt E 1989 Evidence for a close phylogenetic relationship between members of the genera *Frankia*, *Geodematophilus*, and "*Blastococcus*" and emendation of the family Frankiaceae. Syst. Appl. Microbiol. 11, 236–242.

Harriott O T, Khairallah L and Benson D R 1991 Isolation and structure of the lipid envelopes from the nitrogen-fixing vesicles of *Frankia* sp. strain CpI1. J. Bacteriol. 173, 2061–2067.

Hibbs D E and Cromack K 1990 Actinorhizal plants in pacific northwest forests. *In* The Biology of *Frankia* and Actinorhizal Plants. Eds. C R Schwintzer and J D Tjepkema. pp 343–363. Academic Press, San Diego, CA.

Hirsch A M 1992 Developmental biology of legume nodulation. New Phytol. 122, 211–237.

Huss-Dannell K 1990 The physiology of actinorhizal nodules. *In* The Biology of *Frankia* and Actinorhizal Plants. Eds. C R Schwintzer and J D Tjepkema. pp 129–156. Academic Press, San Diego Diego, CA.

Kohls S R, Van Kessel C, Baker D D, Grigal D F and Lawrence D B 1993 Assessment of nitrogen fixation and N cycling by *Dryas* along a chronosequence within the forelands of the Athabasca Glacier, Can. Soil Biol. and Biochem. (In press).

Lambein F and Wolk C P 1973 Structural studies on the glycolipids from the envelope of the heterocyst of *Anabaena cylindrica*. Biochemistry 12, 791–798.

Lamont H C, Silvester W B and Torrey J G 1988 Nile red fluorescence demonstrates lipid in the envelope of vesicles from nitrogen-fixing cultures of *Frankia*. Can. J. Microbiol. 34, 6546–660.

Liu Q and Berry A M 1991 Localization and characterization of pectin polysaccharides in roots and root nodules of *Ceanothus* spp. during intercellular infection by *Frankia*. Protoplasma 163, 93–101.

Lopez M F, Whaling C S and Torrey J G 1983 The polar lipids and free sugars of *Frankia* in culture. Can J. Bot. 61, 2834–2842.

Miller I M and Baker D D 1986. Nodulation of actinorhizal plants by *Frankia* strains capable of both root hair infection and intercellular penetration. Protoplasma 131, 82–91.

Murry M, Daly R and Islaih M 1993 UV induction and isolation of mutants of *Frankia* sp. strain CeSI5 that require combined nitrogen for growth. 9th International Conference on *Frankia* and Actinorhizal Plants Proceedings, Ruapehu NZ (Abstract).

Murry M A, Horne A J and Benemann J R 1984 Physiological studies of oxygen protection mechanisms in the heterocysts of *Anabaena cylindrica* Appl. Environ. Microbiol. 47, 449–454.

Murry M A, Zhang Z and Torrey J G 1985 Effect of oxygen on vesicle formation, acetylene reduction, and oxygen-uptake kinetics in *Frankia* sp. HFPCcI3 isolated from *Casuarina cunninghamiama*. Can. J. Microbiol. 31, 804–809.

Normand P, Cournoyer B, Simonet P and Nazaret S 1992 Analysis of a ribosomal RNA operon in the actinomycete *Frankia*. Gene 111, 119–124.

Normand P, Simonet P and Bardin R 1988 Conservation of *nif* sequence on *Frankia* Mol. Gen. Genet. 213, 238–246.

Parsons R, Silvester WB, Harris S, Gruijters W T M and Buillivant S 1987 *Frankia* vesicles provide inducible and absolute oxygen protection for nitrogenase. Plant Physiol. 83, 728–731.

Prin Y and Rougier M 1987 Preinfection events in the establishment of *Alnus-Frankia* symbiosis: study of the root hair deformation step. Plant Physiol. (Life Sci. Adv.) 6, 99–106.

Racette S and Torrey J G 1989 Root nodule initiation in *Gymnostoma* (Casuarinaceae) and *Shepherdia* (Elaegnaceae) induced by *Frankia* strain HFPGpI1. Can.J.Bot. 67, 2873–2879.

Reddy A Bochenek B and Hirsch A M 1992 A new *Rhizobium meliloti* symbiotic mutant isolated after introducing *Frankia* DNA sequence into a *nodA*::Tn5 strain. Mol. Plant-Microbe Interact. 5, 62–71.

Schwintzer C R, Berry A M and Disney L D 1982 Seasonal patterns of root nodule growth, endophyte morphology, nitrogenase activity and shoot development in *Myrica gale*. Can. J. Bot. 60, 746–657.

Schwintzer C R and Tjepkema J D 1990 The Biology of *Frankia* and Actinorhizal Plants Academic Press, San Diego, CA. 408p.

Silvester W B and Harris S L 1989 Nodule structure and nitrogenase activity of *Coriaria arborea* in response to varying pO_2. Plant and Soil 118, 97–109.

Stevens G A and Berry A M 1988 Cytokinin secretion by *Frankia* sp. HFPArI3 in defined medium. Plant Physiol. 87, 15–16.

Tjepkema J D, Schwintzer C R and Benson D R 1986 Physiology of actinorhizal nodules. Annu. Rev. Plant Physiol. 37, 209–232.

Swensen S M, Mullin B C and Chase M W 1994 Phylogenetic

affinities of datiscaceae based on an analysis of nucleotide sequences from the plastid *rbc*L gene. System Bot *(In press)*.

Torrey J G 1978 Nitrogen fixation by actinomycete-nodulated angiosperms. BioScience 28, 9, 586–592.

Torrey J G and Callaham D 1982 Structural features of the vesicle of *Frankia* sp. CpI1 in culture. Can. J. Microbiol. 18, 749–757.

Twigg P, An C and Mullin B C 1990 Nucleotide sequence of *nifD*, the structural gene coding for the alpha subunit of the MoFe protein of the nitrogenase complex from the actinomycete *Frankia*. *In* Nitrogen Fixation Achievements and Objectives. Eds. PM Gresshoff, LE Roth, G Stacey and W E Newton. p 771 Chapman and Hall, New York.

Van Ghelue M, Lovaas E, Robertsen E, Weber C and Solheim B 1993 Characterization and partial purification of specific molecules involved in the actinorhizal symbiosis between *Alnus glutinosa* and *Frankia* sp Ar13. 9th International Conference on *Frankia* and Actinorhizal Plants Proceedings, Ruapehu NZ (Abstract).

Vikman P-A 1992 The symbiotic vesicle is a major site for respiration in *Frankia* from *Alnus incana* root nodules. Can. J. Microbiol. 38, 779–784.

Wheeler C T, Henson I E and McLaughlin M E 1979 Hormones in plants bearing actinomycete nodules. Bot. Gaz. (Chicago), Suppl. 140, S52–S57.

P.H. Graham, M.J. Sadowsky & C.P. Vance (eds.), Symbiotic nitrogen fixation, 147–152.
© 1994 *Kluwer Academic Publishers.*

Characterization of *Bradyrhizobium japonicum* chromosomal genes involved in the regulation of *hup* gene expression in free-living conditions and in controlling hydrogenase activity

C. Van Soom, J. Vanderleyden and A. P. Van Gool[1]
University of Leuven, F.A. Janssens Memorial Laboratory of Genetics, Willem de Croylaan, 42, B–3001 Heverlee, Belgium. [1]*Corresponding author*

Key words: Bradyrhizobium, hydrogenase, transcriptional regulation

Abstract

Nucleotide sequence analysis of *Bradyrhizobium japonicum* chromosomal DNA, 9 kb downstream of the *hup* genes (hydrogenase structural genes), revealed one incomplete and three complete open reading frames. These were designated as *hypD'*, *hypE*, *hoxX* and *hox A*, respectively, based on the strong homology of the deduced amino acid sequences with those encoded by genes from *Azotobacter vinelandii*, *Escherichia coli*, *Rhodobacter capsulatus*, *Rhizobium leguminosarum* (*hypD'* and *hypE*) and *Alcaligenes eutrophus* (*hoxX* and *hoxA*). Implication of these findings on the regulation of hydrogenase synthesis by hydrogen, oxygen and nickel in free-living *B. japonicum* are discussed. Similarly the nucleotide sequence analysis of a 2.2 kb region downstream the *B. japonicum* hydrogenase structural genes allowed the identification of *hupC*, *hupD* and *hupF* homologous genes forming the distal part of an operon containing the *hup* structural genes *hupS* and *hupL*. A few hundred bp downstream, separated by an AT rich region, an RpoN1/RpoN$_2$ transcribed *hupG* homologous gene was identified. These ORF's show significant homology to genes involved in processing, maturation and functional regulation of the hydrogenase complex, previously described in various hydrogen-oxidizing bacteria.

Introduction

The establishment of micro-aerobiosis in *B. japonicum* bacteroids triggers a respiratory energy generating mechanism driven by a membrane bound hydrogenase recycling hydrogen released during nitrogen fixation. Some strains of *B. japonicum* under free-living microaerobic conditions can exploit this system to support CO_2 fixation and to drive a chemoautotrophic life style.

A substantial amount of genetic data on *B. japonicum* indicates a strong conservation in the genetic make up of the hydrogen uptake system (Hup) in comparison with a diversity of hydrogen oxidizing bacteria (for a review see Pryzbyla et al., 1992). However genetic information on internal and external control mechanisms operating on *B. japonicum* hydrogenase expression and activity is rather scarce. Therefore the present report focuses on the elucidation of *B. japonicum* chromosomal genes involved in the regulation of hydrogenase expression and its environmental control.

Tn5 mutagenesis in the *hup* region of the *B. japonicum* USDA122 DES chromosome identified two regions that are exclusively involved in free-living hydrogenase activity and that are not required for endosymbiotic Hup activity (Lambert et al., 1987). The first region is located immediately downstream of the hydrogenase structural genes *hupSL* while the second region occurs at 9 kb downstream of *hupSL*. The corresponding regions in a USDA122 derivative strain *B. japonicum* CB1809 were sequenced. Genes were identified by ORF analysis and characterized on the basis of nucleotide sequence predicted gene products and homology search of corresponding genes in other hydrogen oxidizing bacteria.

Results and discussion

Sequence analysis and characterization of a 2.2 Kb hupL 3' adjacent region

An *hup* containing cosmid pFAJ1002 was isolated from a *B. japonicum* CB1809 genomic library using a mixture of two synthetic 20-mers. They encoded the first seven amino acids of the small and large hydrogenase subunit (Sayavedro-Soto et al., 1988) respectively. The physical organization of the cosmid DNA appeared to be identical to the *Eco*RI map of the *hup* region in *B. japonicum* USDA122. Together with four additional *hup*-specific clones isolated from a pHC79 genomic library, a 55 kb chromosomal region of *B. japonicum* CB1809, carrying the *hup* structural genes within a 12.9 kb *Eco*RI fragment, was physically characterized. The identity of the physical map of the *hup* region in strains USDA122 and CB1809 points to a high nucleotide conservation and presumably also gene organization.

Here the nucleotide sequence of the region located immediately 3' downstream of the two hydrogenase structural genes in the chromosome of *B. japonicum* CB1809, is reported. The sequence of the first 334 bp was reported by Sayavedra-Soto et al. (1988). The physical and genetic organization of the sequenced region is shown in Figure 1. The double stranded DNA sequencing of a series of overlapping subclones in pUC19 covered part of the 5.9 kb *Hind*III, the entire 0.7 kb *Hind*III and part of the 3.0 kb *Eco*RI fragment (Fig 1). The total sequence containing 2615 base pairs immediately downstream of the *hupL* termination codon has been deposited in the EMBL library under accession number Z21948. Compilation of these sequence data revealed a succession of four open reading frames ORF1, ORF2, ORF3, and ORF4 (Fig. 1). They all are transcribed in the same direction as the hydrogenase structural genes and preceded by a putative ribosome binding site. The 63% overall GC content, the 89.6% GC bias at the third codon position as well as the codon usage in the four ORFs is indicative of *B. japonicum* group III genes (Ramseier et al., 1991). ORF1, ORF2 and ORF3 were not preceded by typical promoter concensus sequences and appeared to be transcriptionally coupled with the *hup-SL* genes. In between ORF3 and ORF4 an AT-rich non-coding region exceeding 200 bp was found upstream from a typical −24/−12 promoter concensus sequence in front of ORF4. Based on their strong nucleotide sequence homology, the ORFs were respectively des-

ignated as *hupC, hupD, hupF* and *hupG* according to the nomenclature of the corresponding genes in *R. leguminosarum* (Table 1). In addition to gene correspondence in *B. japonicum*, the size, molecular mass and amino acid composition of the deduced gene products were compared to those encoded by similar genes occurring in a variety of genetically divergent hydrogen oxidizing bacteria (Table 1). Together with a comparative analysis of amino acid sequence- and structural domain conservation among the homologous gene products, a prediction on the presumed function of the corresponding gene products in *B. japonicum* was made (Table 1).

The number and position of potential transmembrane helices in the ORF1 predicted gene product together with the conservation of a histidine residue at the N-terminal ends of the three transmembrane helices, is characteristic of b-type cytochromes (Friden and Hedestedt, 1990; Widger et al., 1984). This supports the idea that the HupC-like ORF1 gene product functions as an electron carrier between hydrogen and the electron transport chain. A Tn5 mutation located at the "right" border of the 12.9 Kb *Eco*RI fragment, causing a weak Hup$^+$ phenotype both in free-living and symbiotic conditions (Lambert et al., 1987), can be located in *hupC*, indicating the ability of hydrogenase to transfer electrons to alternative electron acceptors. However, recent data indicate that purified heterodimeric hydrogenase efficiently catalyzes H$_2$-ubiquinone-1 oxidoreductase activity in the absence of any detectable cytochromes (Ferber and Maier, 1993) so it seems that the *B. japonicum hupC* gene product is not strictly required for electron transport to ubiquinone. This entails that a possible role in membrane attachment of the hydrophobic *hupC* gene product remains to be considered.

The predicted ORF2 product HupD, has a relatively hydrophobic N-terminal segment that could anchor this protein to the membrane. On the other hand, in *E. coli* the related HyaD and HyaE proteins (Table 1) function as a processing complex of the hydrogenase subunits after insertion of these subunits in the membrane (Menon et al., 1991).

The *hupF* homologous ORF3 *B. japonicum* gene could be the site at which one of the three Tn5 insertions is located that were described by Lambert et al. (1987) and that result in an Hup$^+$ symbiotic and Hup$^-$ free-living phenotype, the other two insertions being located in the *hoxX* homologous gene (cf next session). In *E. coli* the related HypC gene product is a regula-

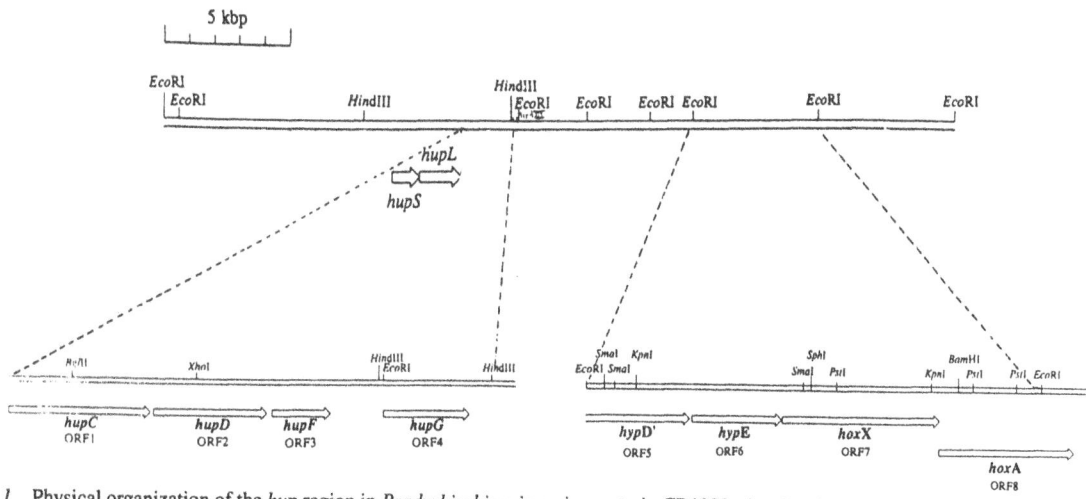

Fig. 1. Physical organization of the *hup* region in *Bradyrhizobium* japonicum strain CB1809, showing the location of the *hup*CDFG genes and *hyp*DE-*hoxXA* genes.

Table 1. Nucleotide and amino acid sequence homology of *hup* genes among hydrogenoxiding bacteria

B. japonicum genes	Gene products size	NS homology to		AA homology % identity	Predicted function
ORF1 *hupC*	244 a.a., 27.8 kD	*R. leguminosarum*[a]	*hupC*	66.1	cytochrome b type electron carrier
		A. vinelandii[b]	ORF 3	60.4	
		R. capsulatus[c]	*hupM*	57.4	
		A. eutrophus[d]	*hoxZ*	53.3	
		E. coli[e]	*hyaC*	43.0	
		W. succinogenes[f]	*hydC*	19.1	
ORF2 *hupD*	21.4 kD	*R. leguminosarum*	*hupD*	64.8	processing of hydrogenase subunits/membrane attachment
		A. eutropus	*hoxM*	45.6	
		E. coli	*hyaD*	40.9	
ORF3 *hupF*	98 a.a., 10.6 kD	*R. leguminosarum*	*hupF*	44.9	regulation of free-living hydrogenase expression
		A. eutrophus	*hoxL*	31.6	
		E coli	*hypC*	23.3	
ORF4 *hupG*	149 a.a., 15.8 kD	*R. leguminosarum*	*hupG*	35.1	processing of hydrogenase subunits
		A. eutrophus	*hupO*	31.3	
		E. coli	*hyaE*	18.2	

References: [a] Hidalgo et al., 1992; [b] Menon et al., 1990a; [c] Richaud et al., 1990; [d] Kortlüke et al., 1992; [e] Menon et al., 1990b; [f] Dross et al., 1992.

tory protein known to be required for the activity of all three hydrogenases.

ORF4 is located on a 2.9 kb fragment that upon deletion in *B. japonicum* results in a Hup⁻ phenotype and the production of unprocessed hydrogenase subunits (Fu and Maier, 1993). In *E. coli* the HupG relat-

ed HyaE protein together with the ORF2 homologous HyaD protein, most likely are involved in processing hydrogenase subunits. A perfect −24/−12 promoter concensus sequence is located 59 bp upstream from the ORF4 initiation codon suggesting that an alternative sigma factor RpoN1 and/or RpoN2 could be involved in the transcription of *hupG* and the more distal gene(s) in the same operon. A similar promoter sequence, characteristic of genes only switched on under specific environmental conditions, is located upstream of the *hupSLCDF* operon (Sayavedro-Soto, 1988) and upstream of the *hoxA* regulatory operon (see section 2 of the Results).

Sequence analysis and characterization of a 5.4 kb region 9 kb downstream of hupSL

From a series of overlapping subclones, containing the region required for free-living hydrogenase activity (Lambert et al., 1987) a 5408 bp sequence was generated by double-stranded DNA sequencing. This sequence revealed four open reading frames (ORF5 through ORF8) adjacent to and downstream from *hupL* (Fig. 1). They were all transcribed from the same strand as the hydrogenase structural genes. With the exception of a 5′ truncation in ORF5, the ORF's were complete and preceded by a putative ribosome binding site. ORF6 and ORF8 did not reveal typical promoter sequences. A potential −24/−12 type promoter was found in front of ORF7. Predictions on the size of the deduced polypeptides are indicated in Table 2. Codon usage, GC bias at the third codon position and average GC% are typical for *B. japonicum* group III genes (Ramseier et al., 1991).

A computer assisted data base search revealed strong homology between the nucleotide sequences of ORF5, ORF6, ORF8, part of ORF7 and genes involved in hydrogen metabolism reported for *E. coli*, *R. capsulatus*, *A. vinelandii*, and *A. eutrophus*. Comparison of the corresponding deduced amino acid sequences indicated functional conservations in the related gene products between different species (Table 2).

The deduced amino acid sequence of ORF5 is homologous to HypD like gene products, found in *E. coli*, *R. capsulatus* and *R. leguminosarum*. A function has not yet been assigned to HypD. The C-X-X-C structural motif at the N-terminal of the ORF5 deduced gene product and homologous HypD proteins in five different species could play a role in binding Fe-S clusters coordinating metal ion transport and/or their insertion into the hydrogenase holoenzyme.

The predicted product of ORF6 is highly homologous to the *R. capsulatus* and *E. coli* HypE protein. A Tn5 insertion in the "left" border of the 5.0 Kb EcoRI fragment (Fig. 1) and abolishing hydrogenase activity in both free-living and endosymbiotic conditions (Lambert et al., 1987), must be located in *hypE*.

The 84 C-terminal amino acids of ORF7 display almost 60% similarity with the C-terminal region of *A. eutrophus* HoxX. For *hoxX* only the 3′ terminal 261 nucleotide sequence immediately upstream of the *hoxA* transcriptional activator sequence has been reported (Eberz and Friedrich, 1991). It has been suggested by Friedrich (personal communication) that HoxX might be a sensor protein that forms a two component sensor/effector couple with HoxA. The occurrence of a potential membrane spanning region in the deduced amino acid sequence of ORF7 together with an amino acid motif characteristic for histidine kinase supports this hypothesis. An HoxX membrane location is very plausible for a sensor protein perceiving environmental stimuli and transducing them to the cell interior via phosphorylation of the HoxA effector to activate its transcriptional regulatory control activity(ies).

Two Tn5 mutations in a 4.9 kb *Eco*RI *B. japonicum* chromosomal fragment were reported by Lambert et al. (1987). These insertions cause a Hup⁻ phenotype only in free-living and not under endosymbiotic conditions. Those appear to be located in ORF7. This points towards a functional role of the deduced ORF7 gene product in the regulation of hydrogenase activity in free-living *B. japonicum*. Therefore we propose to call its corresponding gene *hoxX*. Since these two Tn5 mutants could only be complemented by a 3′ adjacent 5.5 *Eco*RI fragment, it can be assumed that *hoxX* forms an operon with the adjacent and downstream located ORF8 that extends more than 300 nucleotides into the 5.5 kb *Eco*RI fragment. This assumption is supported by the close arrangement of ORF7 and ORF8 and in addition by the absence of promoter consensus sequences in front of ORF8.

The ORF8 deduced gene product displays an extensive homology with *A. eutrophus* HoxA and several other NtrC-family transcriptional regulators such as *R. capsulatus* HupR1 and *E. coli* HydG (see Table 2). All these gene products belong to the two component sensor/effector family and share some structural and functional characteristics. The homology between the different polypeptides is most striking in the central region, where a conserved G-X-X-G-X-G-K-E sequence presumably corresponds to an ATP binding site. In contrast to the conservation reported for other NtrC-like regulators, the amino terminal region displays more variation among HoxA-like regulators. The latter part of NtrC-like regulators has been shown to be involved in their regulatory control activity through a NtrB-mediated phosphorylation of aspartic residue at position. Also the carboxy terminal region appears

Table 2. Nucleotide and amino acid sequence homology of *hypD*, *hypE*, *hoxX* and *hoxA* among hydrogen oxidizing bacteria

B. japonicum genes	Gene product size	NS homology to		AA homology % similarity	Predicted function
hypD'-ORF1 incomplete	347 aa	*A. vinelandii*[a]	ORF8	77.2	Fe-Binding Protein
		R. capsulatus[b]	*hypD*	69.9	
		E. coli[c]	*hypD*	65.5	
		R. leguminosarum[d]	*hypD*	85.7	
hypE-ORF2	321 aa	*R. capsulatus*	HypE	69.9	Free-living hydrogenase activity
		R. leguminosarum	*hypE*	85.7	
		E. coli	*hypE*	65.5	
hoxX-ORF3	566 aa	*A. eutrophus*[e] C-terminal 84 aa	*hoxX*	60.0	NtrB-like sensor protein regulating free-living hydrogenase activity
hoxA-ORF4	484 aa	*A. eutrophus*	*hoxA*	61.9	NtrC-like effector protein of free-living hydrogenase activity
		R. capsulatus[f]	*hupR1*	55.3	
		E. coli[g]	*hydG*	48.7	

References: [a]Chen and Mortenson, 1992; [b]Colbeau et al., 1993; [c]Lutz et al., 1991; [d]Rey et al., 1993; [e]Ebers and Friedrich, 1991; [f]Richaud et al., 1991; [g]Stoker et al., 1989.

less conserved while a helix-turn-helix structural motif, characteristic of DNA-binding proteins, was identified at positions 451 through 470. Together, the preceding data on ORF8 provide strong evidence that it encodes a NtrC-like transcriptional activator involved in the regulation of hydrogenase activity only under free-living conditions while it is not required to establish endosymbiotic hydrogenase activity.

Conclusions

Our results demonstrate that the structure and organization of *hup* genes is conserved among various and genetically distant hydrogen oxidizing bacteria. An identical organization pattern was found with the exception that in *R. leguminosarum*, *hupE* is inserted between *hupD* and *hupF* and in *B. japonicum hupE* is absent. Nevertheless unambiguous assignment of function to these newly uncovered *hup* genes in the *B. japonicum* chromosome requires an elucidation of the effects of non-polar mutations on hydrogenase structure and function.

In *B. japonicum*, free-living *hup* gene expression is regulated at the transcriptional level by nickel, hydrogen and oxygen (Kim and Maier, 1990; Kim et al.,

1991). The −149 to −98 region 5' upstream from the transcriptional start site of the hydrogenase structural genes is essentially required for regulation by all three components (Kim et al., 1991, 1993). A two component sensor/effector would be a suitable candidate for this kind of regulation. Upon activation by a HoxX-like sensor protein an NtrC-like effector could bind to an *hup*SLCDF upstream sequence thereby activating transcription from a −24/−12 promoter. RpoN1/RpoN2 promoter concensus sequences have been identified at the −26 and −14 positions upstream from the determined transcription start site of *hupSL* (Kim et al., 1993). Several observations however suggest a more complex regulatory pathway. First in *B. japonicum* ribulose biphosphate carboxylase is co-ordinately expressed with hydrogenase under free-living conditions but not in nodules (Simpson et al., 1979). Secondly in *A. eutrophus* it has been observed that *hoxA* itself was transcriptionally regulated under the same environmental conditions as the hydrogenase structural genes (Schwartz et al., 1991). Thirdly if HoxX would be the sensor regulating free-living hydrogenase activity in *B. japonicum* this protein should contain a nickel-sensing Ni binding site. No corresponding structural motif could be detected in the ORF7 deduced amino acid sequence. Finally, we can not rule out the presence of genes located downstream of *hoxA* that in addition might be required to establish Hup activity. Also, the role of the *hupF* gene product in this regulatory process is not yet elucidated. The function and regulation of *hoxX* and *hoxA* genes is currently under investigation in our laboratory and should provide further insight into the coordinated regulation of hydrogenase activity in free-living *B. japonicum*.

References

Chen J C and Mortenson L E 1992 Biochim. Biophys. Acta 1131, 199–202

Colbeau A et al. 1993 Mol. Microbiol. 8, 15–29.

Dross F et al. 1992 Eur. J. Biochem. 206, 93–102.

Eberz G and Friedrich B 1991 J. Bacteriol. 173, 1845–1854.

Ferber D M and Maier R J 1993 FEMS Microbiol. Lett. 110, 257–264

Friden H and Hedestedt L 1990 Mol. Microbiol. 4, 1045–1056.

Fu C and Maier R J 1993 J. Bacteriol. 175, 295–298.

Hidalgo E et al. 1992 J. Bacteriol. 174, 4130–4139.

Kim H and Maier R J 1990 J. Biol. Chem. 265, 18729–18732.

Kim H et al. 1991 J. Bacteriol. 173, 3993–3999.

Kim H et al. 1993 Arch. Microbiol. 160, 43–50.

Kortlüke C et al. 1992 J. Bacteriol. 174, 6277–6289.

Lambert G R et al. 1987 Appl. Environ. Microbiol. 53, 422–428.

Lutz S et al. 1991 Mol. Microbiol. 5, 123–135.

Menon N K et al. 1991 J Bacteriol. 173, 4851–4861.

Menon A L et al 1990a Gene 96, 67–74.

Menon N K et al. 1990 J. Bacteriol. 172, 1969–1977.

Pryzbyla A E et al. 1992 FEMS Microbiol. Rev. 88, 109–156.

Ramseir T M et al. 1991 Arch. Microbiol. 156, 270–276.

Rey L et al. 1993 Mol. Microbiol. 8, 471–481.

Richaud P et al. 1990 FEMX Microbiol. Rev. 87, 413–418.

Richaud P et al. 1991 J. Bacteriol. 173, 5928–5932.

Sayavedra-Soto L A et al. 1988 Proc. Natl. Acad. Sci. USA 85, 8395–8399.

Schwartz E et al. 1991 3rd Intern. Conf. on Hydrogenases, Troia, Portugal.

Simpson F B et al. 1979 Arch. Microbiol. 123, 1–8.

Stoker K et al. 1989 J. Bacteriol. 171, 4448–4456.

Van Soom C et al. 1993 Mol. Gen. Genet. 239, 235–245.

Van Soom C et al. 1993 J. Mol. Biol. 234, 508–512.

Widger W R et al. 1984 Proc. Natl. Acad. Sci. USA 81, 647–678.

P.H. Graham, M.J. Sadowsky & C.P. Vance (eds.), Symbiotic nitrogen fixation, 153–158.

Characterization of *Rhizobium galegae* by REP-PCR, PFGE and 16S rRNA sequencing

Ingrid Huber and Sonja Selenska-Pobell[1]
Department of Genetics, University of Bayreuth, D-95440 Bayreuth, Germany. [1]*Corresponding author*

Key words: digoxygenin (DIG)-sequencing, Pulsed Field Gel Electrophoresis (PFGE), Repetitive Extragenic Palynodromic (REP)-PCR, *R. galegae,* 16S rRNA

Abstract

Repetitive Extragenic Palindromic (REP)-PCR and Pulsed Field Gel Electrophoresis (PFGE)-fingerprinting were used for characterization of different bulgarian isolates of *R. galegae* subspecies *officinalis* effectively nodulating *Galegae officinalis* plants. Products of PCR-amplifications of total DNA of the studied strains generated by single REP-primers had very characteristic patterns when separated electrophoreticaly in NuSieve 3:1 agarose. These patterns were found to be related to one another for the analysed *R. galegae (of.)* straims. Individual fingerprints for every particular strain within the investigated group of natural *R. galegae (of.)* isolates were also obtained by PFGE of their total DNA, digested with the rare cutting restriction endonuclease *Spe*I. On the basis of the length of the fragments obtained by PFGE, the minimum genome size of *R. galegae (of.)* was estimated as 6 ± 0.5 Mbp. Different parts of 16S rRNA genes of the studied *R. galegae (of.)* isolates were sequenced by nonradioactive DIG technique, using the direct blotting electrophoresis system. Two regions containing nucleotides distinguishing *Galegae* rhizobia on the level of species and subspecies were determined by comparison of the sequenced regions of *R. galegae (of.)* 16S rRNA to the already published 16S rRNA sequences of the other subspecies of *R. galegae (or.)* effective on *Galegae orientalis* host plant, and also to the available corresponding sequences of *R. leguminosarum,* *R. meliloti,* *A. tumefaciens* and *A. vitis.*

Introduction

The prospect of more extensive releases of rhizobia and especially of their improved derivatives requires development of techniques for monitoring of their fate and their effects in natural conditions. With new isolates, which are perspective for deliberate releases, it is very important to establish appropriate identification procedures for their characterization and identification.

The use of a pair of REP primers in PCR amplifications was shown to be a promising technique for molecular taxonomy of soil bacteria including *Rhizobiaceae* (De Bruijn, 1992). In some cases short unspecific "arbitrary" primers (AP) were successfully used for PCR with identification purposes (Caetano-Anolles et al., 1991; Williams et al., 1990). Another fingerprinting technique which was also successfully used for characterization of several *Rhizobium* species is PFGE (Corich et al., 1991; Sobral et al., 1991). The most powerful approach in molecular classification of bacteria, however, is 16S rRNA sequence comparative analysis (Weizenegger et al., 1992; Willems and Collins, 1993; Yanagi and Yamasato, 1993; Young et al., 1991; Young, 1992). Moreover, the recent fast development of taxon-specific oligonucleotide probes complementary to particular highly specific regions of 16S rRNA provides an important tool for studying bacterial community structures in natural environments and for in situ detection of bacteria of special interest (Hahn et al., 1990; Stackebrandt et al., 1991).

The species *Rhizobium galegae* has two subspecies *R. galegae (of.)* and *R. galegae (or.),* which are very host specific and effectively nodulate only their corresponding host plants - *Galega officinalis* and *Galega orientalis,* respectively (Lindström, 1989). In this work we present the results of molecular genetical analyses of several natural bulgarian isolates of *R. galegae (of.).* Both *R. galegae* symbiotic systems are supposed to have good perspectives for forage production in Europe, where the host plants can grow very well (Varis, 1986).

We are presenting REP and PFGE fingerprints of the studied *R. galegae (of.)* strains. These fingerprints could be used for strain identification on the level of bacterial genome. We present also *R. galegae* 16S rRNA species- and subspecies- specific sequences which could be used for the preparation of probes for in situ detection of these bacteria in soil or plant tissue samples.

Materials and methods

Bacterial strains, media and growing conditions

Bacterial strains used are listed in Table 1. Natural isolates of *R. galegae (of.)*, were obtained as described by Vincent (1970) from nodules of *Galegae officinalis* plants. All strains were grown in YM or TY medium (Corich et al., 1991) at 28°C.

Isolation of total bacterial DNA, oligonucleotide primers and PCR-conditions

Total bacterial DNA was isolated according to Masterson et al. (1985). The oligonucleotide primers used were supplied by MWG-BIOTECH. Their sequences were as follows: REP₁-primer: 5'-IIIICGICGICACIGGC-3' and REP₂-primer: 5'-ICGICTTATCIGG CCTAC-3'(De Bruijn, 1992). The reactions were carried out in 20 μL volume and they contained as follows: 10 ng total bacterial DNA, 200 μM of each dNTP, 100 pmol of the REP₁- or REP₂-primer, 2 U of Taq polymerase (USB) with the corresponding buffer from USB. The amplifications were performed with a OmniGene thermocycler. For both primers, the cycles used were: 1 cycle at 95°C for 5 min, 45 cycles at 94°C for 1 min, at 36°C for 1 min, and at 72°C for 2 min; and 1 extension cycle at 72°C for 10 min. The products of reactions were analysed electrophoretically in 3% NuSieve agarose 3:1.

PFGE

PFGE experiments were carried out using the Bio-Rad CHEFF-DR™ System. Bacteria were grown in TY medium to an adsorbance (A₆₀₀=0,15). The cells were washed, closed in FMC-Biozym InCert agarose and treated according to the prescriptions of the firm. Digestion of the genomic DNA with 30 U of the *Spe*I

enzyme was done in 100 μL plugs at 37°C. The electrophoresis was done in 1% Pulsed Field Agarose (USB) in 0.5×TBE buffer (Sambrook et al., 1989). Running time was 20 to 26 hours with pulse times ramping from 3 to 15 s, 10 to 35 s, 5 to 40 s, 40 to 80 s and 60 to 120 s.

DIG-Sequencing of 16S rRNA:1 500 bp sequences of the 16S rRNA genes of *R. galegae* strains were amplified in total bacterial DNA using conservative eubacterial primers: 5'-TATA**GCGGCCGC**AGAGTTTGATTYMTGGCTC-AG-3' and 5'-ATAT**GCGGCCGC**AGAAAGGAGGT-GATCC-3', corresponding to possitions 8–27 and 1528–1540 bp of the *E. coli* 16S rRNA gene (Neefs et al., 1991) and creating *Not*I endonuclease recognition site (5'**GC/GGCCGC**3'). The PCR products were sequenced directly or after cloning into the pBluescript II SK⁺ vector (STRATAGENE). The GATC-sequence direct blotting apparatus of MWG-BIOTECH, Munich, Germany and a DIG nonradioactive kit from Boeringer were used as described by Pohl and Feger (1992). A cycle sequencing procedure was applied with the following parameters: 1 cycle at 95°C for 5 min, 25 cycles 36 s at 95°C, 36 s at 45°C and 124 s at 72°C.

Resulting 16S rRNA sequences were compared with those available from the literature (Willems and Collins, 1992) and Gene Bank Nucleotide Sequence Database.

Results

REP-PCR-fingerprinting

In contrast to De Bruijn (1992) who successfully applied the pair of REP₁ and REP₂ for development of specific fingerprints of some soil bacteria including rhizobia, we used these primers separately. The REP patterns of the studied rhizobial strains generated by REP₁ or REP₂ primers are presented in Figure 1. The *R. galegae (of.)* REP₁ fingerprints are characteristic for every strain, while that generated by REP₂ are more related.

PFGE-fingerprinting

As shown in Figure 2, six of the analysed *R. galegae (of.)* isolates have their individual *Spe*I pattern. The optimal fingerprints were obtained running the electrophoresis with 5 to 40 s ramping pulses for 26

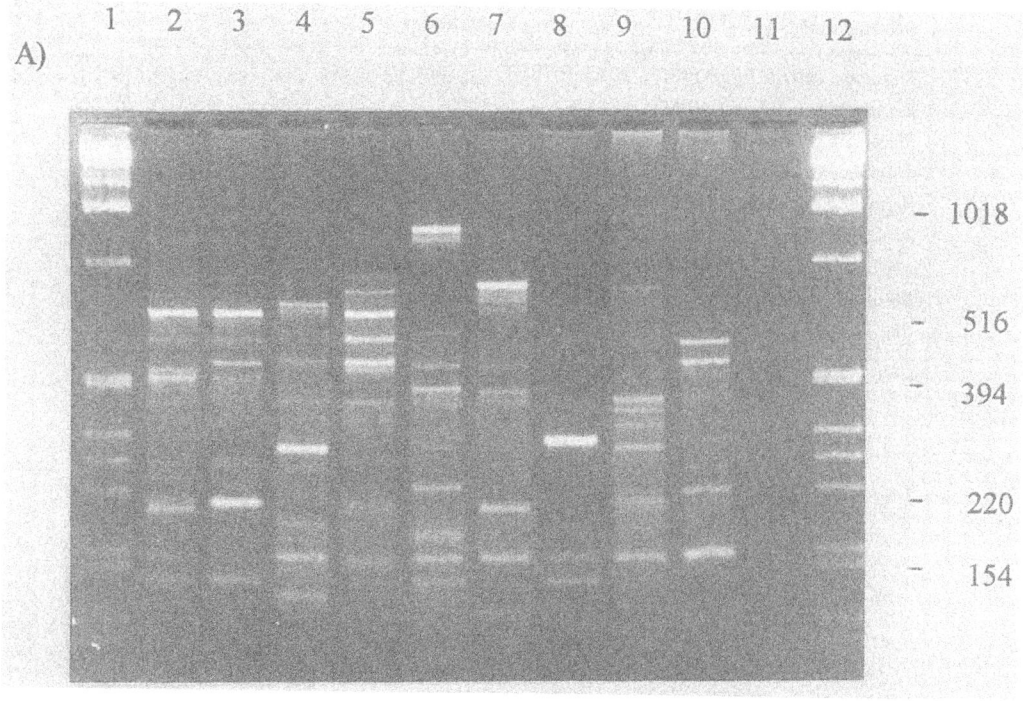

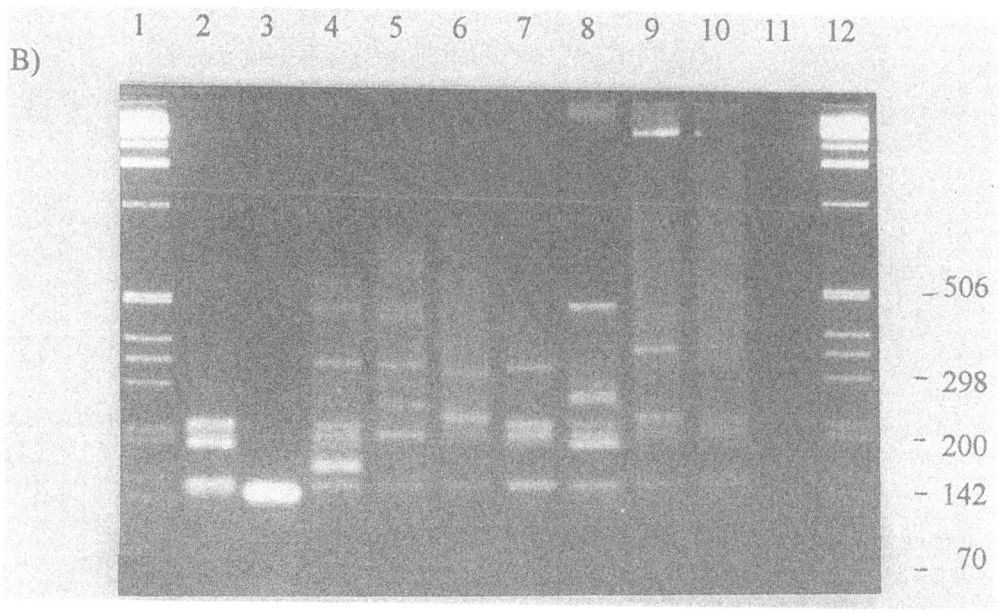

Fig. 1. REP-PCR fingerprint patterns of different rhizobial strains: 1 and 12-kilobase ladder (BRL); 2-*R. galegae* NBIMTC 2246; 3-*R. galegae* NBIMTC 2247; 4-*R. galegae* NBIMTC 2248; 5-*R. galegae* NBIMTC 2249; 6-*R. galegae* NBIMTC 2250; 7-*R. galegae* 2251; 8-*R. galegae* BG-9; 9-*R. meliloti* 114; 10-*R. leguminosarum* VF39; 11-reaction mixture without adding of DNA. **A)** Fingerprints generated by **REP**₁-primer; **B)** Fingerprints generated by **REP**₂-primer.

Table 1. Bacterial strains used

Bacterial strain	Source or reference
Rhizobium galegae (of.) NBIMTC 2246	NBIMTC*, Sofia, Bulgaria
Rhizobium galegae (of.) NBIMTC 2247	" " " "
Rhizobium galegae (of.) NBIMTC 2248	" " " "
Rhizobium galegae (of.) NBIMTC 2249	" " " "
Rhizobium galegae (of.) NBIMTC 2250	" " " "
Rhizobium galegae (of.) NBIMTC 2251	" " " "
Rhizobium galegae (of.) BG-9	Institute of Molecular Biology, Sofia, Bulgaria
Rhizobium meliloti 114	Selenska-Trajkova et al. (1990)
Rhizobium leguminosarum VF39	Hynes et al. (1985)

* - National Bank for Industrial Microorganisms and Tissue Cultures - Sofia 1113, Bulgaria.

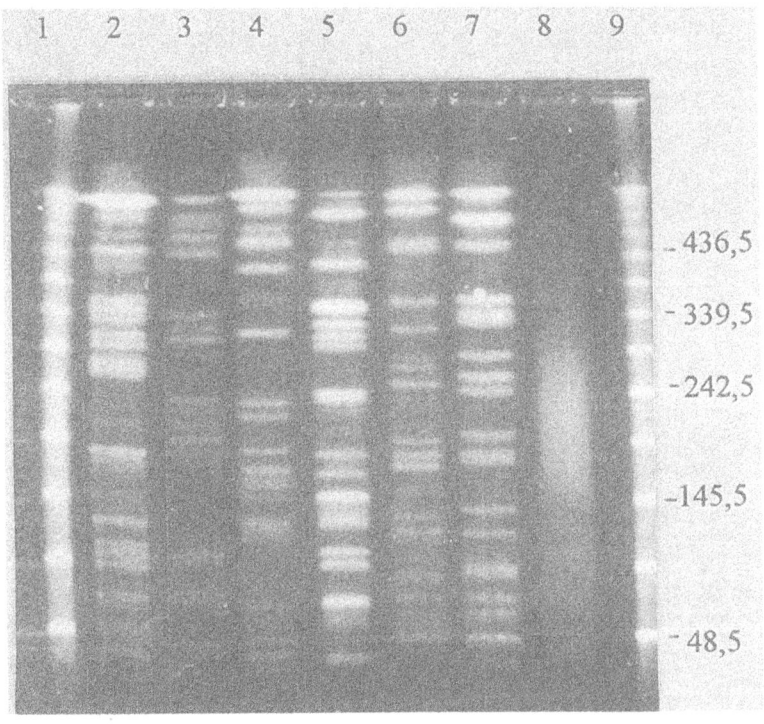

*Fig. 2. Spe*I PFGE-fingerprints of the studied *R. galegae (of.)* strains. 1 and 9 - λ-concatamers; 2-NBIMTC 2246; 3-NBIMTC 2247; 4-NBIMTC 2248; 5-NBIMTC 2249; 6-NBIMTC 2250; 7-NBIMTC 2251; 8-BG-9.

hours. DNA-fragments in range of 50 to 1500 kbp were obtained by PFGE by using appropriate time pulses. For example, ramping pulses from 60 to 120 s separate very large fragments, with a size of 800 to 1500 kbp. On the basis of the lengths of DNA fragments, resulting after *Spe*I digestions, it was possible to estimate the minimal genomic size of *R. galegae (of.)* as 6 ± 0.5 Mbp.

16S rRNA-sequencing

As sequencing primers we used the 5'DIG-labelled 5'-CGGTGAAATGCGTAG-3',- 5'-ACGCGNARAACC-TTA-3' and 5'-CAGAAAGGAGGTGATCC-3' oligonucleotides, which correspond to the highly conserved positions 688 to 702, 968 to 982 and 1540 to 1528 of the *E. coli* 16S rRNA (Neefs et al., 1991). We succeeded in sequencing the variable regions V_6, V_7 and V_8 of

V₆:

5'-CCCGG - - - **CT**<u>AGC</u> - <u>TACA</u> - GAGA-3'   *R. galegae (of.)*

5'-CCCGG - - - **AC**<u>AGC</u> - <u>TACA</u> - GAGA-3'   *R. galegae (or.)*

5'-CTGTG.-.-.- ACCGC.-.CACG:-:GAGA-3'   *A. vitis*

5'-CGGGG - - TTTGGG CAGTG - GAGA-3'   *A. tumefaciens* (biovar 2)

5'-CCCGG - - - CTACT - TGCA - GAGA-3'   *R. leguminosarum (trifolii)*

5'-CCGAT - CGCGGAT - ACGA - GAGA-3'   *R. meliloti*

V₈:

5'- GC - AGC - GA<u>AGG</u>AG - CGAT<u>CCC</u> - GAGCTA-3'   *R. galegae (of.)*

5'- GC - AGC - **GG**<u>AGG</u>AG - CGAT<u>CCC</u> - GAGCTA-3'   *R. galegae* (or.)

5'- GC.-.AGC.-.GAGACCG.-.CGAGGTC.-.GAGCTA-3'   *A. vitis*

5'- GC - AGC - GAGACAG - CGATGTC - GAGCTA-3'   *A. tumefaciens* (biovar 2)

5'- GC - AGC - GAGCACG - CGAGTGT - GAGCTA-3'   *R. leguminosarum (trifolii)*

5'- GC - AGC - GAGACCG - CGAGGTC - GAGCTA- 3'   *R. meliloti*

Fig. 3. Primary structure of the species- and subspecies- specific 16S rRNA sequences from the V₆ and V₈ regions of *R. galegae (of.)* compared to the corresponding sequences of *R. galegae (or.), A. vitis, A. tumefaciens, R.meliloti* and *R. leguminosarum.* Species-specific sequences are underlined, and subspecies mismatches are given with fet letters.

the strains NBIMTC 2246, 2249 and 2251. Resulting sequences were completely identical for these strains. When a comparative analysis with the corresponding parts of *R. galegae (or.)*, *R. meliloti*, *R. leguminosarum* (biovar *trifolii*), *A. tumefaciens* (biovar 2) and *A. vitis* 16S rRNA genes was done, species- and subspecies-specific sequences in V₆ and V₈ regions were determined (see Fig. 3).

Discussion

The results of REP-PCR amplifications confirm the opinion of De Bruijn (1992) that this technique could be applied for molecular genetic characterization of rhizobia. We show here that these primers, when used not as a pair but separately, generate very characteristic fingerprints and could differentiate *R. galegae (of.)* strains (see Fig.1). The REP-PCR generated fingerprints, however, may be changed by varying concentrations of DNA, primers, dNTP or Taq polymerase in the reaction mixtures. Moreover, the results obtained with Taq polymerases from different firms were also

not well reproducible even when all other parameters of the reaction were the same. That is why we used another fingerprinting technique - PFGE, which gives more reproducible results. As rhizobia have a G±C content of about 65%, we used a rare cutting enzyme *Spe*I (5'-A/**CTAG**T-3'). The recognition site of this enzyme contains the tetranucleotide "**CTAG**", which was shown to be very rare in bacteria with high G+C content (Mc Clelland et al., 1987). The *Spe*I PFGE-fingerprints of the studied *R. galegae (of.)* isolates were very specific (see Fig. 2) and could be used for identification of every particular strain with one exception - the strain BG-9, the PFGE analysis of which is still not successful. The same isolate, however, has very characteristic REP patterns which share common bands with some of the other analysed strains and on the basis of this one can conclude that it belongs to the studied group of bacteria.

The estimated genome size of the *R. galegae (of.)* strains of 6±0.5 Mbp corresponds to that estimated by the same method for *R. meliloti* (Sobral et al., 1991). This is in agreement with the fact that both species contain megaplasmids with similar size (about 1500 kbp), which are involved in their symbiotic functions (Selenska-Trajkova et al., 1990 a,b).

The prospect of application of *R. galegae* in agriculture makes important the development of techniques for their direct monitoring in the environment. The species- and subspecies-specific 16S rRNA sequences determined in this work could be used for such purposes.

Acknowledgements

We thank Ms. Maria Kraut for her highly qualified help at the final stages of our work. This work was financially supported by a grant No. 6496–1053–12936 to Prof W. Klingmüller by Bayerisches Staatsministerium für Landesentwicklung und Umweltfragen.

References

Corich V, Giacomini A, Ollero F J, Squartini A and Nuti M F 1991 Pulse-field electrophoresis in contour-clamped homogenous electric fields (CHEF) for the fingerprinting of *Rhizobium* spp. FEMS Microbiol. Lett. 83, 193–198.

De Bruijn F 1992 Use of repetitive (repetitive extragenic palindromic and Enterobacterial intergeneric consensus) sequences and the polymerase chain reaction to fingerprint the genomes of *Rhizobium meliloti* isolates and other soil bacteria. Appl. Environ. Microbiol. 58, 2180–2187.

Caetano-Anolles G, Bassam B J and Gresshoff P M 1991 DNA amplification fingerprinting using very short arbitrary oligonucleotide primers. Biotechnology 9, 553–557.

Hahn D, Kester R, Starrenburg M J C and Akkermans A D L 1990 Extraction of rRNA from soil for detection of *Frankia* with oligonucleotide probes. Arch. Microbiol. 154, 329–335.

Hynes M F, Bricksch K and Priefer U 1988. Melanin production encoded by a cryptic plasmid in a *Rhizobium leguminosarum* strain. Arch. Microbiol. 150, 326–332.

Lindström K 1989 *Rhizobium galegae*, a new species of legume root nodule bacteria. Int. J. Syst. Bacteriol. 39, 365–367.

Masterson R V, Prakash R K and Atherly A G 1985 Conservation of symbiotic nitrogen fixation gene sequences in *Rhizobium japonicum* and *Bradyrhizobium japonicum*. J. Bacteriol. 163, 21–26.

Mc Clelland M, Jones R, Patel Y and Nelson M 1987 Restriction nucleases for pulsed field mapping of bacterial genomes. Nucl. Acid Res. 15, 5985–6005.

Neefs J M, Van de Peer Y, De Rijk P, Goris A and De Wachter R 1991 Compilation of small ribosomal subunit RNA sequences. Nucleic Acid Research 19, Supplement 1987–2009.

Pohl T M and Feger G M 1992 The DBE-DIG sequencing system. Colloquium, The Newsletter for the Molecular Biologist 2, 17–18, Boehringer Mannheim, Biochemica.

Sambrook J, Fritsch E F and Maniatis T 1989 Molecular Cloning: A Laboratory Manual. 2nd edn. Cold Spring Harbor Laboratory Press, New York

Selenska-Trajkova S, Radeva G and Markov K 1990a Comparison between *Rhizobium galegae*, and *Rhizobium meliloti* plasmid contents. Lett. Appl. Microbiol. 11, 123–126.

Selenska-Trajkova S, Radeva G, Gigova L and Markov K 1990b Localization of nif genes on large plasmids in *Rhizobium galegae*. Lett. Appl. Bacteriol 11, 73–76.

Sobral B W S, Honeycut R J and Atherly 1991 The genomes of the family *Rhizobiaceae*: Size, stability and rarely cutting restriction endonucleases. J. Bacteriol 173, 704–709.

Stackebrandt E, Witt D, Kemmerling C, Kroppenstedt R and Liesack W 1991 Designation of *Streptomycete* 16S and 23S rRNA-based target regions for oligonucleotide probes. Appl. Environ. Microbiol. 57, 1468–1477.

Weizenegger M, Neumann M, Stackebrandt E, Weiss N and Ludwig W 1992 *Eubacterium alactolyticum* phylogenetically groups with *Eubacterium limosum*, *Acetobacterium woodii* and *Clostridium barkeri*. System. Appl. Microbiol. 15, 32–36.

Willems A and Collins M D 1993 Phylogenetic analysis of Rhizobia and Agrobacteria based on 16S rRNA gene sequences. Int. J. Systematic Bacteriol. 43, 305–313.

Williams J G K, Kubelik A R, Livak K J, Rafalski J A and Tingey S V 1990 DNA polymorphism, amplified by arbitrary primers are useful as genetic markers. Nucl. Acids Res. 22, 6531–6535.

Vincent J M 1970 A Manual for the Practical Study of Root-nodule Bacteria. International Biological Program Handbook 15, Blackwell Scientific Publications, Oxford.

Varis E 1986 Goat's rue (*Galega orientalis* Lam) a potential pasture legume for temperate conditions. J. Agric. Sci. Finland 58, 83–101.

Yanagi M and Yamasato K 1993 Phylogenetic analysis of the family *Rhizobiaceae* and related bacteria by sequencing of 16S rRNA gene using PCR and DNA sequencer. FEMS Microbiol. Lett. 107, 115–120.

Young J P W 1992 Phylogenetic classification of nitrogen-fixing organisms. *In* Biological Nitrogen Fixation. Eds. G Stacey, R H Burris and H J Evans. pp 43–86. Chapman and Hall, New York

Young J P W, Downer H L and Eardly B D 1991 Phylogeny of the phototrophic *Rhizobium* strain BTAi1 by polymerase chain reaction-based sequencing of a 16S rRNA gene segment. J. Bacteriol. 173, 2271–2277.

P.H. Graham, M.J. Sadowsky & C.P. Vance (eds.), Symbiotic nitrogen fixation, 159–164.
© 1994 *Kluwer Academic Publishers.*

An hypothesis for the role of malic enzyme in symbiotic nitrogen fixation in soybean nodules

David A. Day[1], Rosanne G. Quinnell[1] and Fraser J. Bergersen[2]
[1]*Division of Biochemistry and Molecular Biology, Faculty of Science, Australian National University, Canberra, ACT 0200 and* [2]*CSIRO, Division of Plant Industry, Canberra, ACT 2601, Australia*

Key words: bacteroid, malic enzyme, nitrogen fixation, poly-β-hydroxybutyrate

Abstract

Recent measurements of respiration and nitrogen fixation by isolated soybean bacteroids, incubated with oxyleghaemoglobin in a flow chamber under steady-state conditions, have shown that while addition of malate stimulated respiration, the effect on N_2 fixation depended on the concentration of malate and oxygen. At low malate concentrations, N_2 fixation was stimulated, but at higher malate concentrations (more than 0.5 mM at less than 60 nM O_2) N_2 fixation was inhibited and carbon diverted to poly-β-hydroxybutyrate formation. These results are interpreted in terms of the redox poise of pyridine nucleotides and the relative rates of poly-β-hydroxybutyrate synthesis and tricarboxylic acid cycle operation. Soybean bacteroids contain both NAD- and NADP-linked malic enzymes which have very different affinities for malate, and thereby have the capacity to alter the NAD(P)H/NAD(P) ratios in the bacteroid in response to varying malate concentrations. It is suggested that in vivo the rate of delivery of malate to the bacteroid must be carefully regulated to optimise N_2 fixation.

Introduction

Nitrogen fixation in legume nodules requires a constant supply of carbon substrates to the bacteroids to provide ATP and NAD(P)H for nitrogenase activity, and to maintain oxygen tensions within the nodule at acceptable levels. There is now overwhelming evidence that the main form of this carbon is supplied as organic acids, principally malate, although other forms may play minor roles (for recent reviews see Day and Copeland, 1991; Streeter, 1991; Vance and Heichel, 1991.)

Sucrose supplied to the nodule from the shoot is degraded in the first instance by sucrose synthase located, in soybean, mainly in the inner cortex cells close to vascular bundles and in the arrays of uninfected cells of the inner infected zone (Gordon, 1992; Zammit and Copeland, 1993). There is also evidence that at least some of the glycolytic enzymes, which further degrade hexoses to phosphoenolpyruvate (PEP), are also predominately localised in the uninfected cells of mature soybean nodules (Zammit et al., 1992). The PEP produced is carboxylated by PEP carboxylase to form oxaloacetic acid (OAA) which can then be reduced to malate via malate dehydrogenase (MDH); the activity of both of these enzymes is elevated in nodule cells. Thus organic acid production can take place to some extent outside of the infected cells which have the ability to import malate via a plasma membrane dicarboxylate carrier (Li and Day, 1991). Regardless of where it is produced, it is obvious that malate is a major carbon substrate for bacteroids and its metabolism is of key importance to symbiotic; nitrogen fixation. Malic enzyme in the bacteroid is likely to be involved in this metabolism.

Malic enzyme (ME) catalyses the oxidative decarboxylation of malate to pyruvate; (Fig. 1). Two forms of ME are found in bacteria and plant cells: NADP-linked (EC1.1.1.40) and NAD-linked (EC1.1.1.38). In plants the NADP form is cytosolic while the NAD form is localized to the mitochondrion. In bacteria such as *E. coli*, the two enzymes co-exist in the cytoplasm but probably perform different functions (Hanson and Juni, 1975). Both forms of ME have been found in *Bradyrhizobium japonicum* (Copeland et al., 1989; Kimura and Tajima, 1989) and *Rhizobium meliloti* (Driscoll and Finan, 1993). Mutants of *R. meliloti* lacking NAD-ME

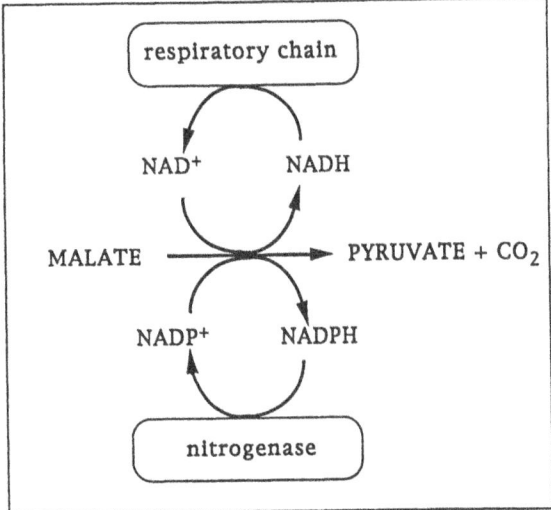

Fig. 1. Proposed roles of NAD- and NADP-linked malic enzymes in bacteroids. It is proposed that NADH produced is used as a substrate for the respiratory electron transport chain while NADPH is used, directly or indirectly, as reductant for nitrogenase.

are fix$^-$, confiming that this enzyme is involved in nitrogen fixation (Driscoll and Finan, 1993).

Malic enzyme, in concert with MDH and pyruvate dehydrogenase, allows both OAA. and acetyl-CoA to be produced within the bacteroid from the oxidation of malate (Fig. 2) NADH produced can be oxidised by the respiratory chain to support oxidative phosphorylation; NADPH is not oxidised by the respiratory chain of *B. japonicum* bacteroids but could be used directly to reduce nitrogenase (Copeland et al., 1989). These putative roles of ME in bacteroids are summarised in Figure 1. In this paper, the properties of soybean ME and isolated bacteroids are reviewed and incorporated into a model for malate metabolism during nitrogen fixation. Some aspects of this model have been presented previously (Bergersen and Turner, 1993).

Materials and methods

Soybeans (*Glycine max*[L.] Merr. cv Lincoln or Bragg) inoculated with *Bradyrhizobium japonicum* (strains CB 1809/USDA 136 or USDA 110) were grown as described previously (Bergersen and Turner, 1993; Copeland et al., 1989). The methods used for bacteroid isolation, leghaemoglobin purification, flow-chamber experiments, uptake of malate, and ME assays have also been published previously (Bergersen and Turner, 1993; Copeland et al., 1989; Udvardi et al., 1988). Mal-

ic enzymes were isolated and partially purified according to Quinnell and Day (1993).

Results and discussion

Malate oxidation during N$_2$ fixation by isolated bacteroids

Experiments in which bacteroids isolated from soybean were incubated with leghaemoglobin under limited oxygen supply in a flow-through chamber, have shown that the metabolism of malate and the concomitant fixation of N$_2$ is strongly dependent on the quantities of malate supplied to the bacteroid (Bergersen and Turner, 1990a,b, 1992, 1993). At O$_2$ concentrations of 20–60 nM, which approximate those estimated in whole nodules (Layzell et al., 1993), external concentrations of malate (or succinate) up to 0.5 mM stimulated N$_2$ fixation above endogenous rates (Table 1; Bergersen and Turner, 1993). Higher concentrations of malate in the external medium, on the other hand, depressed the rate of N$_2$ fixation (Table 1); subsequent removal of malate from the medium stimulated N$_2$ fixation while endogenous reserves of carbon were utilised (Bergersen and Turner, 1991). If the O$_2$ concentration was decreased, then the concentration of external malate required to inhibit N$_2$ fixation also decreased (Bergersen and Turner, 1990b, 1992). It was further shown that during diminished nitrogen fixation at higher malate concentrations, carbon was directed away from the TCA cycle towards the synthesis of poly-β-hydroxybutyrate (PHB) which was utilised during the subsequent withdrawal of malate and recovery of N$_2$ fixation (Bergersen and Turner, 1990b).

Kinetics of malic enzyme and malate transport in bacteroids

In all of the above experiments, addition of malate stimulated respiration and caused immediate release of CO$_2$ (Table 1), suggesting the participation of ME. NAD- and NADP- linked ME have been separated and partially purified from soybean bacteroids and their kinetics studied (Table 2). NADP-ME had a substantially higher affinity for malate than did its NAD-linked counterpart, while the V$_{max}$ for the two enzymes was similar, on a bacteroid protein basis. Malate uptake by isolated soybean bacteroids showed an apparent Km of 10–15 μM with a V$_{max}$ two to three times that of the two ME (Table 2). Malate can be accumulated against a

Table 1. Effects of malate concentration on nitrogen fixation and respiration of isolated soybean bacteroids in flow-chamber reactions with oxyleghaemoglobin. Oxygen consumption, nitrogen fixation and CO_2 output were measured during steady-states of 15–30 min duration, as described by Bergersen and Turner (1993). Except for 5 mM malate, data are from three similar experiments with overlapping oxygen concentrations, conducted over a period of about one month

Malate concentration (mM)	O_2 concentration (nM)	O_2 uptake	CO_2 released	NH_3 produced
		(nmol min^{-1}mg^{-1} protein)		
0[a]	9.3 (0.06)**	5.2 (0.03)	7.3 (0.1)	1.7 (0.1)
	11.5 (0.2)	5.7 (0.05)	5.8 (0.1)	1.0 (0.1)
	46.7 (0.7)	4.6 (0.1)	6.1 (0.01)	1.1 (0.1)
0.05	9.2(0.05)	5.9 (0.03)	9.7 (0.1)	1.5 (0.03)
	10.8 (0.1)	6.6 (0.03)	8.8 (0.1)	2.0 (0.1)
	46.6 (0.5)	7.6 (0.04)	10.7 (0.4)	2.3 (0.03)
0.5	8.2 (0.01)	4.8 (0.01)	12.1 (0.01)	0.7 (0.09)
	15.1 (0.01)	8.2 (0.01)	21.7 (0.03)	1.5 (0.03)
	51.2 (2.2)	13.2 (0.09)	25.7 (0.8)	2.8 (0.1)
5.0	13.2 (0.2)	10.2 (0.03)	13.2 (0.7)	0.4 (0.04)
	39.4 (1.5)	11.4 ((0.03)	26.3 (0.7)	0.04 (0.03)

[a] Activity supported only by endogenous PHB. [b] Values in parentheses are S.E.M.

Table 2. Kinetic parameters of malic enzymes and malate uptake by soybean bacteroids

Enzyme	Km$_{app}$ (mM)	Vmax (nmol min^{-1} mg^{-1})
NADP-ME	0.16	10–20
NAD-ME	2.5	8–15
Malate transporter	0.01–0.02	20–30

concentration gradient in isolated bacteroids (Udvardi et al., 1988). In short-term uptake experiments where metabolism was negligible, even 10 μM malate in the medium was accumulated to 1–2 mM inside the bacteroid (Udvardi, 1989), sufficient to give full activity of NAD-ME. However, in steady-state experiments such as those described above, the internal concentration will depend on the rate of oxidation which in turn will depend on the rate of respiration and hence the concentration of O_2.

Regulation of carbon metabolism in bacteroids

Our interpretation of the results obtained in the steady-state measurements of malate utilisation and nitrogen fixation described above, is summarised schematically in Figure 2 (see Bergersen and Turner, 1993). The fate of the malate, and the rate of N_2 fixation, depends on the redox poise of pyridine nucleotides within the bacteroids, the different kinetics of the two ME, and possible competition between PHB synthesis and citrate synthase for acetyl-CoA on the one hand, and between PHB synthesis and nitrogenase for NADPH on the other. When malate supply to bacteroids is low, NADP-ME activity is dominant (although some NAD-ME activity may also be evident), the NADH/NAD$^+$ ratio is maintained at a level which permits the TCA cycle to operate, PHB synthesis is in check and NADPH is available for nitrogenase (Fig. 2A). When malate supply to bacteroids is higher (> 0.5 mM at 60 nM O_2 in the above experiments), NAD-ME becomes fully active and, because low oxygen restricts the rate of electron transport through the respiratory chain, the NADH/NAD$^+$ ratio increases to the point where TCA cycle enzymes (such as MDH and isocitrate and oxo-

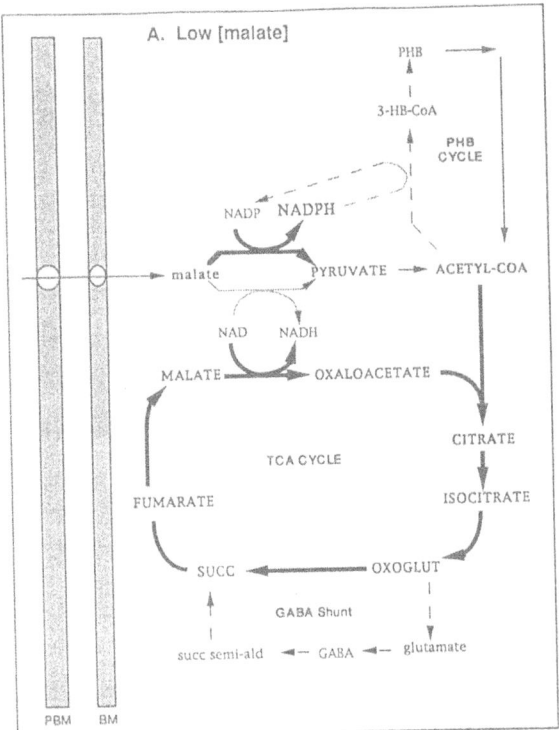
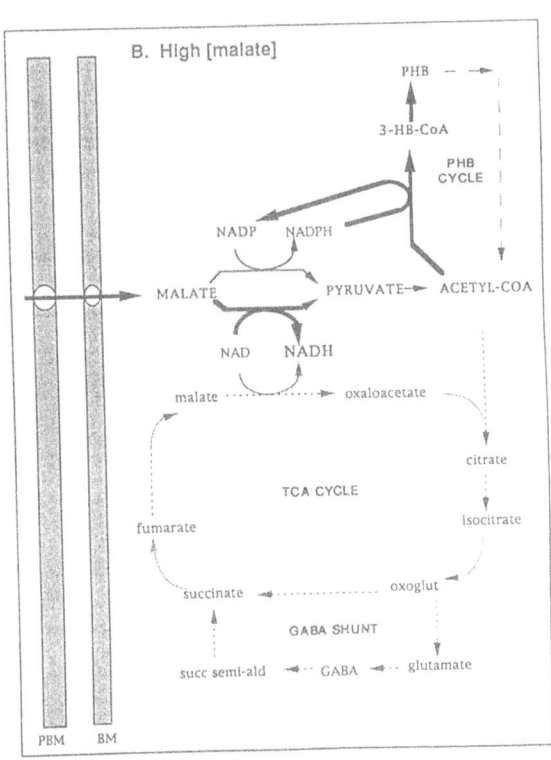

Fig. 2. Proposed pathways of carbon metabolism in bacteroids in the presence of low (A) and high (B) concentrations of external malate in the presence of limiting O₂ (see text and Bergersen and Turner 1993 for experimental details). **A:** At low concentrations of malate, the high affinity NADP-ME is fully active and NADPH is available for nitrogenase. The lower affinity NAD-ME is only partially active, if at all. ME activity is coordinated with pyruvate dehydrogenase and the tricarboxylic acid (TCA) cycle, and the NADH produced is oxidised by the electron transport chain. Nitrogen fixation is stimulated and very little carbon is directed towards poly-β-hydroxybutyrate (PHB) synthesis; however, previously synthesised PHB may be utilised as endogenous substrate. **B:** When higher concentrations of malate are supplied, NAD-ME becomes fully active and NADH accumulates faster than it can be oxidised, inhibiting TCA cycle operation and causing acetyl-CoA to accumulate. This stimulates PHB synthesis which reoxidises the NADPH formed by NADP-ME, thus restricting nitrogenase. Subsequent lowering of the malate concentration allows the stored PHB to be used as endogenous substrate. Bold lines and arrows indicate the most active pathways; dotted and broken lines indicated repressed pathways. PBM: peribacteroid membrane with its carrier; BM: bacteroid membrane with its carrier.

glutarate dehydrogenases) are inhibited. Under these conditions, acetyl-CoA will accumulate and be used in PHB synthesis (Fig. 2B). Consequently, NADPH will be oxidised (Fig. 2B) and nitrogenase activity inhibited. It should be noted that a high NADH/NAD⁺ ratio also inhibits NAD-ME (Quinnell and Day, unpubl.), but not to the extent that malate oxidation by MDH is inhibited. Measurements of PHB synthesis during oxidation of malate by soybean bacteroids have confirmed that redirection of carbon into storage does occur when malate supply is non-limiting (Bergersen and Turner, 1990b).

In legumes other than soybean, whose bacteroids do not accumulate PHB, the model depicted in Figure 2 will obviously not hold. Malate metabolism by these bacteroids needs to be investigated in more detail.

Regulation of malate supply in vivo

Clearly the schemes outlined in Figure 2 are speculative at this stage and their proof or otherwise awaits a more detailed knowledge of the enzymes involved. Nonetheless, they do explain the data at hand and demonstrate that efficient N₂ fixation requires the coordinated regulation of the rate of malate supply, the synthesis of PHB, and the operation of the bacteroid respiratory chain. The level of oxygen within the infected zone will be one important factor, but the rate of synthesis of malate in the plant cytosol and its subsequent delivery to the bacteroid will also be important. Two phosphorylation events involving plant protein kinases may be involved at this level.

The rate of malate synthesis in the nodule cytoplasm will depend to a large extent on the activity of PEP carboxylase (see Introduction). It has recently been shown that the nodule isoform of this enzyme, like its leaf counterparts, is regulated by phosphorylation via a protein kinase, the phosphorylated form being much less sensitive to feedback inhibition by malate and consequently more active (Schuller and Werner, 1993). It has also been demonstrated that the rate of malate uptake across the peribacteroid membrane (PBM) of symbiosomes isolated from soybean is stimulated by phosphorylation of PBM proteins via a protein kinase (Ouyang et al., 1991). Thus both synthesis and delivery of malate may be controlled by protein phosphorylation. The relationship between the two protein kinases is unknown but the PBM enzyme is apparently membrane bound (Weaver et al., 1991) and at least some of the PEP carboxylase (and presumably its protein kinase) is located in uninfected cells, suggesting that the kinases are different. The two events may be coordinated, however, by the rate of mitochondrial ATP synthesis in the infected zone and that in turn will be dependent, like bacteroid metabolism, on substrate supply and O_2 levels. Plant cytoplasmic ATP concentrations will also affect the operation of the PBM ATPase which may play a role in metabolite transport to the bacteroid (Udvardi et al., 1991).

Interestingly, while bacteroids contain substantial ME activity, nodule mitochondria have very little (Day and Mannix, 1988; Rawsthorne and La Rue, 1986). All other plant mitochondria examined to date have high levels of NAD-ME, suggesting that nodule mitochondrial malate metabolism is modified to suit the demands of that particular tissue. Apart from C_4 photosynthetic tissues, NAD-ME is thought to allow anaplerotic operation of mitochondria in plant tissues. Mitochondria from the nodule infected zone also lack non-phosphorylating electron transport components (Day and Mannix, 1988; Rawsthorne and La Rue, 1986), and it may be that the energy demands of the infected tissue preclude operation of these ancillary mitochondrial enzymes.

Summary

Organic acids, especially malate, are important sources of carbon for bacteroids in legume nodules. Studies with isolated bacteroids have shown that the rate at which malate is supplied, together with the concentration of O_2, regulates pathways of carbon metabolism and the rate of N_2 fixation in the endosymbiont. It is probable that malic enzyme activity within the bacteroid plays an important role in this regulation.

Acknowledgements

D A Day acknowledges the support of the Australian Research Council; R G Quinnell is the recipient of an Australian Postgraduate Award.

References

Bergersen F J and Turner G L 1990a Bacteroids from soybean root nodules: respiration and N_2 fixation in flow-chamber reactions with oxyleghaemoglobin. Proc. R. Soc. Lond. B 238, 295–320.

Bergersen F J and Turner G L 1990b Bacteroids from soybean root nodules: accumulation of poly-β-hydroxy butyrate during supply of malate and succinate in relation to N_2 fixation in flow-chamber reactions. Proc. R. Soc. Lond. B 240, 39–59.

Bergersen F J and Turner G L 1992 Supply of O_2 regulates O_2 demand during utilization of reserves of poly-β-hydroxy butyrate in N_2-fixing soybean bacteroids. Proc. R. Soc. Lond. B 238, 295–320.

Bergersen F J and Turner G L 1993 Effects of concentrations of substrates supplied to N_2-fixing soybean bacteroids in flow-chamber reactions. Proc. R. Soc. Lond. B 251, 95–102.

Copeland L, Quinnell R G and Day D A 1989 Malic enzyme activity in bacteroids from soybean nodules. J. Gen. Micrbiol. 135, 2005–2011.

Day D A and Copeland L 1991 Carbon metabolism and compartmentation in nitrogen-fixing legume nodules. Plant Physiol. Biochem. 29, 185–201.

Day D A and Mannix, M 1988 Malate oxidation by soybean mitochondria and the possible consequences for nitrogen fixation. Plant Physiol. Biochem. 26, 567–573

Driscoll B T and Finan, T M 1993 NAD^+-dependent malic enzyme of *Rhizobium meliloti* is required for symbiotic nitrogen fixation. Molecular Microbiology 7, 865–873.

Gordon A J 1992 Carbon metabolism in the legume nodule. *In* Carbon Partitioning Within and Between Organisms. Ed. C J Pollock, J F Farrar and A J Gordon. pp 133–162. BIOS. Oxford, UK.

Hanson E J and Juni E 1975 Isolation of mutants of *E. coli* lacking NAD- and NADP-linked malic enzyme activities. Biochem. Biophys. Res. Comm. 65, 559–566.

Layzell D B, Diaz Del Castillo L, Hunt S, Kuzma M, Van Cauwenberghe O and Oresnik I 1993 The regulation of oxygen and its role in regulating nodule metabolism. *In* Current Horizons in Nitrogen Fixation. Eds R Palacios, J Moira and W E Newton. pp 393–403. Kluwer Academic Publishers, Dordrecht.

Kimura I and Tajima S 1989 Presence and characteristics of NADP-malic enzyme in soybean nodule bacteroids. Soil Sci. Plant Nutr. 35, 271–279.

Li Y and Day D A 1991 Permeability of isolated infected cells from soybean nodules. J. Expt. Bot. 243, 1325–1329.

Ouyang L-J, Whelan J, Roberts D M, Weaver C D and Day D A 1991 Protein phosphorylation stimulates the rate of malate uptake across the peribacteroid membrane of soybean nodules. FEBS Letters 293, 188–190.

Quinnel R G and Day D A 1993 Purification of NAD-malic enzyme from *Bradyrhizobium japonicum*. *In* Current Horizons in Nitrogen Fixation. Eds. R Palacios, J Moira and W E Newton. pp 560. Kluwer Academic Publishers, Dordrecht.

Rawsthorne S and La Rue T A 1986 Preparation and properties of mitochondria from cowpea nodules. Plant Physiol. 81, 1092–1096.

Schuller K A and Werner D 1993 Phosphorylation of soybean nodule phosphoenolpyruvate carboxylase in vitro decreases sensitivity to inhibition by L-malate. Plant Physiol.

Streeter J G 1991 Transport and carbon metabolism and nitrogen in legume nodules. Adv. Bot. Res. 18, 129–187.

Udvardi M K 1989 Membrane transport processes and regulation of symbiotic nitrogen fixation. PhD thesis. Australian National University.

Udvardi M K, Price G D, Gresshoff P M and Day D A 1988 A dicarboxylate transporter on the peribacteroid membrane of soybean nodules. FEBS Lett. 231, 36–40.

Udvardi M K, Lister, D L and Day D A 1991 ATPase activity and anion transport across the peribacteroid membrane of isolated soybean symbiosomes. Arch. Microbiol. 156, 362–366.

Vance C P and Heichel G H 1991 Carbon in nitrogen fixation: limitation or exquisite adaptation? Annu. Rev. Plant Physiol. Plant Molec. Biol. 42, 373–392.

Weaver C D, Crombie B, Stacey G and Roberts D M 1991 Calcium-dependent phosphorylation of symbiosome membrane proteins from soybean nodules. Evidence for phosphorylation of nodulin-26. Plant Physiol. 95, 222–227.

Zammit, A and Copeland, L 1993 Immunocytochemical localisation of nodule-specific sucrose synthase in soybean nodules. Aust. J. Plant Physiol. 20, 25–32.

Zammit A, Copeland L, Miller C and Craig S 1992 Immunocytochemical localisation of NAD-dependent glyceraldehyde-3-phosphate dehydrogenase. in soybean nodules. Physiol. Plant. 84, 549–554.

P.H. Graham, M.J. Sadowsky & C.P. Vance (eds.), Symbiotic nitrogen fixation, 165–169.
© 1994 *Kluwer Academic Publishers.*

Eastern Canadian soybean field trials of rhizobial strain NS 1 in two commercial carriers

Richard E. Sanders

Applied Microbe Consultants, Inc., 21 Dixon Court, Truro, Nova Scotia, Canada B2N 4H5

Key words: inoculant, nitrogen, NS1, *Rhizobium*, soybean

Abstract

Rhizobial strain NS 1 was isolated in 1987 from a field in Nova Scotia with no known history of soybean as part of a program to develop regionally-specific legume inoculants for Atlantic Canada. In 1989–90 strain NS 1, in preliminary small-plot field trials utilizing a non-commercial inoculant carrier, gave seed yields 25% greater than the most popular commercial strain, 532C, in five site-years of tests in Atlantic Canada. In 1991 strain NS 1, in two commercial carriers, was tested in both small-plot and farm-level field trials at five sites extending from Ontario to Prince Edward Island. In the small-plot tests, NS 1 gave an average of 10% higher seed yield than strain 532C in both the Grip (TM) and the Nitragin (TM) commercial carriers. In the farm-level field trial, NS 1 gave a 15% greater seed yield than strain 532C in the Nitragin (TM) carrier.

Introduction

The introduction of a leguminous plant into a region for systematic cultivation should be accompanied by the development of a corresponding rhizobial inoculant to maximize harvest yields (Burton, 1982; Buttery et al., 1992; FAO 1983, 1984; Vincent 1970, 1982). Over the last ten years, early-maturing varieties of soybean have been introduced into Atlantic Canada (MacLeod and Goit, 1985). I report here the isolation and partial characterization of rhizobial strain NS 1, which has supported higher soybean seed yields than the most popular Canadian inoculant strain, 532C (also designated 61A152) (Buttery et al., 1992; Hume and Shelp, 1990; Hume et al., 1990 ; Ravuri and Hume, 1992; Wiersma and Orf, 1992), in both non-commercial and commercial carriers on soybean cultivars grown in Eastern Canada.

Materials and methods

Isolation and initial screening of strain NS 1

Strain NS 1 was isolated from a nodule located on the root system of a soybean plant (cv. Maple Isle) grown from a surface-sterilized seed in soil obtained from a field in Nova Scotia with no known history of soybean, using standard bacteriological techniques (Kelley and Post, 1989; Vincent, 1970). Strain NS 1 was greenhouse-tested on several cultivars of soybean cultivated in Atlantic Canada in growth pouches (Wacek and Brill, 1976), soil cores (Vincent 1970, 1982), and an artificial soil system (consisting of a mixture of equal volumes of peat, sand, perlite, vermiculite, and brick clay), using the best commercially available strain, 532C, as a positive control. Growth pouches were evaluated for early nodule formation. The evaluation criterion for the soil cores and the artificial soil system was leaf colour at approximately day 75 after planting. Strain NS 1 out-performed strain 532C in all greenhouse tests.

Preliminary small-plot field trials with non-commercial inoculant

In 1989–1990, NS 1 was tested in preliminary small-plot field trials using a sterile, neutralized peat carrier prepared with standard protocols (Vincent, 1970). The inoculant preparations of strains NS 1 and 532C were adjusted to contain approximately five times ten to the eighth colony forming units per gram by the addition of sterile peat. In the field, inoculant was applied at a rate of two grams of inoculant to one kg of seed.

Based on an estimated five thousand seeds per kg, each seed received approximately two times ten to the eight rhizobia, or approximately twice the Canadian Standard for inoculants for large-seeded legumes such as soybean (Anonymous, 1988; FAO, 1984). Nitracoat (TM) was used as a sticker in accordance with the label instructions for this product.

All field tests were done in accordance with accepted agronomic practice (Vincent, 1970) on land with no known history of soybean. Background levels of nodulation were typically less than one nodule per plant. A hand-pushed V-belt seeder was used to plant the field trials. Between inoculant treatments, the parts of the seeder contacting the soil were bathed in 100% Javex (TM), and the seeder's V-belt was wiped with 70% ethanol. Contamination of un-inoculated guard plots located between the main plots was not significant.

Statistical design consisted of four replications in a split plot experiment with inoculants as the main plots and cultivars as the subplots. Seed yields were subject to the general linear models procedure (SAS, 1985) at the 5% level of significance. When there was a significant main effect of cultivar or inoculant, a Duncan's Multiple Range Test was applied for comparison of the means.

Small-plot and farm-scale field tests with two commercial inoculants

The 1991 field tests compared strains NS 1 and 532C in the Nitragin (TM) and Grip (TM) commercial inoculants. The Nitragin (TM) preparation used a non-sterile peat and was manufactured by Lipha Tech, Inc. The Grip (TM) preparation used a sterile peat and was manufactured by Titre, Inc.

The commercial inoculants of strains NS 1 and 532C were prepared by: Dr. S. Smith, Director of Agricultural Research and Development, Lipha Tech, Inc., 3101 West Custer Avenue, Milwaukee, WI 53209; and, Mr. D. Sutherland, President, Titre, Inc., 361 Rothiemay Road, Ryegate, MO, 59074. These preparations were mailed directly to cooperators in Ontario (Dr. D. Hume, Professor, Crop Science Department, Guelph University, Guelph, ON) and Quebec (Dr. D. Smith, Professor, Plant Science Department, McGill University Montreal, PQ) and to me in Nova Scotia. I distributed portions of the commercial preparations to the cooperators in Prince Edward Island (Dr. John MacLeod, Soil and Water Project Leader, Agriculture Canada Research Station, Charlottetown, PEI), New Brunswick (Mr. David Walker at the Agriculture

Canada Research Station, Fredericton, NB), and Nova Scotia (the Crop Development Institute in Truro, NS).

The protocol for the 1991 small-plot field tests was identical to the protocol described above for the 1989–1990 field trials except that the commercial inoculants were utilized in accordance with their label instructions in place of the non-commercial inoculant preparations. The Nitragin (TM) label rate was 3.34 gm inoculant per kg of seed, with a minimum guaranteed content of two times ten to the eight *Bradyrhizobium japonicum* cells per gram of inoculant. The Grip (TM) label rate was 1.04 g inoculant per kg of seed, with a minimum guaranteed content of two times ten to the ninth *B. japonicum* per gram of inoculant. Based on an estimated five thousand seeds per kg, the Nitragin (TM) preparation would have delivered at least 1.34 times ten to the fifth rhizobia per seed, and the Grip (TM) preparation would have delivered at least $4.16 \cdot 10^5$ rhizobia per seed.

The two-acre strip field trial was carried out on the farm of Mr. Peter Bunnet in Petitcodiac, NB, under the supervision of the cooperator in New Brunswick.

Commercial inoculant titres

Inoculant quality and stability were estimated by determining the number of colony forming units in the preparations from both companies sent to Nova Scotia. The inoculants were titred using standard protocols (Kelley and Post, 1989; Vincent, 1970) upon receipt, after six months storage in ambient laboratory conditions, and after six months storage at four degrees Celsius. All preparations at all times contained approximately ten to the ninth colony forming units per gram.

Results

In 1989–1990, rhizobial strain NS 1 was tested at five sites in Atlantic Canada using non-commercial peat inoculant preparations. The seed yield data from these experiments are presented in Table 1. Strain NS 1 gave a 25% higher seed yield than the most popular Canadian inoculant strain, 532C.

In 1991, the field performance of strain NS 1 was directly compared with strain 532C in two commercial inoculant preparations at five sites in Eastern Canada. The seed yield data from these experiments are presented in Table 2. Strain NS 1 gave 10% higher seed yield than 532C when the data from all sites and both inoculant preparations were averaged.

Table 1. 1989–1990. Summary of soybean seed yield data for five site-years using non-commercial inoculant preparations of strains NS 1 and 532C

Treatment	Seed yield (kg per ha @ 14% moisture)					
	1989		1990		Average	
	Truro, NS	Harring-ton, PEI	Truro, NS	Harring-ton, PEI	Woodstock, NB	
No Nitrogen/ [1]						
No Inoculant	2192c	2269c	2300c	1662c	1860c	2121c
200 kg N/ha	3181a	3246b	3982a	2708a	2314abc	3121a
NS 1	2643b	3667a	3722a	2335bc	2934a	3141a
532C	2823ab	3170b	2133c	1672c	1949c	2457bc

Strain NS 1 gave a 25% higher seed yield than strain 532C in five site-years of small-plot field tests in 1989–1990 using a non-commercial neutralized, sterile peat carrier.

Two cultivars, Maple Isle and KG 20, were used at all sites in all tests. The averaged data from all sites in both years for these two cultivars is presented. There were no strain by cultivar interactions.

Controls included an uninoculated treatment receiving no nitrogen fertilizer and uninoculated treatment receiving 200 kg/ha of fertilizer nitrogen in the form of ammonium nitrate (hand spread in three applications at two weeks, six weeks and ten weeks after seeding).

[1] a-c Means within the same column followed by the same letter are not significantly different (p=0.05) according to Duncan's Multiple Range Test.

Discussion

Superior legume-*Rhizobium* partnerships may rationally be developed when the genetic basis of agronomically important symbiotic traits of legumes and rhizobia is understood (Buttery et al., 1992; Denarie et al., 1992; Phillips et al., 1992; Triplett and Sadowsky, 1992). In the long term, it should be possible to construct regionally-adapted soybean cultivars which are exclusively nodulated by highly effective and competitive rhizobia that flourish in local soils. However, the genetic engineering of more efficient legume-*Rhizobium* symbioses can prove to be complicated (Leigh and Coplin, 1992; Sanders et al., 1978, 1981), and additional studies are necessary before agronomically significant results are consistently obtained. In the short term, screening of wild type rhizobial strains on important cultivars of legumes remains the most common method of superior legume inoculant development (Ravure and Hume 1992; Vincent, 1982). Employing this traditional method of inoculant development, strain NS 1 was identified during screening of indigenous and exotic wild type rhizobial strains for ability to support high seed yields of soybean in Atlantic Canada.

A simple interpretation of the data presented in Tables 1 and 2 is that rhizobial strain NS 1 can support higher soybean harvest seed yields than the most popular commercial strain, 532C, in fields in Eastern Canada which have no history of soybean. Caution should be exercised in more complex interpretations of the data in Tables 1 and 2, since the precise number of rhizobia that were seed-applied was not determined. The experiments reported here were designed to deliver a minimum of one hundred thousand viable rhizobia to each soybean seed, in accordance with the Canadian standard for soybean inoculants (Anonymous 1988; FAO, 1984). However, since these experiments were conducted, it has been reported that seed yield of soybean grown in Canada increases significantly when the number of rhizobia applied per seed is increased above ten to the fifth (Hume and Blair, 1992). Consequently, interpretation of soybean harvest seed yield data should be guarded unless the exact number of seed-applied rhizobia is known.

Table 2. Summary of soybean seed yield field trial data from five sites using two commercial inoculant preparations

| Treatment | Seed yield (kg per ha @ 14% moisture) | | | | | Average |
	Alma ONa	Rothsay ONb	Montreal PQcd	Harrington PEIe	Petitcodiac NBf	
No Nitrogen/ No Inoculant	2192	2331	3336d	1756	2218	2367
200 kg N/ha	3727	2917	3884c	2195	—	3056
532C-N	2471	2769	3203d	1949	2487	2576
NS 1-N	3345	2903	3609cd	2003	2891	2950
532C-G	2318	3185	3350d	2127	—	2745
NS 1-G	2938	2995	3681cd	1980	—	2899

Strain NS 1 gave a 10% higher seed yield than strain 532C in five-site years of small-plot and farm-level field tests carried out in 1991 using two commercial carrier preparations.

"N" or "G" after the strain designation indicates that the commercial inoculant is the Nitragin (TM) or Grip (TM) preparation, respectively. Controls are as described in the legend to Table 1. A small-plot field test at Woodstock, NB, was not harvested due to wet conditions. Data from a small plot field test at Truro, NS were not analyzed due to problems with plant populations.

a
   Alma, Ontario
      Cultivar: Maple Glenn
      Type of test: small-plot
      Statistical analysis: CV = 11.655%; LSD (0.05) = 492 kg/ha

b
   Rothsay, Ontario
      Cultivar: Maple Glenn
      Type of test: small-plot
      Statistical analysis: CV = 8.075%; LSD (0.05) = 344 kg/ha

cd
   Montreal, Quebec
      Cultivars: Maple Isle, Libra, KG 30
      Type of test: small-plot
      Statistical analysis: Means within a column followed by the same letter are not significantly different (p=0.05) according to Duncan's Multiple Range Test.

e
   Harrington, Prince Edward Island
      Cultivars: Maple Isle, KG 20, OT89-16
      Type of test: small-plot
      Statistical analysis: e.s.e. = 41.1

f
   Petitcodiac, New Brunswick
      Cultivar: Maple Amber
      Type of test: two-acre (approx.) strip
      Statistical analysis: none

Acknowledgements

Major funding for the 1991 Eastern Canadian field trials of strain NS 1 was provided by the Prince Edward Island Atlantic Livestock Feed Initiative to Applied Microbe Consultants, Truro, NS. Additional support was provided by the New Brunswick Livestock Feed Initiative to Richard Sanders and by the cooperators at each site.

Support for the initial screening and preliminary field trials of strain NS 1 was provided by grants from the Canada/Prince Edward Island Atlantic Livestock Feed Initiative, the Canada/New Brunswick Atlantic Livestock Feed Initiative and the Canada/Nova Scotia Atlantic Livestock Feed Initiative. Technical assistance was provided by the co-operators in New Brunswick and Prince Edward Island and the Nova Scotia Agricultural College.

The idea to develop regionally-specific legume inoculants was the suggestion of Dr. David Patriquin, Department of Biology, Dalhousie University, Halifax, NS. In addition, Dr. Patriquin kindly provided space in his laboratory for me to write the initial grant proposal to support this work to the Canada/Nova Scotia Livestock Feed Initiative. This grant was later administered by Dr. C. Caldwell, Plant Science Department, Nova Scotia Agricultural College, Truro, NS. Dr. Caldwell also provided the suggestion to focus initial efforts for inoculant development on soybean.

The statistical analysis of the field data for years 1988–1990 was done by Dr. Gary Atlin and Dr. Ralph Martin, both of the Plant Science Department of the Nova Scotia Agricultural College, Truro, Nova Scotia.

References

Anonymous 1988 Fertilizers act and fertilizers regulation, amendment SOR/88–353. Canada Department of Agriculture, Ottawa, ON.

Burton J C 1982 New developments in inoculating legumes. *In* Recent Advances in Biological Nitrogen Fixation. Ed. N S Subba Rao. pp 380–405. Edward Arnold, London, England.

Buttery B R, Park S J and Hume D J 1992 Potential for increasing nitrogen fixation in grain legumes. Can. J. Plant Sci. 72, 323–349.

Denarie J, Debelle F and Rosenberg C 1992 Signalling and host range variation in nodulation. Annu. Rev. Microbiol. 46, 497–531.

FAO 1983 Technical handbook on symbiotic nitrogen fixation. Food and Agriculture Organization of the United Nations, Rome, Italy. 105 p.

FAO 1984 Legume inoculants and their use. Food and Agriculture Organization of the United Nations, Rome, Italy. 61 p.

Kelley S G and Post F J 1989 Basic microbiology techniques. Star, Belmont, CA, USA. pp 31–47.

Hume D H and Blair D H 1992 Effect of numbers of *Bradyrhizobium japonicum* applied in commercial inoculants on soybean seed yield in Ontario. Can. J. Microbiol. 38, 588–593.

Hume D J, Blair D H, Feindel D E and Beeraraghavaiah R 1990 Performance of *Bradyrhizobium japonicum* strain 532C in new soybean fields in Ontario. Can. J. Plant Sci. 70, 319–324.

Hume D J and Shelp F 1990 Superior performance of the Hup⁻ *Bradyrhizobium japonicum* strain 532C in Ontario soybean field trials. Can. J. Plant Sci. 70, 661–666.

Leigh J A and Coplin D L 1992 Exopolysaccharides in plant-bacterial interactions. Annu. Rev. Microbiol. 46, 307–347.

MacLeod J A and Goit J B 1985 Soybean Production in the Atlantic Provinces. Publication No. 117, Atlantic Provinces Agriculture Services Co-ordinating Committee, Truro, NS, Canada.

Phillips D A and Teuber L R 1992 Plant Genetics of Symbiotic Nitrogen Fixation. *In* Biological Nitrogen Fixation. Eds. G Stacey, H J Evans and R W Burris. pp 625–647. Chapman and Hall, New York, USA.

Ravure V and Hume D J 1992 Performance of a superior *Bradyrhizobium japonicum* and a selected *Sinorhizobium fredii* strain with soybean cultivars. Agron. J. 84, 1051–1056.

Sanders R E, Carlson R W and Albersheim P 1978 A *Rhizobium* mutant incapable of nodulation and normal polysaccharide secretion. Nature 271, 240–242.

Sanders R E, Raleigh E and Signer E 1981 Lack of correlation between extracellular polysaccharide and nodulation ability in *Rhizobium*. Nature 292, 148–149.

SAS Institute, Inc. 1985 SAS Users Guide: Statistical Analysis Institute Inc., Cary, NC.

Triplett E W and Sadowsky M J 1992 Genetics of competition for nodulation of legumes. Annu. Rev. Microbiol. 46, 399–428.

Vincent J M 1970 A Manual for the practical Study of the Root-Nodule Bacteria. Blackwell Scientific, Oxford and Edinburgh, UK. 159 p.

Vincent J M 1982 Nitrogen Fixation in Legumes. Academic Press, Sydney, Australia. 283 p.

Wacek T J and Brill W 1976 Simple, Rapid Assay for Screening Nitrogen-Fixing Ability in Soybean. Crop Sci. 16, 519–523.

Wiersma J V and Orf J H 1992 Early maturing soybean nodulation and performance with selected *Bradyrhizobium japonicum* strains. Agron. J. 84, 449–458.

P.H. Graham, M.J. Sadowsky & C.P. Vance (eds.), Symbiotic nitrogen fixation, 171–176.
© 1994 *Kluwer Academic Publishers.*

Nitrogen fixation efficiency of cold-adapted rhizobia on sainfoin (*Onobrychis viciifolia*): Laboratory and field evaluation *

D. Prévost[1], L.M. Bordeleau[1], R. Michaud[1], C. Lafrenière[2], J. Waddington[3] and V.O. Biederbeck[3]

[1]*Research Station, Agriculture Canada, Sainte-Foy, Québec G1V 2J3,* [2]*Experimental Farm, Agriculture Canada, Kapuskasing, Ontario P5N 2Y3 and* [3]*Research Station, Agriculture Canada, Swift Current, Saskatchewan, Canada S9H 3X2*

Key words: arctic rhizobia, cold adaptation, nitrogenase, dry matter yield

Abstract

The ability of cold-adapted arctic rhizobia (from *Astragalus* and *Oxytropis*) to promote growth of sainfoin was investigated by measuring dry matter yield under both controlled and natural conditions. Under controlled conditions simulating spring and summer temperatures, the efficiency of two arctic strains (N10 and N31) was similar to that of two temperate strains (SM2 and 116A15) for two harvests. After the second harvest, the strain efficiency determined as the regrowth of plants maintained at a high temperature regime of 23°C/10°C (summer) was in the following order : N10 > SM2 = N31 > 116A15. However, under a low temperature regime of 14°C/3°C (spring), shoot dry weights and nitrogenase activities were significantly higher with plants nodulated by the arctic strains. In field tests, significant differences among treatments in sainfoin dry matter yield were observed only the year after seeding and not at all sites. For instance, at two sites in western Canada, no significant differences were found in a four year study while another study showed a significantly higher efficiency for one arctic strain at one site. In eastern Canada, the two sites showed higher efficiency of the arctic strain over the temperate and it was more significant at the coldest site. Even though several factors (environmental and plant related) can interfere with the efficiency of the strains, our results show that the cold adaptation of arctic strains is reflected on the regrowth of sainfoin. Thus, it can be advantageous to inoculate sainfoin with cold-adapted rhizobia.

Introduction

In temperate regions, low temperature is one of the major climatological limitations to legume growth. This factor is known to adversely affect both partners of the legume-*Rhizobium* symbiosis by inhibiting nodule development (Sutton, 1983), N_2-fixing efficiency (Heichel and Vance, 1983) and nodulating competitiveness among rhizobial strains (Rice and Olsen, 1988). Consequently, the enhancement of the nitrogen-fixing efficiency at low temperatures would be helpful in sustaining legume growth during cold phases of the growing season. Maximization of symbioses under cool conditions can be obtained by selecting host plant genotypes (Hardwick, 1983) and rhizobial strains (Lip-

sanen and Lindstrom, 1986) that are cold-adapted. In a previous study (Prévost et al., 1987a), cold-adapted rhizobia isolated from *Astragalus* and *Oxytropis* spp. indigenous to the high Arctic were shown to form an effective symbiosis with the temperate forage legume sainfoin (*Onobrychis viciifolia*). The association of arctic rhizobia-sainfoin was used in subsequent studies to establish the advantages of using cold-adapted strains with a temperate legume. Nitrogenase activity of sainfoin nodulated with arctic rhizobia was greater at 5°C and 10°C than that with sainfoin temperate rhizobia (Prévost et al., 1987b). Moreover, at 9°C, arctic rhizobia were more competitive than temperate rhizobia to form nodules on sainfoin (Prévost and Bromfield, 1991). Even though sainfoin has potential in western Canada as a pasture or hay crop because it is non-bloating, relatively drought tolerant and winter-hardy (Hanna et al., 1979), it is not very popular due

_____
* Contribution no 82 of Agriculture Canada Research Station, Sainte Foy, Québec Canada

to its lack of persistence (Krall and Delanay, 1982). This undesirable agronomic trait may be explained by a relatively low growth rate and the inefficiency of the N_2-fixing system to meet the plant's N requirement (Hume et al., 1985). The objective of the present study was to evaluate the symbiotic efficiency of arctic strains, defined as their potential to improve yield of sainfoin, under controlled conditions simulating the growing season and under natural conditions in field tests conducted in eastern and western Canada.

Materials and methods

Origin of rhizobia and inocula

Two arctic rhizobial strains N10 (from *Oxytropis maydelliana*) and N31 (from *Astragalus alpinus*) and two temperate strains SM2 (Ag. Canada) and 116A115 (commercial strain, Nitragin) from sainfoin were used in this study (Prévost et al., 1987a). Liquid inocula were prepared by growing cells at 25°C in yeast-extract mannitol broth (Vincent, 1970) (10^9 cells mL^{-1}). Peat base inoculant (>10^7 cells g^{-1}) was produced for the arctic strains or obtained from Liphatec (Nitragin, Milwaukee) for the commercial inoculant.

Experiments under controlled conditions

Four seedlings from surface-sterilized dehulled seeds of sainfoin cv Nova were grown in each sterilized 13 cm diameter pot containing an autoclaved mixture of 50% (v/v) vermiculite and perlite and fed with a plant nutrient solution supplemented with 10 mg L^{-1} N-NO$_3$ (Prévost et al., 1987a). After emergence, plants were inoculated with 10 mL of liquid inoculum. The treatments were the two arctic strains N10 and N31 and the two temperate strains SM2 and 116A15, with six replicates (pots) in a complete randomized block design. Growing conditions were set to simulate the Canadian Climate Normals for Saskatchewan. Plants were grown at 14°C (350 μE m^{-2} s^{-1}, 14-h day) and 3°C (10-h night) (springtime temperatures) for 3 wk, after which temperatures were increased to 23°C (16-h day) and 10°C (8-h night) (summertime temperatures). The sainfoin was harvested twice (after 12 and 20 wk of growth) at 40% bloom. After the second harvest, plants from three replicates were maintained at the high temperature regime of 23°C/10°C for 8 wk and those from the three other replicates were returned to the low temperature regime of 14°C/3°C for 12 wk at which time a

third harvest was taken. Shoot dry weights were determined for each harvest and nitrogenase activity was evaluated at different times during regrowth for the third harvest by the acetylene reduction assay (Turner and Gibson, 1980). Intact plants from each temperature regime were incubated in growth cabinets (one cabinet per temperature regime) in sealed jars with 10% acetylene and assayed at each day and night temperatures. Data were subjected to analyses of variance.

Field trials

Hulled seeds of sainfoin cv Nova were sown in 1.5 × 5m plots at the recommended seeding rate with the appropriate P and K fertilization according to soil tests and peat-base inoculants were used at double recommended rate to increase the chances of nodulation from the rhizobial inoculant. The two sites in western Canada (Saskatchewan) were seeded in rows in irrigated plots, in 1987 and 1990, at Swift Current (50°17'N, 107°45'W) on a Swinton loam soil and at Melfort (52°N, 104°W) on a nonirrigated Black loam soil. At Swift-Current, there was a starting N-fertilization (15 kg ha^{-1}) and the herbicide Treflan was applied before seeding to control annual weed grasses. The treatments were the two arctic strains N31 and N10, a commercial inoculant and an inoculated control arranged in a randomized complete block design with six replicates. The two sites in eastern Canada were broadcast seeded in 1990 and 1991 at Kapuskasing, Ontario (49°24'N, 82°26'W) on a clay soil and at St-David, Québec (46°48'N, 71°23'W) on a sandy loam soil. Plots were hand weeded after sainfoin emergence. The treatments were the arctic strain N31 and a commercial inoculant; each with two levels of nitrogen fertilization applied only at seeding (0 and 50 kg/ha N-NH$_4$NO$_3$) and arranged in a randomized complete block design with four replicates. Above-ground dry matter yield was determined at 40% bloom for two harvests every year starting the year after seeding. In a few plots, roots were randomly sampled for the determination of strain nodule occupancy as described by Prévost and Bromfield (1991).

Results and discussion

Laboratory experiments

Under controlled conditions (Table 1), arctic and temperate rhizobia showed the same symbiotic efficiency

Table 1. Growth response of sainfoin inoculated with arctic and temperate strains of rhizobia, under controlled conditions

Rhizobial strain	Shoot dry matter yield (g pot^{-1}) of sainfoin			
	1st harvest at	2nd harvest at	3rd harvest at	
	14°/3°C, 23°C/10°C	23°/10°C	14°C/3°C or 23°C/10°C	
Arctic N10	1.15 a^z	5.26 a	3.29 a	3.13 a
N31	1.02 a	4.11 a	2.93 a	2.16 b
Temperate 116A15	1.53 a	5.38 a	0.90 b	1.41 c
SM2	1.33 a	4.25 a	0.87 b	2.21 b
CV (%)	16.4	17.7	9.4	12.4

^{z}Means within a column followed by the same letter are not significantly different at $p < 0.05$ according to Duncan's multiple range test.

as measured by dry matter yield on sainfoin at the first harvest. These results suggest that a 3-wk period under low temperatures may be too short to reflect the cold adaptation of arctic strains as demonstrated previously on sainfoin grown for 12 wk at a root temperature of 9°C (Prévost and Bromfield, 1991). Under optimal conditions, sainfoin is slow to develop an efficient N_2-fixing system since it was observed that fully-formed nodules appeared after 34 days and that root growth was greater than top growth during the first 50 days (Hume and Withers, 1985). In our experiment, the nodulation, hence the establishment of an effective symbiosis may have been stimulated only when the high temperatures (23°C/10°C) were applied resulting in a similar efficiency for both types of strains for the two first harvests, as previously reported under optimal growth conditions (Prévost et al., 1987a). In the present study, the starting temperatures of 14°C/3°C may have caused a stress which delayed the growth since the yield of the first harvest was only 25% of that of the second harvest whereas in the previous study (Prévost et al., 1987a), yield of sainfoin nodulated by these strains was similar for the first two harvests.

Simulated spring and summer temperatures were applied after the second harvest in order to determine the effect of temperature on regrowth of well-established plants. At 14°C/3°C, the regrowth of sainfoin nodulated by the arctic strains was threefold that of plants nodulated by the temperate strains. The increased biomass of sainfoin grown at 14°C/3°C nodulated by the arctic strains compared with the temperate strains at the third harvest indicates that at low temperatures the arctic strains are more efficient for

nitrogen fixation than the temperate strains. Under high temperatures (23°C/10°C), strain efficiencies did not respond according to their origin, the highest yield being obtained with the arctic strain N10 and the lowest with the temperate strain 116A15 (Table 1). For all strains, the yield of the third cut was lower than that obtained for the second harvest. Regrowth processes after harvest draw on root carbohydrate reserves which are also used for N_2-fixation. Krall and Delaney (1982) reported that the low level of carbohydrate storage in sainfoin may contribute to its persistence problem.

Nitrogenase activity (Fig 1) during regrowth for the third harvest followed the same trend as yield (Table 1). Under both day and night temperatures of the cool regime, plants nodulated by the two arctic strains showed the highest nitrogenase activities while at the high temperatures regime (23°C and 10°C), only slight differences among strains were detected. Furthermore, nitrogenase activities with arctic strains were higher at 14°C/3°C than at 23°/10°C, indicating a major change in nitrogen metabolism at low temperature. In a previous experiment with plants grown under optimal temperatures, nitrogenase activity was also higher with plants nodulated by the arctic strains when tested at low temperatures (Prévost et al., 1987b). This cold adaptation for nitrogen fixation may be due to a better substrate uptake to support nitrogenase activity or to the use of a different substrate by the bacteroids of the arctic strains. These factors may be influenced also by the different structure of nodules of sainfoin formed by the two types of strains; the bacteroids of the arctic strain N31 are spherical and included in low number (1–3) in each membrane envelope whereas those of

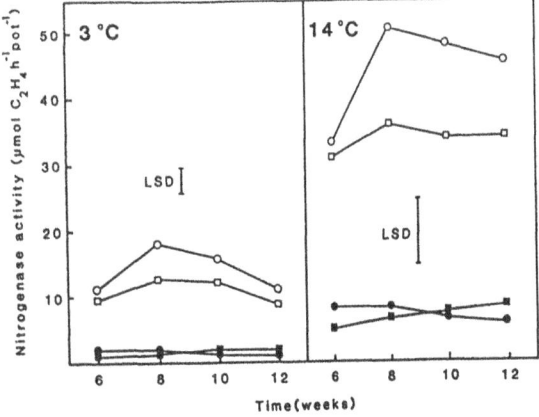

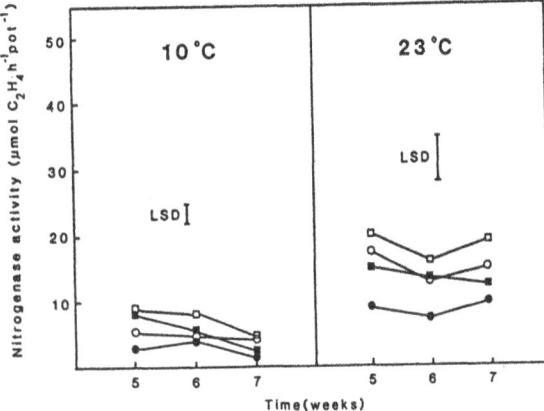

Fig. 1. Nitrogenase activity of sainfoin plants nodulated by the arctic strains N31 (○) and N10 (□) and the temperate strains SM2 (■) and 116A15 (●) during regrowth for the third harvest under a low (14°C/3°C) (upper part) and a high (23°C/10°C) (bottom part) temperature regimes.

the temperate strain SM2 are elongated and included in large number (3–9) (Prévost et al., 1989). It has been reported that succinate and other C_4-dicarboxylic acids support nitrogenase activity (Mc Rae et al., 1989); however, in free-living cells of strains N31 and SM2, the effect of low temperature on succinate transport system was similar for both strains, indicating that the succinate carrier is not adapted to cold (Bigwanesa et al., 1993). Future studies should investigate the effect of temperature on bacteroid activities.

Field trials

In western Canada, the 1987 seedings (results not shown) did not allow us to conclude that there was any beneficial effect of inoculation on sainfoin yield over four years, whatever the rhizobial strain and the site. Inconsistant responses of sainfoin to inoculation

have been reported in many field tests in this area (Hanna et al., 1979). For the 1990 seeding at Swift Current (Table 2), no significant differences were observed at the first harvest in 1991.

At the second harvest and for the two 1992 harvests, the arctic strain N31 was the most effective; other inoculants were similar to the uninoculated control. This strain effect did not manifest at Melfort (Table 2) where the benefit of inoculation was shown only at the first harvest of 1992. It seems that weed control, specially once sainfoin was established, and drought at Melfort were the major problems that may partially explain the poor second cut yields in 1992. In fact, productive life of sainfoin is related to competition from weeds and the crop gives poor yields in very dry conditions (Smoliak et al., 1981).

In eastern Canada, the 1991 seedings were a failure at both sites due to drought during the establishment year at St-David and winterkill at Kapuskasing. However, for the 1990 seedings, marked differences between the arctic strain N31 and the commercial inoculant were observed at the two sites (Table 3). At St-David in 1991, where only one harvest was taken due to drought, the N-fertilized treatments gave the highest yields. In 1992, the effect of N-fertilization was not maintained and yields of sainfoin inoculated with the arctic strain N31 were higher although not significantly different from those of commercial inoculant.

At the second harvest, the arctic strain treatments yielded up to twice as much as the commercial inoculant. At Kapuskasing in 1991, there was no effect of fertilization and sainfoin nodulated by the arctic strain N31 produced 2 and 3 times more dry matter yield than the commercial inoculant for the first and second harvests respectively. However, sainfoin plants did not survive over winter at this site.

Variations in symbiotic efficiencies obtained from field tests could be due to many factors. First, agronomic practices for seeding and weeding differed among eastern and western sites. Second, it was difficult to determine if there was a competition problem among rhizobial strains. The presence of indigenous ineffective rhizobia was evident at all sites since sainfoin inoculated with soil dilutions from eastern sites showed tiny ineffective nodules as did uninoculated sainfoin grown in western sites. Nodule occupancy in eastern sites revealed that nodulation was mainly due to the introduced strain (data not shown). Also, the different edaphic conditions between eastern and western sites may influence the growth and persistence of both partners of the symbiosis. Finally, other factors such as

Table 2. Growth response of sainfoin inoculated with arctic rhizobia and commercial inoculant, seeded in 1990 at two sites in western Canada

Treatment	Dry matter yield (kg ha^{-1}) of sainfoin			
	1991		1992	
	1	2	1	2
Swift Current				
Arctic strain N10	3282 a[z]	1120 b	386 b	408 b
Arctic strain N31	3543 a	1691 a	592 a	703 a
Commercial inoculant	3562 a	1255 b	484 b	523 ab
Uninoculated	3187 a	1177 b	456 b	461 b
CV (%)	12.3	14.0	17.9	34.4
Melfort				
Arctic strain N10	2979 a	1472 a	2867 a	204 a
Arctic strain N31	2899 a	1289 a	2818 a	234 a
Commercial inoculant	2970 a	1259 a	2903 a	271a
Uninoculated	2898 a	1312 a	2465 b	189 a
CV(%)	8.3	13.9	6.1	37.7

[z]Means within a column followed by the same letter are not significantly different at $p < 0.05$ according to Duncan's multiple range test.

Table 3. Growth response of sainfoin inoculated with arctic rhizobia and commercial inoculant, seeded in 1990 at two sites in eastern Canada

Treatment	Dry matter yield (kg ha^{-1}) of sainfoin			
	1991		1992	
	1	2	1	2
St-David				
Arctic strain N31 (50N)[z]	1969 a[y]	_[x]	3238 a	3286 a
Arctic strain N31 (0N)	1272 b		4036 a	3276 a
Commercial inoculant (50N)	2072 a		2699 a	1777 b
Commercial inoculant (0N)	1272 b		2644 a	2282 b
CV (%)	16.7		27.6	21.6
Kapuskasing				
Arctic strain N31 (50N)	3295 a	2120 a	_[w]	
Arctic strain N31 (0N)	2791 a	2354 a		
Commercial inoculant (50N)	862 b	1360 b		
Commercial inoculant (0N)	998 b	1258 b		
CV (%)	20.8	19.5		

[z]50N and 0N = 50 and 0 kg ha^{-1} of N-NH$_4$NO$_3$ at seeding.

[y]Means within a column followed by the same letter are not significantly different at $p < 0.05$ according to Duncan's multiple range test.

[x]Regrowth too poor for a second harvest.

[w]No winter survival.

temperature, moisture regimes and length of the growing period may play a significant role on crop growth (Heichel, 1987). For instance, the efficiency of the arctic strain N31 was best at Kapuskasing, the coldest site in the tests with a monthly average temperature of 4.2°C in spring (April-May) and under irrigation at Swift Current, the warmest site (9.3°C). However, this strain did not show any positive effect at the non-irrigated site of Melfort where spring is also cold (5.8°C).

Nevertheless, our results show that under all conditions, arctic rhizobia were better, or as efficient as temperate rhizobia. Under controlled low temperatures and particular field conditions, regrowth of sainfoin nodulated by the arctic strains gave 1.5 to 3 times more shoot dry matter yield than those nodulated by temperate rhizobia. The fact that the arctic strain N10 was less efficient than strain N31 in the field may be due to some site specific factors or may indicate that traits other than cold adaptation may be involved. We conclude that the cold adaptation of arctic strains is reflected in the regrowth of sainfoin. Thus, it could be advantageous to inoculate sainfoin with arctic rhizobia.

Acknowledgements

We would like to thank J Geissler and M P Schellenberg from Swift Current, R Horton and M Hiltz from Melfort, L Guillemette from Kapuskasing, A Chabot and L Lambert from Sainte-Foy for their collaboration.

References

Bigwanesa P C, Prévost D, Bordeleau L M and Antoun H 1993 Effect of temperature on succinate transport by an arctic and a temperate strain of rhizobia. Can. J. Microbiol. (In press).

Hanna M R, Cooke D A, Smoliak S, Goplen B P and Wilson D B 1979 Sainfoin for western Canada. Publication 1470, Agriculture Canada, 18p.

Hardwick R C 1983 Genetic variation in yield processes of grain legumes and their responses to low temperatures. *In* Temperate Legumes (Physiology, genetics and nodulation). Eds. D G Jones and D R Davies. pp 55–76. Pitman Books Limited, London, UK.

Heichel G H 1987 Legumes as a source of nitrogen in conservation tillage systems. *In* The Role of Legumes in Conservation Tillage Systems. Ed. J F Power. pp 29–35. Soil Conserv. Soc. Am. Ankeny, Iowa.

Heichel G H and Vance C P 1983 Physiology and morphology of perennial legumes. *In* Nitrogen Fixation Vol. 3 (Legumes). Ed, W J Broughton. pp 99–143. Clarendon Press, Oxford.

Hume L J and Withers N J 1985 Nitrogen fixation in sainfoin (*Onobrychis viciifolia*) 1. Responses to nitrogen nutrition. N. Z. J. Agric. Res. 28, 325–335.

Hume L J, Withers N J and Rhoades D A 1985 Nitrogen fixation in sainfoin (*Onobrychis viciifolia*) 2. Effectiveness of the nitrogen-fixing system. N. Z. J. Agric. Res. 28, 337–348.

Krall J M and Delaney R H 1982 Assessment of acetylene reduction by sainfoin and alfalfa over three growing seasons. Crop Sci. 22, 762–766.

Lipsanen P and Lindstrom K 1986 Adaptation of red clover rhizobia to low temperatures. Plant and Soil 92, 55–62.

Mc Rae D G, Miller R W, Berndt W B and Joy K 1989 Transport of C_4-Dicarboxylates and amino acids by *Rhizobium meliloti* bacteroids. Mol. Plant-Microbe Interact. 2, 273–278.

Prévost D, Antoun H and Bordeleau L M 1987a Symbiotic effectiveness of arctic rhizobia on a temperate forage legume sainfoin (*Onobrychis viciifolia*). Plant and Soil 104, 63–69.

Prévost D, Antoun H and Bordeleau L M 1987b Effects of low temperatures on nitrogenase activity in sainfoin (*Onobrychis viciifolia*) nodulated by arctic rhizobia. FEMS Microbial. Ecol. 45, 205–210.

Prévost D, Bordeleau L M and Antoun H 1989 Effet des souches arctiques de *Rhizobium* sur la structure des nodules du sainfoin (*Onobrychis viciifolia*) et de légumineuses arctiques (*Astragalus* et *Oxytropis* spp). Can. J. Bot. 67, 3164–3168.

Prévost D and Bromfield E S P 1991 Effect of low root temperature on symbiotic nitrogen fixation and on competitive nodulation of *Onobrychis viciifolia* (sainfoin) by strains of arctic and temperate rhizobia. Biol. Fertil. Soils 12, 161–164.

Rice W A and Olson P E 1988 Root temperature effects on competition for nodule occupancy between two *Rhizobium meliloti* strains. Biol. Fertil. Soils 6, 137–140.

Smoliak S, Bjorge M, Penny D, Harper A M and Horricks J S 1981 Alberta forage manual. Ed Alberta Agriculture, Alberta, 85 p.

Sutton W D 1983 Nodule development and senescence. *In* Nitrogen Fixation Vol. 3 (Legumes). Ed. W J Broughton. pp 144–212. Clarendon Press, Oxford.

Turner G L and Gibson A H 1980 Measurement of nitrogen fixation by indirect means. *In* Methods for Evaluating Biological Nitrogen Fixation. Ed F J Bergersen. pp 328–362. John Wiley and Sons, New-York, NY.

Vincent J M 1970 A Manual for the Practical study of Root-Nodule Bacteria. IBP Handbook no 15. Blackwell Scientific Publications, Oxford. 164 p.

P.H. Graham, M.J. Sadowsky & C.P. Vance (eds.), Symbiotic nitrogen fixation, 177–180.

Survival of *Bradyrhizobium japonicum* in pig slurry used as carrier for soil inoculation

G. Ciafardini and G. C. Turtura

Department of Animal, Plant and Environmental Sciences, Agriculture Faculty of the Molise University, Via Cavourn. 50, 86100 Campobasso, Italy and Institute of Agriculture Microbiology, Agriculture Faculty of Bologna University, Via Filippo Re 6, I40126 Bologna, Italy

Key words: Bradyrhizobium japonicum, pig slurry, Rhizobia soil inoculation

Abstract

Pig slurry is a highly polluting animal effluent whose discharge directly into waterways is prohibited by law in many countries. It is usually collected during the winter in special storage lagoons where it undergoes partial purification and then in the spring it is used to irrigate soybean or corn. This paper reports on the survival of *Bradyrhizobium japonicum* USDA 122 in pig slurry used as a carrier to inoculate soil and soybean plants at the V_3 stage of development. The survival of *B. japonicum* was determined in water (control), in slurry from a storage lagoon (treated), and in newly produced pig slurry from a pigsty (untreated). Sample analysis at time zero and at 2-h intervals showed that the survival of *B. japonicum* in the two types of slurry remained unchanged during the initial 24 h of contact at 20°C. However, after 48 h cell mortality increased significantly reaching 40% in the untreated slurry and 15% in the treated slurry. The three treatments were enriched with *B. japonicum* and after 1 h applied with a tractor-pulled tank at a rate of 100 m^3 ha^{-1} on *B. japonicum*-free clay-loam soil to cover inoculate cv. Evans soybean. The field trials showed the presence of abundant nodulation (ca. 40-60 nodules/plant) on all plants inoculated with the bacterium applied with water (control) as well as with the two types of pig slurry. In addition, microbiological analyses of the soil during the first three weeks after surface inoculation showed no significant differences in the survival of the microsymbiont applied with the pig-slurry carrier. Pig slurry applied a few hours after enrichment with the rhizobium may prove to be a valid method to enrich soybean-planted soil with both fertilizing elements and *B. japonicum*.

Introduction

Bradyrhizobium japonicum is a gram-negative soil bacterium that induces the formation of nodules on soybean (*Glycine max* (L.) Merr.) root. When grown in soil with few or no effective *B. japonicum*, soybeans demonstrate poor nitrogen fixation resulting from the small number of nodules produced. To obtain sufficient nodulation, soybeans are usually inoculated at sowing with suitable peat-based inoculants of *B. japonicum*. Previous work (Ciafardini and Barbieri 1984a, 1987; Ciafardini and Lombardo, 1991) has demonstrated the possibility of inducing nodulation on soybean roots by cover inoculating the microsymbiont in the soil with the irrigation water when the plants have reached the three-node (V_3) phenological stage (Fehr and Caviness, 1977). Cover inoculation of V_3 stage

plants, previously seed inoculated, doubled the nodulation by producing on secondary roots numerous nodules that were about one month younger than those on taproots obtained by seed inoculation (Ciafardini and Barbieri, 1987). However, not all farms have river or ground water sufficient for irrigation, whereas livestock farms frequently use pig or stall slurry for irrigation. Being highly polluting, pig slurry cannot be discharged directly into waterways. In fact, it is collected during the winter in special storage lagoons and used in summer to irrigate corn or soybean. This paper presents a study of the survival of *B. japonicum* in pig slurry used as a carrier to soil inoculate the microsymbiont on soybean.

Materials and methods

Laboratory tests

The pig slurry used for this study came from a pigsty with a lagoon-type purification system, which consists essentially of a solid-liquid separator and a storage lagoon (Ciafardini and Barbieri, 1982a, b, 1984). The chemical components of the pig slurry were evaluated according to Cottenie (1980) for nitrogen and phosphorus and according to the Consiglio Nazionale delle Ricerche (CNR-IRSA) (1985) for the other elements. The aqueous phase and suspended solids were evaluated, respectively, by drying 100 mL of whole pig slurry at 90°C and the solid fraction alone by centrifuging 100 mL of whole pig slurry at 7000g. Survival tests were conducted using *Bradyrhizobium japonicum* USDA strain 122. The bacterium was grown in yeast extract mannitol (YEM) broth (Vincent, 1970) at 28°C in 3-L flasks with stirring at 20 rpm. After 5 days, 2.7 L of the bacterial suspension (2×10^7 cells mL^{-1}) was centrifuged. The supernatant was discarded and the bacterial cells resuspended in 270 mL of water. Thirty mL of bacterial suspension was placed with stirring in to 3-L flasks containing 2.7 L of the following three carriers: 1) new untreated pig slurry coming directly from the pigsty, 2) treated pig slurry taken from the storage lagoon, 3) water (control). Tests were conducted at 20°C with three replications per treatment. Ten mL, samples were taken with an autosampler at the start of the test (time 0) and at 2-h intervals up to 48 h, and analyzed immediately (daytime samples) or the following morning after storage at 4°C (nighttime samples). Samples were analyzed with the most probable number (MPN) method, inoculating cv Weber soybean in growth pouches (Weaver et al., 1972). Following statistical analysis these data were summarized and plotted at 4-h intervals.

Field trials

The field trials were conducted on *B. japonicum*-free clay-loam soil planted with cv Evans soybean. The trial was conducted by delivering the microsymbiont via the new and treated pig slurry carriers used for the previous test, plus an uninoculated control (water without the bacterium). A randomized block design with three repetitions was used. Fifty m^2 plots were inoculated when the soybean plants were in the V$_3$ stage of development. The microsymbiont was grown in YEM broth. After 6-days' growth, it was centrifuged and

Table 1. Chemical and physical characteristics of the pig slurries used as inoculant carriers

Element	Pig slurry	
	Untreated	Treated
	(% of dry matter in the slurry)	
N	6.70	4.09
P	3.01	0.14
Ca	7.42	5.27
Mg	0.44	0.77
K	1.28	1.89
Na	0.29	0.84
Fe	0.62	0.15
	(µg/g of dry matter in the slury)	
Cu	978.00	259.00
Zn	3360.00	436.00
Mn	666.00	27.00
Cd	3.00	<0.01
Pb	3.09	<0.10
Cr	18.58	6.80
Ni	16.00	11.30
Hg	<0.01	<0.01
As	1.85	<.10
	(% of pig slurry)	
Moisture	97.31	99.78
Total suspended solids	2.50	0.57

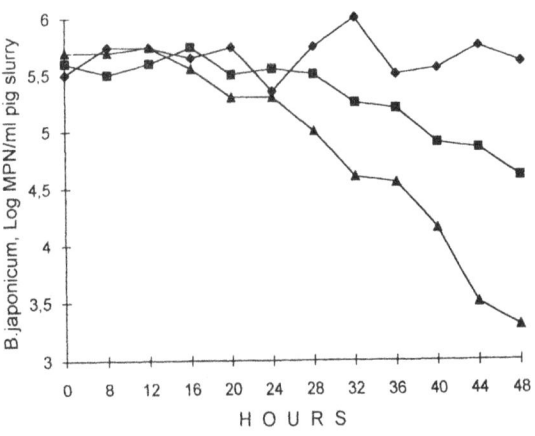

Fig. 1. Survival of *Bradyrhizobium japonicum* USDA 122 in different carriers (♦ water; ■ treated pig slurry; ▲ untreated pig slurry).

resuspended in water to obtain 1×10^{11} cells mL^{-1}. One litre of concentrated inoculum was delivered in 100 m^3 ha^{-1} of carrier. The carriers were enriched with *B. japonicum* by incorporating sufficient inoculant into the wagon tanks to provide the amount of water suffi-

Table 2. Survival in the soil of *Bradyrhizobium japonicum* USDA 122 after soil-inoculation using pig slurry as carrier

Carrier	*Bradyrhizobium japonicum* USDA 122 (MPN 10^{-2} g dry soil^{-1})	
	10 days after inoculation	21 days after inoculation
Untreated pig slurry	8.7±0.9	7.3±0.3
Treated pig slurry	14.2±1.2	13.1±0.5
Water	15.8±1.5	13.7±0.7
Uninoculated control	—	—

Table 3. Nodulation obtained with *Bradyrhizobium japonicum* USDA 122 soil-inoculated using pig slurry as carrier

Carrier	Nodulation[a]	
	No. of nodules per plant	mg of nodules per plant[b]
Untreated pig slurry	40±5	850±40
Treated pig slurry	52±8	1010±150
Water	60±7	1280±110
Uninoculated control	—	—

[a] $\overline{x}$ ± S.E. [b] Dry weight

cient to treat one plot. After allowing 1 h of contact, the field was irrigated with the carrier treatments. Survival of *B. japonicum* in the soil was evaluated after 10 and 21 days by sampling five soybean plants per plot. The soil was freed from the roots and analyzed with the MPN method to estimate the population of *B. japonicum*. Nodulation was evaluated, at the time of pod filling, by sampling five plants per plot and determining the number and biomass of the nodules.

Results

The chemical and physical analyses of the pig slurry indicated a greater concentration of heavy metals in the untreated slurry from the pigsty. In particular this type of slurry is richer in both suspended solids and heavy metals such as Cu, Zn, Mn, Cd, Pb and Cr (Table 1). The survival of *B. japonicum* suspended in the two types of pig slurry did not vary significantly during the first 24 h. After 48 h of contact the viability of the microsymbiont varied significantly both in the untreated as well as the treated slurry, decreasing by 40% and 15%, respectively (Fig. 1).

In the field trials, the soybean treated with the two types of slurry showed no symptoms of phytotoxicity.

In addition, soil inoculation led to the appearance of nodules along the secondary roots and adventitia.

Soil analyses showed a lower presence of *B. japonicum* in the plots inoculated with untreated pig slurry; however, the viability of the symbiont remained nearly constant for a period of 21 days (Table 2). We observed no *B. japonicum* in soil samples from the control plots that were treated with symbiont-free water (Table 2). Nodulation was seen in all soybean plants inoculated with *B. japonicum*. The number and biomass of the nodules, however, was lower in plots inoculated with untreated pig slurry (Table 3).

Discussion

Slurry kept in the storage lagoons has a lower polluting load than untreated slurry coming from the pigsty. In fact, it shows a lower concentration of heavy metals, which are removed from the liquid phase during passage through the solid-liquid separator of the purification system (Ciafardini and Barbieri, 1982a, b, 1984b) (Table 1). This type of slurry is typically used in farming to irrigate fields in the spring. The high survival rate of *B. japonicum*, for at least 24 h in this type of slurry, makes it especially interesting as a carrier to deliver the microsymbiont to soybean. In fact, 24 h of microsymbiont survival is more than sufficient time to spread even large quantities of *B. japonicum*-enriched pig slurry without appreciable lose of viability of the bacterium because of prolonged contact with the carrier (Fig. 1). The fewer numbers of *B. japonicum* and nodules per plant observed in field plots treated with untreated pig slurry (Tables 2, 3) may be explained by both the lower survival rate of the microsymbiont in this type of slurry—which continues to exert its toxicity (Fig. 1) even at ground level—and by the higher levels of suspended solids, that may form a thin surface crust on the soil. This may trap many bradyrhizobia at the surface leading to their subsequently destruction by both drying and by light. Because pig slurry treated by the storage-lagoon system contains only small amounts of suspended solids, it can be delivered to the field with normal watering-gun systems. *B. japonicum* liquid inoculant can be added to the irrigation system with the standard dispenser used to spread liquid chemical fertilizers.

References

Ciafardini G and Barbieri C 1982a An ecological study of pig slurry in storage lagoons. Agrochimica 26, 243–245.

Ciafardini G and Barbieri C 1982b Dynamics of some biological parameters in the purification of pig slurry in storage lagoons. Agrochimica 26, 246–253.

Ciafardini G and Barbieri C 1984a Un nuovo sistema di inoculazione della soia: la distribuzione del simbionte in copertura. Inf Agr.19, 67–69.

Ciafardini G and Barbieri C 1984b Effects of copper and zinc on the microflora of soil fertilized with pig slurry. Agrochimica 28, 133–139.

Ciafardini G and Barbieri C 1987 Effects of cover inoculation of soybean on nodulation, nitrogen fixation and yield. Agron. J. 79, 645–648.

Ciafardini G and Lombardo G M 1991 Nodulation, dinitrogen fixation and yield improvement in second-crop soybean cover-inoculated with *Bradyrhizobium japonicum*. Agron. J. 83, 622–625.

Consiglio Nazionale delle Ricerche (CNR-IRSA) 1985 Metodi analitici per fanghi. Quaderno n. 64 fascicolo 10: metalli pesenti. Rome, Italy.

Cottenie A 1980 Soil and plant testing as a basis of fertilizer recommendation. FAO-Soil Bulletin 38/2, 96-100.

Fehr W R and Caviness C E 1977 Stages of soybean development. Coop. Ext. Serv. Iowa State Univ. Special Rep. 80, Ames.

Vincent J M 1970 A Manual for the practical Study of Root-Nodule Bacteria. Int. Biol. Prog. Handb. 16, Blackwell Scientific Publications, Oxford, England.

Weaver R W and Frederick L R 1972 A new techniques for Most Probable Number count of rhizobia. Plant and Soil 36, 219–222.

P.H. Graham, M.J. Sadowsky & C.P. Vance (eds.), Symbiotic nitrogen fixation, 181–188.
© 1994 *Kluwer Academic Publishers.*

Plasmid DNA content of several agronomically important *Rhizobium* species that nodulate alfalfa, berseem clover, or *Leucaena*

Fawzy M. Hashem and David Kuykendall[1]
Soybean and Alfalfa Research Laboratory, USDA/ARS, 10300 Baltimore Avenue, Beltsville, MD 20705–2350, USA. [1]*Corresponding author*

Key words: extrachromosomal inheritance, legume, nitrogen fixation, nodulation, symbiosis

Abstract

Plasmids control the effective symbiotic properties of legume-nodulating *Rhizobium* species. Our long-term goal is to genetically improve inoculant strains using plasmid transfer, therefore, in this study, we examined the plasmid content of eight agronomically-important strains of *Rhizobium meliloti* from alfalfa (*Medicago sativa* L.), six strains of *R. leguminosarum* biovar *trifolii* from berseem clover (*Trifolium alexandrinum* L.), and thirteen strains of *Rhizobium* sp. (*Leucaena*) from *L. leucocephala* (Lam.) Dewit, isolated in Egypt. Most of the *R. meliloti* strains had an approximately 800-MDa cryptic megaplasmid; USDA strain 1021a and ARC strains 1, 103, B3, and B7 also contained a smaller symbiotic megaplasmid that hybridized with *nifHD*. These Sym plasmids varied in size from 430–615 MDa. Both strain ARC 104, with a single megaplasmid greater than 1600 MDa, and strain USDA1011, which had six plasmids ranging from 1000 to < 50 MDa, failed to hybridize with the *nif* gene probe. *R. leguminosarum* bv. *trifolii* USDA strains 2101, 2128, 2129 and 2131 each had a 500-MDa *nif* megaplasmid plus a 220 and a 150-MDa plasmid, whereas ARC strains 100 and 101 each had a 200-MDa *nif* plasmid and six other plasmids ranging in size from 75 to 310 MDa. *Rhizobium* sp. (*Leucaena*) strains fell into five different plasmid profile groups, most with a *nif*-carrying plasmid of 200, 230 or 290 MDa. Plasmid characterization will be useful for plasmid transfer studies aimed at developing improved *Rhizobium* inoculant strains. This is the first study to document diversity in plasmid profiles among inoculant *Rhizobium* strains for alfalfa and berseem clover in Egypt. It is also the first report describing plasmids and megaplasmids of *Rhizobium* sp. (*Leucaena*) strains isolated in Egypt.

Introduction

Plasmids, extrachromosomal DNA molecules present in a wide variety of bacterial species, control many different phenotypic characteristics (Beringer and Hirsch, 1984). Those in *Rhizobium* species carry genes controlling symbiotic functions (Banfalvi et al., 1981; Djordjevic et al., 1982; Hirsch et al., 1980; Hooykaas et al., 1982), which are clustered on a single large plasmid termed "pSym" (Banfalvi et al., 1981, 1985; Barbour et al., 1985; Heron and Pueppke, 1984; Hynes et al., 1986; Krol et al., 1982; Rosenberg et al., 1981). Two or more plasmids that carry genes controlling symbiotic functions have been found in certain strains of *R. leguminosarum* biovar *trifolii* (Hynes and McGregor, 1990), of *R. etli* (Brom et al., 1992) and of *R. meliloti* (Hynes et al., 1986). *Rhizobium* plasmid DNA molecules represent a significant part of the genome, or between 25% and 50% of the total DNA (Martinez et al., 1990; Prakash and Atherly, 1986). Because there is high variability, plasmid profile analysis has proven to be a useful tool for differentiating *Rhizobium* strains and studying their ecology (Brockman and Bezdicek, 1989; Buendia-Claveria et al., 1989; Glynn et al., 1985; Harrison et al., 1989a,b; Hartmann and Amarger, 1991; Shishido and Pepper, 1990; Weaver et al., 1990 ; Young and Wexler, 1988; Zahran, 1992).

Differences in the symbiosis-controlling plasmids of *R. leguminosarum* have been noted (Hombrecher et al., 1981; Young and Wexler, 1988). *R. leguminosarum* bv. *trifolii* and *phaseoli*, and *R. fredii*, contain large plasmids varying in size from 100 to 300 MDa that carry genes involved in nodulation (*nod*) and symbiotic nitrogen fixation (*fix*), as well as the nitrogenase structural genes (*nifHDK*) (Berryhill et al., 1985; Beynon et al., 1980; Barbour et al., 1985; Barbour and Elkan, 1989; Harrison et al., 1988; Mozo et al., 1988). In *R. meliloti, nod* and *nif* genes are carried on megaplasmids that range in size from 400-1000 MDa (Banfalvi et al., 1981, 1985; Farrai et al., 1983; Hynes et al., 1986; Rosenberg et al., 1981).

In addition to symbiotic plasmid(s), *Rhizobium* strains may carry 1–10 other plasmids (Thurman et al., 1985) which range in size from about 30 to more than 1000 MDa (Baldani and Weaver, 1992; Chen et al., 1993 ; Heron and Pueppke, 1984; Mozo et al., 1988; Selenska-Trajkowa et al., 1990). These plasmids are highly stable and are assumed to have beneficial roles in the soil environment (Weaver et al., 1990). This study examined the plasmid profiles and identified the *nif*-hybridizing plasmids of twenty-seven strains of *Rhizobium* species nodulating alfalfa, berseem clover, or *Leucaena*. Our long-term goal is to transfer plasmids between strains in order to generate new combinations of desired traits for genetic improvement of inoculant strains.

Materials and methods

Bacterial strains and growth conditions

The bacterial strains studied are listed in Table 1. *Rhizobium leguminosarum* bv. *trifolii* and *R. meliloti* strains are used in legume inoculant production both in Egypt and in the U.S.; strains of *Rhizobium* sp. (*Leucaena*) were isolated in Egypt. For plasmid analysis, all *Rhizobium* strains were grown in HP medium described by Hynes et al. (1985), except that they were not subcultured in TY medium, at 30°C for 16–22 h. The *Escherichia coli* and *Pseudomonas aeruginosa* reference strains were grown in LB medium at 37°C for 6–8 h. All cultures were grown in an incubator shaker (100 rpm) and were in early log phase [optical density (A_{575}) was less than 0.375].

Plasmid profiles

Plasmid profiles were determined on horizontal agarose gels using an in-well lysis method (Eckhardt, 1978; Hynes and McGregor, 1990; Hynes et al., 1985; Kuykendall and Hengen, 1988; and Plazinski et al., 1985) with some modifications. Gels were prepared in TBE buffer (89 m*M* Tris-borate, 2 m*M* EDTA, pH 8.0) with 0.75% (w:v) agarose and 1% (w:v) SDS.

0.2 mL of culture was transferred to a sterile Eppendorf tube on ice, then five volumes of 0.3% (w:v) N-lauroylsarcosine in TBE buffer were added to the culture and mixed gently by tapping. The tube was kept on ice for at least 10 min. The cells were pelleted by centrifugation in a benchtop microcentrifuge at 17,000 rpm at 4°C for 5 min and the supernatant was removed as completely as possible without disturbing the pellet. Pellets were resuspended in 25 μL of lysis solution containing lysozyme (0.2 mg/mL), RNase (10 μg/mL), and 10% (w:v) sucrose in TBE, then immediately loaded into the wells in the gel. Five volts were applied for 30–45 min, or until turbidity disappeared, then 85 volts for 8 h. Gels were stained in deionized water containing 1 μL of a 1% (w:v) ethidium bromide stock solution per 100 mL for 1 h at room temperature, then destained in deionized water for at least 1 h at room temperature before photographing on a UV transilluminator.

Molecular weights for plasmids were estimated from the known molecular weights listed in Table 2, essentially as described by Shishido and Pepper (1990).

Hybridization analysis

Southern transfers of DNA from gels to nylon membranes were performed using vacuum blotting. The hybridization protocol was as described by Sambrook et al. (1989) using the *nifHD* probe pCHK12, which consists of a 3.7-kb *Eco*RI fragment from *Rhizobium meliloti* 102F34 cloned into pUC9 (Baldani and Weaver, 1992)

Results

Plasmid profiles and plasmid characterization

Modified in-well lysis and electrophoresis produced plasmid profiles (Figures 1–3) that revealed variation in plasmid content of the *Rhizobium* strains analysed in this study (Table 3). Plasmid sizes ranged from greater than 1600 MDa to less than 50 MDa. Each strain had between one and seven plasmids, except for two *Rhizobium* sp. (*Leucaena*) strains, DS 65 and DS 78, which apparently lacked plasmids.

Based on the number and sizes of plasmids, as well as *nif* homology, it was possible to distinguish seven of the eight *Rhizobium* strains isolated from alfalfa; strains ARC 1 and ARC 103 were indistinguishable (Table 3). The *R. meliloti* strains studied had six differ-

Table 1. Rhizobium strains

Strain[a]	Parent host	Origin	Source
Rhizobium	*Medicago sativa*		
meliloti	subsp. *sativa*		
USDA 1011		Maryland, USA	P. van Berkum, USDA
USDA 1021a		N. Dakota, USA	''
USDA 1025		Arizona, USA	''
ARC 1		Giza, Egypt	S. Abdel-Wahab, ARC
ARC 103		''	''
ARC 104		Bensweif, Egypt	''
ARCB3		Bahrain	''
ARC B7		''	''
R. leguminosarum	*Trifolium*		
biovar *trifolii*	*alexandrinum*		
USDA 2101	Not known	P. van Berkum, USDA	
USDA 2128	Tunisia	''	
USDA 2129	''	''	
USDA 2131	''	''	
ARC 100	Egypt	S. Abdel-Wahab, ARC	
ARC 101	''	''	
Rhizobium sp.	*Leucaena*		
(*Leucaena*)	*leucocephala*		
TAL 583		New Guinea	H. Keyser, NifTAL
DS 3		Giza, Egypt	D. Swelim, ARC
DS 9		''	''
DS 43		Minia, Egypt	''
DS 65		Cairo, Egypt	''
DS 78		''	''
DS 91		Giza, Egypt	''
DS 93		''	''
DS 129		''	''
DS 137		''	''
DS 144/1		''	''
DS 158		Ismalia, Egypt	''

[a]ARC = Egyptian Agricultural Research Center; USDA = Beltsville Collection;
DS = isolated in Egypt by Diaa Swelim; TAL = NifTAL, Maui, Hawaii.

ent geographical origins (Table 1) and fell into seven different plasmid groups (Table 3, Fig. 1). Although strains ARC B3 and B7 had the same geographical origin, their plasmid profiles differed.

The *R. leguminosarum* bv. *trifolii* strains examined fell into two groups (Fig. 2). The plasmid profiles of the bacterial strains within each group were indistinguishable. The two Egyptian strains constituted one group, while the four USDA berseem clover strains made up

the second group. All of the strains, even those within the same group, were distinguishable by RFLPs, restriction fragment length polymorphisms (unpubl.).

The thirteen strains of *Rhizobium* isolated from *L. leucocephala* fell into five groups according to plasmid profile. The strains varied in the number (two to five plasmids) and in size (75 MDa to about 800 MDa) of the plasmids they contained (Table 3, Fig. 3). Plasmids were not detected in strains DS 65 and DS 78. Differ-

Table 2. Reference strains and plasmids

Strains/plasmid	Plasmid sizes (MDa)	Reference or source
R. leguminosarum biovar *phaseoli*		
DB1	583, 248, 166	Berryhill et al., 1985
DB1ES[c]	583, 248, 225[a], 166	E. Schroder, Univ. of Puerto Rico
Escherichia coli		
PRC357 (pTP116)	143	Hansen and Olsen, 1978
Pseudomonas aeruginosa		
PAO2(pMG5)	280	Hansen and Olsen, 1978
Rhizobium fredii		
USDA 205	~ 1000[a], 248, *127*[b] 83	Heron and Peuppke, 1984 and this study

[a]The estimated sizes of these plasmids were determined by extrapolation from a curve derived from the known molecular weights of the other reference plasmids.
[b]We determined that this plasmid carried *nif* genes.
[c]This strain was evidently a spontaneous variant of DB1.

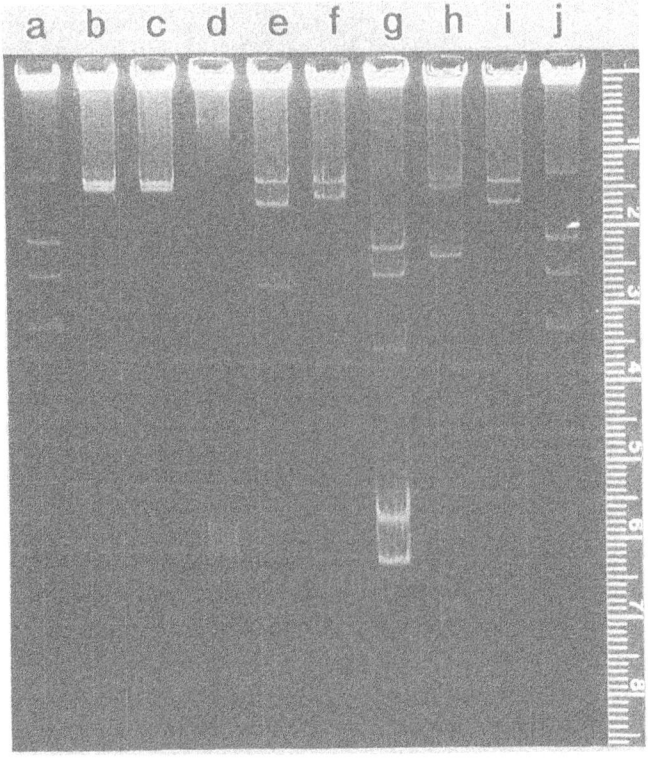

Fig. 1. Plasmid profiles of *R. meliloti* strains: a and j) reference strain *R. fredii* USDA 205; b) ARC 1; c) ARC 103; d) ARC 104; e) ARC B3; f) ARC B7; g) USDA 1011; h) USDA 1021a; and i) USDA 1025.

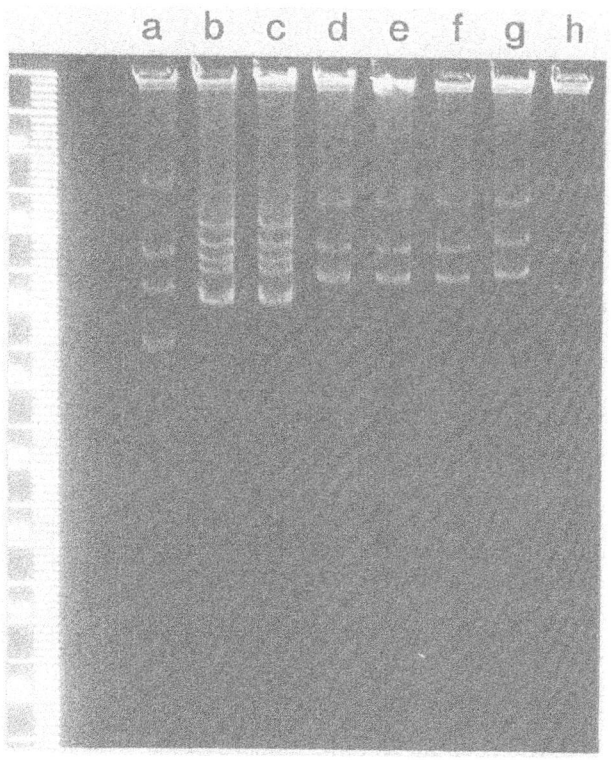

Fig. 2. Plasmid profiles of *R. leguminosarum* biovar *trifolii* strains: a and h) reference strain *R. fredii* USDA 205; b) ARC 100; c) ARC 101; d) USDA 2101; e) USDA 2128; f) USDA 2129; and g) USDA 2131.

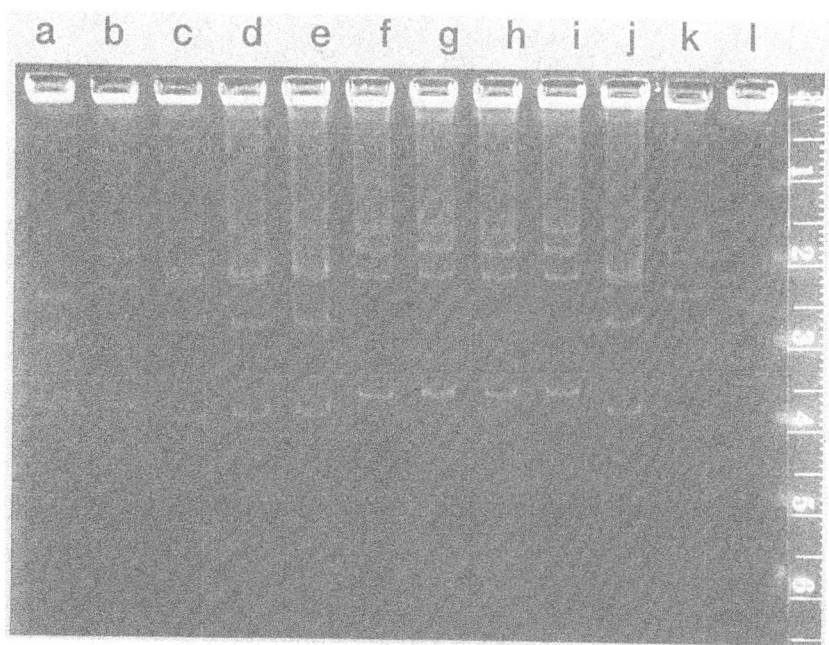

Fig. 3. Plasmid profiles of *Rhizobium* sp. *(Leucaena)* strains: a) reference strain *R. fredii* USDA 205; b) DS 9; c) DS 91; d) D6 93: e) DS 144/1; f) DS 3; g) DS 124; h) DS 129; i) DS 137; j) DS 43; k) DS 158; and l) TAL 583.

Table 3. Plasmid profiles, estimated molecular weights, and *nif*[+] Sym plasmids of *Rhizobium* strains

Strain	Plasmid number, code, and size (MDa)[a]							
	NO.	a	b	c	d	e	f	g
R.meliloti								
USDA 1011[b]	6	1000	200	140	60	<50	<50	
USDA 1021[a]	5	1000	800	**600**[c]	185	155		
USDA 1025	2	700	**430**					
ARC 1	2	800	**615**					
ARC 103	2	800	**615**					
ARC 104[b]	1	>1600						
ARC B3	3	800	**430**	120				
ARC B7	2	800	**500**					
R. leguminosarum								
biovar *trifolii*								
USDA 2101	3	**500**	220	150				
USDA 2128	3	**500**	220	150				
USDA 2129	3	**500**	220	150				
USDA 2131	3	**500**	220	150				
ARC 100	7	310	245	**200**	180	125	115	75
ARC 101	7	310	245	**200**	180	125	115	75
Rhizobium sp.								
(*Leucaena*)								
TAL 583	2	**200**	100					
DS 3	4	800	470	**290**	90			
DS 9	4	800	470	**290**	90			
DS 43	5	800	325	**290**	170	75		
DS 65	none detected							
DS 78	none detected							
DS 91	5	800	325	**290**	170	75		
DS 93	5	800	325	**290**	170	75		
DS 124	4	800	470	**290**	90			
DS 129	4	800	470	**290**	90			
DS 137	4	800	470	**290**	90			
DS 144/1	5	800	325	**290**	170	75		
DS 158	3	800	430	**230**				

[a]Plasmid sizes are estimates made from more than one gel.

[b]Plasmids in these two strains did not hybridize with the *nifHD* gene probe.

[c]Boldface type and underlining indicates plasmids that hybridized to the *nifHD* gene probe, which are Sym plasmids.

ing plasmid profile groups apparently correlated with geographical origins (Table 1). The reference strain TAL 583 did not have any megaplasmids, but did have a 100 and a 200-MDa plasmid.

Location of symbiotic genes on plasmids

Only one plasmid band per strain hybridized with radiolabeled *nifHD* probe when the agarose gels used to visualize plasmid bands (Figures 1–3) were subjected to Southern blot analysis (not shown). The huge megaplasmid of *R. meliloti* strain ARC 104 did not hybridize with the probe nor did any of the plasmids of strain USDA 1011.

R. leguminosarum biovar *trifolii* strains USDA 2101, 2128, 2129, and 2131 had a 500-MDa megaplasmid that was identified as the *nif*-carrying symbiotic plasmid, but in strains ARC 100 and 101, a 200-MDa plasmid had *nifHD* homology. The *R. meliloti nif*-carrying megaplasmids ranged in size from 430 to 615 MDa. Eleven *Rhizobium* sp.(*Leucaena*) strains from Egypt evidently carried the same 290-MDa *nif*-hybridizing plasmid, except for strain DS 158 which had a 230-MDa *nif* plasmid. Strain TAL 583 had *nif* on a 200-MDa plasmid. All Egyptian strains, except strains DS 65 and DS 78, had *nif* genes localized on either a 230-MDa or a 290-MDa plasmid and also had cryptic 800-MDa megaplasmids, however strains varied in their content of several other plasmids.

Discussion

This investigation is the first to document genetic diversity among inoculant *Rhizobium* strains for alfalfa and berseem clover in Egypt. Also, this is the first report identifying plasmids and megaplasmids of *Rhizobium* strains isolated from *L. leuocephala* in Egypt. Studies examining diversity among legume root nodule bacteria have shown that natural rhizobial populations are comprised of a multitude of strains. Variation in the plasmid profiles of individual *Rhizobium* species has been reported (Cadahia et al., 1986; Glynn et al., 1985; Mozo et al., 1988; Young and Wexler, 1988). Even within a given serogroup, considerable variation in plasmid profile of field isolates of *Rhizobium leguminosarum* bv. *vicaeae* has been documented (Brockman and Bezdicek, 1989). The results of the present study clearly document diversity of plasmid profiles among 27 strains of the several different *Rhizobium* species that nodulate berseem clover, *Leucaena*, or alfalfa. Therefore, plasmid profile analysis was shown to be a valuable tool for demonstrating genetic variation among strains of different *Rhizobium* species.

Since plasmid profiles can be more discriminating than insertion sequence (IS) fingerprints (Hartmann and Amarger, 1991), one goal of this investigation was to compare the plasmid profiles and to identify the *nif* plasmids of several agronomically important *Rhizobium* species. Plasmid profiles were correlated with but were generally less discriminating than RFLP analysis (unpubl.). This study provided some indication that geographic origin and plasmid profile may be correlat-

ed; and, to further explore this phenomenon, we plan to examine additional strains.

The information obtained in this study on plasmid content and identification of *nif*-carrying plasmids of several *Rhizobium* species will be useful in plasmid transfer studies. The transfer of the symbiosis-controlling plasmid from a very effective symbiont to another which is adapted to a particular geographic region, such as one that is semi-arid, could result in an improved inoculant strain. Plasmid transfer can also be used to determine whether some plasmids control bacterial phenotypes essential to geographic adaption to stress conditions such as drought, high temperature, or salinity. If so, it would be desirable to transfer such plasmids to other, symbiotically superior *Rhizobium* strains. Plasmid characterization could thus be used ultimately for the genetic improvement of symbiotic nitrogen fixation under climate stress conditions.

Acknowledgements

We thank Michael Behler for technical assistance, Eduardo Schroder and Mary Voll for suggestions, and Gary Bauchan for reading the manuscript. We especially thank Diaa Swelim, Samir Abdel-Wahab, Harold Keyser, David Berryhill and Peter van Berkum for providing *Rhizobium* strains.

References

Baldani J I and Weaver R W 1992 Survival of clover rhizobia and their plasmid-cured derivatives in soil under heat and drought stress. Soil Biol. Biochem. 24, 737–742.

Banfalvi Z, Kondorosi E and Kondorosi A 1985 *Rhizobium meliloti* carries two megaplasmids. Plasmid 13, 129–138.

Banfalvi Z, Sakanyan V, Konez C, Kiss A, Dusha Z and Kondorosi A 1981 Location of nodulation and nitrogen fixation genes on a high molecular weight plasmid of *Rhizobium meliloti*. Mol. Gen. Genet. 184, 318–325.

Barbour W M and Elkan G H 1989 Relationship of the presence and copy number of plasmids to exopolysaccharide production and symbiotic effectiveness in *Rhizobium fredii* USDA 206. Appl. Environ. Microbiol. 55, 813–818.

Barbour W M, Mathis J N and Elkan G H 1985 Evidence for plasmid- and chromosome-borne multiple *nif* genes in *Rhizobium fredii*. Appl. Environ. Microbiol. 50, 41–44.

Beringer J E and Hirsch P R 1984 The role of plasmids in microbial ecology. *In* Current perspectives in Microbial Ecology. Eds. M J Klug and C A Reddy. pp 63–70. American Society for Microbiology, Washington, DC, USA.

Berryhill D L, Schroder M B and Obermiller T L 1985 Plasmid contents of commercial *Rhizobium leguminosarum* biovar *phaseoli* strains. Research report No. 105. Agricultural Expirimental Station, North Dakota State University, Fargo, ND.

Beynon J L, Beringer J E and Johnston A W B 1980 Plasmids and host range in *Rhizobium leguminosarum* and *Rhizobium phaseoli*. J. Gen. Microbiol. 120, 421–429.

Brockman F J and Bezdicek D F 1989 Diversity within serogroups of *Rhizobium leguminosarum* biovar *vicaeae* in the Palouse region of eastern Washington as indicated by plasmid profiles, intrinsic antibiotic resistance and topography. Appl. Environ. Microbiol. 55, 109–115.

Brom S, Garcia de los Santos A, Stepkowsky T, Flores M, Davila G, Romero D and Palacios R 1992 Different plasmids of *Rhizobium leguninosarum* bv. *phaseoli* are required for optimal symbiotic performance. J. Bacteriol. 174, 5183–5189.

Buendia-Claveria A M, Chamber M and Ruiz-Sainz J E 1989 A comparative study of the physiological characteristics, plasmid content and symbiotic properties of different *Rhizobium fredii* strains. System. Appl. Microbiol. 12, 203–209.

Cadahia E, Leyva A and Ruiz-Argueso T 1986 Indigenous plasmids and cultural characteristics of rhizobia nodulating chickpeas (*Cicer arietinum* L.). Arch. Microbiol. 146, 239–244.

Chen H, Gartner E and Rolfe B G 1993 Involvement of genes on a megaplasmid in the acid-tolerant phenotype of *Rhizobium leguminosarum* biovar *trifolii*. Appl. Environ. Microbiol. 59, 1058–1064.

Djordjevic M A, Zurkowski W and Rolfe B G 1982 Plasmids and stability of symbiotic properties of *Rhizobium trifolii*. J. Bacteriol. 151, 560–568.

Eckhardt T 1978 A rapid method for the identification of plasmid deoxyribonucleic acid in bacteria. Plasmid 1, 584–588.

Farrai T, Vincze E, Banfalvi Z, Kiss G B, Randhawa G S and Kondorosi A 1983 Localization of symbiotic mutations in *Rhizobium meliloti*. J. Bacteriol. 153, 635–643.

Glynn P, Higgins P, Squartini A and O'Gara F 1985 Strain identification in *Rhizobium trifolii* using DNA restriction analysis, plasmid DNA profiles and intrinsic antibiotic resistance. FEMS Microbiol. Lett. 30, 177–182.

Hansen J B and Olsen R H 1978 Isolation of large bacterial plasmids and characterization of the P2 incompatibility group plasmids pMG1 and pMG5. J. Bacteriol. 135, 227–236.

Harrison S P, Jones D G, Schunmann P H D, Forster J W and Young J P W 1988 Variation in *Rhizobium leguminosarum* biovar *trifolii* Sym plasmids and the association with effectiveness of nitrogen fixation. J. Gen. Microbiol. 134, 2721–2730.

Harrison S P, Jones D G and Young J P W 1989 *Rhizobium* population genetics: genetic variation within and between populations from diverse locations. J. Gen. Microbiol. 135, 1061–1069.

Harrison S P, Young J P W and Jones D G 1989 *Rhizobium* population genetics: host preference and strain competition effects on the range of *Rhizobium leguminosarumm* biovar *trifolii* genotypes isolated from natural populations. Soil Biol. Biochem. 21, 981–986.

Hartmann A and Amarger N 1991 Genotypic diversity of an indigenous *Rhizobium meliloti* field population assessed by plasmid profiles, DNA fingerprinting, and insertion sequence typing. Can. J. Microbiol. 37, 600–608.

Heron S H, Pueppke S G 1984 Mode of infection, nodulation specificity, and indigenous plasmids of eleven fast-growing *Rhizobium japonicum* strains. J. Bacteriol. 160, 1061–1066.

Hirch P R, van Montagu M, Johnston A W B, Brewin N J and Schell J 1980 Physical identification of bacteriocinogenic, nodulation and other plasmids in strains of *Rhizobium leguminosarum*. J. Gen. Microbiol. 120, 403–412.

Hombrecher G, Brewin N J and Johnston A W B 1981 Linkage of genes for nitrogenase and nodulation ability on plasmids in *Rhizobium leguminosarum* and *Rhizobium phaseoli*. Mol. Gen. Genet. 182, 133–136.

Hooykaas P J J, Snijdewint F G M and Schilperoort R A 1982 Identification of the Sym plasmid of *Rhizobium leguminosarum* strain 1001 and its transfer to and expression in other rhizobia and *Agrobacterium tumefaciens*. Plasmid 8, 73–82.

Hynes M F and McGregor N F 1990 Two plasmids other than the nodulation plasmid are necessary for formation of nitrogen-fixing nodules by *Rhizobium leguminosarum*. Mol. Microbiol. 4, 567–574.

Hynes M F, Simon R, Muller P, Niehaus K, Labes M and Puhler A 1986 The two megaplasmids of *Rhizobium meliloti* are involved in the effective nodulation of alfalfa. Mol. Gen. Genet. 202, 356–362.

Hynes M F, Simon R and Puhler A 1985 The development of plasmid-free strains of *Agrobacterium tumefaciens* by using incompatibility with a *Rhizobium meliloti* plasmid to eliminate pAtC58. Plasmid 13, 99–105.

Krol A J M, Hontelez J G J and van Kammen A 1982 Only one of the large plasmids in *Rhizobium leguminosarum* strain PRE is stongly espressed in the endosymbiotic state. J. Gen. Microbiol. 128, 1839–1847.

Kuykendall L D and Hengen P N 1988 Microbial genetics of legume root nodulation and nitrogen fixation. *In* Recent advances in Symbiotic Nitrogen Fixation. Ed. N S Subba Rao. pp 71–112. Oxford Press, New Delhi, India.

Martinez E, Romero D and Palacios R 1990 The *Rhizobium* genome. Crit. Rev. Plant Sci. 9, 59–93.

Mozo T, Cabrera E and Ruiz-Argueso T 1988 Diversity of plasmid profiles and conservation of symbiotic nitrogen fixation genes in newly isolated *Rhizobium* strains nodulating sulla (*Hedysarum coronarium* L.). Appl. Environ. Microbiol. 54, 1262–1267.

Plazinski J, Cen Y H and Rolfe B G 1985 General method for the identification of plasmid species in fast-growing soil microorganisms. Appl. Environ. Microbiol. 48, 1001–1003.

Prakash R K and Atherly A G 1986 Plasmids of *Rhizobium* and their role in symbiotic nitrogen fixation. Int. Rev. Cytol. 104, 1–24.

Rosenberg C, Boistard P, Denarie J and Casse-Delbart F 1981 Genes controlling early and late functions in symbioses are located on a megaplasmid in *Rhizobium meliloti*. Mol. Gen. Genet. 184, 326–333.

Sambrook J, Fritsch E F and Maniatis T 1989 Molecular cloning: a laboratory manual. Cold Spring Harbor Laboratory, Cold Spring Harbor, NY.

Selenska-Trajkowa S, Radewa G and Markov K 1990 Comparison between *Rhizobium galegae* and *Rhizobium meliloti* plasmid contents. Lett. Appl. Microbiol. 10, 123–126.

Shishido M and Pepper L L 1990 Identification of dominant indigenous *Rhizobium meliloti* by plasmid profiles and intrinsic antibiotic resistance. Soil Biol. Biochem. 22, 11–16.

Thurman N P, Lewis D M and Jones D G 1985 The relationship of plasmid number to growth, acid tolerance and symbiotic efficiency in isolates of *Rhizobium trifolii*. J. Appl. Bacteriol. 58, 1–6.

Weaver R W, Wei G R and Berryhill D L 1990 Stability of plasmids in *Rhizobium phaseoli* during culture. Soil Biol. Biochem. 22, 465–469.

Young J P W and Wexler W 1988 Sym plasmid and chromosomal genotypes are correlated in field populations of *Rhizobium leguminosarum*. J. Gen. Microbiol. 134, 2731–2739.

Zahran H H 1992 Characterization of root-nodule bacteria indigenous in the salt-affected soils of Egypt by lipopolysaccharide, protein and plasmid profiles. J. Basic Microbiol. 32, 279–287.

P.H. Graham, M.J. Sadowsky & C.P. Vance (eds.), Symbiotic nitrogen fixation, 189–202.
© 1994 Kluwer Academic Publishers.

International FAO/IAEA programmes on biological nitrogen fixation

Gudni Hardarson
FAO/IAEA Programme, Agency's Agriculture Laboratory, A-2444 Seibersdorf, Austria

Key words: Azolla, isotope dilution, legumes, ^{15}N, nitrogen fixation, *Rhizobium*

Abstract

The Food and Agriculture Organization of the United Nations and the International Atomic Energy Agency have through their Joint FAO/IAEA Division in Vienna and the Agriculture Laboratory in Seibersdorf, Austria coordinated international programmes on biological nitrogen fixation (BNF) in developing countries for more than two decades. The main objectives of these programmes have been to enhance nitrogen fixation in various cropping systems and to develop and optimize the ^{15}N isotope dilution method to quantify N_2 fixation in leguminous crops. In connection to the various FAO/IAEA programmes fellowships and training courses are being offered.

Experiments, conducted as part of Coordinated Research Programmes, are performed in a number of countries simultaneously and, therefore, produce results applicable to a wide range of environmental conditions. Due to this replication of experiments, solutions to problems can often be found in one season. Another feature of the FAO/IAEA Coordinated Research Programmes is the transfer of research techniques to developing countries through the Research Co-ordination Meetings and workshops. Volunteer experts in the field of study help the FAO/IAEA staff with this task.

The BNF programmes initiated some twenty years ago focused on grain legumes and the development of ^{15}N methodology to measure BNF. These programmes were followed by others in which BNF in forage, pasture legumes, *Azolla* and tree legumes was quantified. The most recent programmes emphasize the enhancement of nitrogen fixation in some grain legume species through genetic improvement. Selected results from the above programmes will be presented.

Introduction

In 1964, the Food and Agriculture Organization of the United Nations and the International Atomic Energy Agency established a Joint FAO/IAEA Division responsible for food and agricultural research and development involving isotopes and radiation techniques. The Joint FAO/IAEA Division is located in Vienna and its associated laboratory is approximately 40-km south of Vienna at Seibersdorf. The programmes of the Joint FAO/IAEA Division operate through six disciplinary Sections and corresponding Units at the Laboratory: Soil Fertility, Irrigation and Crop Production, Plant Breeding and Genetics, Animal Production and Health, Agrochemicals and Residues, Insect and Pest Control and Food Preservation.

Coordinated Research Programmes (CRPs)

At present, approximately 400 universities and research institutes in FAO and IAEA Member States participate in some 40 Co-ordinated Research Programmes (CRP). Each CRP aims to identify and resolve problems affecting agricultural production in developing countries. The research conducted by CRP participants is directed towards increasing crop and livestock production. Scientists from 10–15 institutes participate in each programme as funded contractors from developing countries or as cost-free agreement holders from developed countries or International Institutes. Researchers in these programmes form teams with links that sometime continue long after the programmes are over. Each programme lasts about 5 years.

The specific aims of each CRP are defined by the various Sections of the Joint FAO/IAEA Division, which coordinate the programmes. All participants from developed and developing countries are encouraged to collaborate and produce results that will benefit the inhabitants of FAO/IAEA Member States. The programmes very often provide incentive to work as results have to be reported each year to the research group. Technologies developed must be affordable by the farmers in those countries and be sustainable.

Research Co-ordination Meetings, funded by the FAO/IAEA, are held to monitor the progress of research, share ideas and responsibilities, to plan future work and finally to report on results and recommend future work. These meetings help focus and clarify objectives, for the researchers themselves as well as for the FAO/IAEA. The results and conclusions are usually published by the FAO/IAEA and individual participants are encouraged to publish their own results separately.

Projects that are funded through a CRP receive an annual lump sum contract. Usually the amount awarded to each contractor is relatively small (US$ 5000/year) but this is usually enough to conduct some important and often difficult research. Scientists in developing countries receiving support from the FAO/IAEA programme often obtain additional support from their institutes or other external sources.

Other activities coordinated by the Joint FAO/IAEA Division are the so called Technical Co-operation Projects (TCP). The main objectives of these projects are to improve the physical laboratory infrastructure, training of scientists and the transfer of technology through expert services to Member States.

International programmes on biological nitrogen fixation

In spite of the abundance of nitrogen in the atmosphere it is one of the most limiting factors for crop growth, and nitrogen fertilizer represents one of the major costs of crop production. BNF is a feasible alternative for farmers in both developing and developed countries. The main advantages are: reduced cost of crop production, less groundwater pollution, enhanced protein production - as most legumes have higher protein content than cereals, more residual N for succeeding crops, and the build up of soil fertility.

Great emphasis has been placed on BNF in the FAO/IAEA Programme due to the importance of this process in developing countries and the unique use of ^{15}N to quantify N_2 fixation.

There are several methods to quantify N_2 fixation by leguminous, crops; e.g. 1) the N difference or N balance method (Peoples et. al., 1989); 2) acetylene reduction assay (Bergersen, 1980); 3) xylem solute technique (Peoples et al., 1989); and 4) ^{15}N methodologies. The FAO/IAEA Programme has used many of these methods but has emphasized the development and use of the ^{15}N isotope dilution method to quantify biological nitrogen fixation (Hardarson, 1990; Hardarson and Danso, 1993). The work by McAuliffe et al. (1958) on the use of ^{15}N to quantify nitrogen fixation by isotope dilution paved the way for development of the "A-value" method (Fried and Broeshart, 1975) and the use of the ^{15}N isotope dilution method under field conditions.

Fertilization of grain legumes, 1972–1977

^{15}N isotope methods were tested and further developed during the first FAO/IAEA programme on BNF (1972–1977), whose main objective was to study fertilization of grain legumes (Table 1). The ^{15}N isotope dilution method involves the growth of N_2 fixing and non-fixing reference plants in soil labelled with ^{15}N enriched inorganic or organic fertilizers. It is based on differential dilution in the plants of ^{15}N tracer by soil and fixed nitrogen (Fried and Middelboe, 1977; Mc Auliffe et al., 1958). A N_2 fixing plant will have a lower ^{15}N enrichment as compared to the non-fixing reference plant due to assimilation of unlabelled N_2 from the air.

Without N_2 fixation both will have the same enrichment. This methodology provides an integrated measurement of the amount of fixed N_2 accumulated by a crop over the growing season. Calculation of % N derived from the atmosphere (%Ndfa) can be made by the following equation:

$$\%Ndfa = \left(1 - \frac{\%Ndff_F}{\%Ndff_{NF}} \right) \times 100$$

where $\%Ndff_F$ and $\%Ndff_{NF}$ are % N derived from fertilizer or tracer by fixing (F) and non-fixing (NF) plants, respectively. In the above equation both fixing and non-fixing plants receive the same amount and enrichment of ^{15}N.

A modification of the isotope dilution method is the so called "A-value" method (Fried and Broeshart, 1975), where the F and NF crops receive different amounts of fertilizer or tracers. In this case a second

Table 1. Coordinated Research Programmes on biological nitrogen fixation which have been conducted by the Soil Fertility, Irrigation and Crop Production Section of the Joint FAO/IAEA Division

Programme Short Title (FAO/IAEA Project Officer)	Duration	Participating countries
a) Fertilization of grain legumes, (Nethsinghe)	1972–1977	Australia[1], Brazil, Egypt, Ghana, Greece, Hungary, India, Peru, Romania, Senegal, Sri Lanka, UK[1], USA[1].
b) Grain legumes[2], (Danso)	1979–1983	Argentina, Australia[1], Bangladesh, Brazil, Egypt, France[1], Germany[1], Greece, India, Kenya, Mexico, Nigeria, Pakistan, Romania, Senegal, Sri Lanka, Syria, UK[1], USA[1].
c) Multiple cropping, (Bole, Kalinin)	1980–1985	Bangladesh, Ghana, Indonesia, Panama, Tanzania, Thailand, Turkey, USA[1], Zambia.
d) Pasture[3], (Danso)	1983–1988	Brazil, China, Colombia[1], Cyprus, Ethiopia[1], Greece, Iceland, India, Kenya, Malaysia, New Zealand[1], Spain[1], Sri Lanka, Sudan, Syria[1], Switzerland[1], UK[1], Uruguay, USA[1].
e) Azolla[2], (Kumarasinghe/Eskew)	1984–1989	Austria[1], Bangladesh, Belgium[1], Brazil, China, Hungary, Indonesia, Pakistan, Philippines, Sri Lanka, Sweden[1], Thailand, USA[1].

(*continued next page*)

equation has to be used:

$$\%\text{Ndfa} = 100 \left(1 - \frac{\%\text{Ndff}_F}{n\%\text{Ndff}_{NF}} \right) + \%\text{Ndff}_F \left(\frac{1}{n} - 1 \right)$$

where n is the relative amount of fertilizer applied, i.e. n = amount of N fertilizer applied to the F crop divided by the amount applied to the NF crop (Hardarson et al.,

1991). This method was originally presented using the "A-value"concept, which involves slightly different equations (Fried and Broeshart, 1975).

Table 1. continued

Programme Short Title (FAO/IAEA Project Officer)	Duration	Participating countries
f) Common bean in Latin America, (Hardarson)	1986–1991	Brazil, Chile Colombia[1], Guatemala Mexico, Peru, USA[1].
g) Grain legumes in Asia[4] (Danso)	1987–(1994)	Australia[1], Bangladesh, China, Pakistan, Philippines, India[1] Malaysia, Sri Lanka, Thailand, Vietnam.
h) Tree legumes, (Danso)	1989–(1995)	Chile, Cameroon Canada[1], China, Ghana, Malaysia, Nigeria Pakistan, Philippines, Senegal, Sri Lanka, Zaire.
i) Micorbial ecology, (Hardarson)	1992–(1997)	Brazil, Egypt, Germany[1], Ireland[1], Mexico, The Netherlands[1], Pakistan, Philippines, Romania, Switzerland[1], Thailand, UK[1].

[1]Cost free Agreement holders from developed countries or International Institutes.

[2]Funded by the Swedish International Development Authority.

[3]Funded by the Government of Italy.

[4]Funded by the United Nations Development Programme.

Grain legumes, 1979–1983

The second FAO/IAEA programme on BNF (Table 1) was on grain legumes (1979–1983). The main objectives of this programme were to use the ^{15}N methodology to quantify nitrogen fixation in different grain legumes in symbiosis with various rhizobial strains. Also the effect of agronomic and environmental factors on N_2 fixation were investigated. Examples of the results obtained are presented in Figures 1–4. Figure 1 illustrates the % N derived from the atmosphere (%Ndfa) in several grain legume species when tested under field conditions (Danso, personal communication). The results are averages from investigators participating in the programme (Table 1). There were large differences between the grain legume species, with some being very effective, others intermediate and some, such as the common bean, rather poor. Although grain legumes had previously been found to be variable in their N_2 fixation effectiveness no previous study had investigated this in so many countries. This is an unique aspect of many of the CRPs.

The time course of nitrogen fixation in some grain legume species were also studied as part of this and the following CRPs (Fig. 2a–d). Surprising differences in the time course of fixation were observed between species; for instance groundnut was effective only during the last part of the growing season, soybean fixed effectively during the 20 day span of the reproductive stage and faba bean was very effective over most of

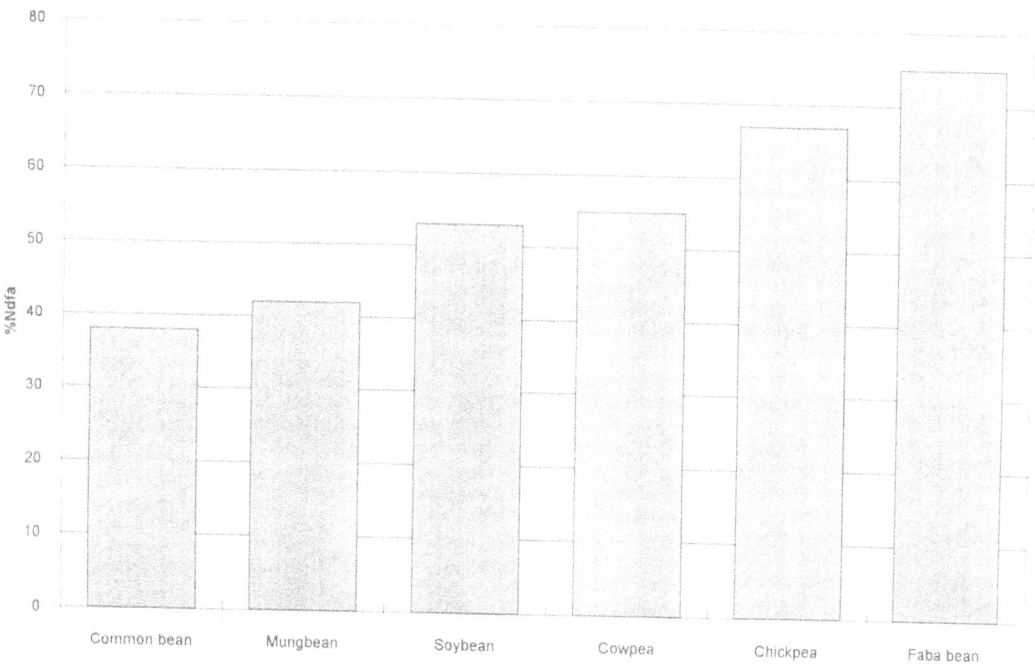

Fig. 1. Percent nitrogen derived from atmosphere (%Ndfa) by several grain legume species tested under field conditions. The results are averages from several investigators in countries mentioned in Table 1. (Danso, personal communications).

the growing season. This type of data is useful when N_2 fixation is to be enhanced in a particular legume crop. It should for example be possible to increase early nodulation and fixation in common bean, groundnut and soybean.

Other agronomical factors such as the effect of P fertilizer and rhizobial inoculation method were also investigated in this CRP (Figs 3 and 4). Such findings are important for setting future research priorities locally and for developing appropriate technologies for farmers in a specific region.

Multiple cropping, 1980–1985

The third CRP programme, whose main emphasis was on multiple cropping (Table 1), included the effect of nitrogen fixing legumes in cropping systems. The results of this programme have been summarized (IAEA, 1986). The uptake of nitrogen by cereals after a legume crop compared with continuous cereal cropping was increased from 5 to 40 kg N ha^{-1} (Fig. 5). The amount of fertilizer N which would have to be applied to the cereals to obtain the same increase in total N uptake ranged from 20 to 150 kg N ha^{-1}.

A large portion of the increase in N yield after a legume crop is likely to be due to the saving of soil nitrogen as illustrated in the study by Senaratne and

Hardarson (1988) (Fig. 6). This experiment included preceding crops of faba bean, pea and barley as well as a fallow treatment. After incorporation of organic matter (excluding grain) from the preceding crop into the soil two successive cereals were used to evaluate N derived from residue and soil. It was clear that a large part of the N in the succeeding crops was derived from soil which had been used less by the preceding legume compared to the cereal. Only a small portion of the N was derived from the legume residue.

Pasture, 1983–1988

This CRP was initiated after a consultant's meeting which was held in 1980. The proceedings of this meeting were published by the IAEA (1983). A critical problem was how to apply the ^{15}N methodology, previously used with grain legumes, to a perennial system and how to select reference crops for mixed pastures. Experiments were conducted at the Seibersdorf Laboratory to find the best method of ^{15}N application and to determine whether a non-fixing crop grown in mixture with the legume could be used as a reference (Danso et al., 1988; Hardarson et al., 1988). Very little N was found to be transferred from the fixing to the non-fixing companion crop and therefore, such a reference crop could be used as a control. Sole crops of forage or

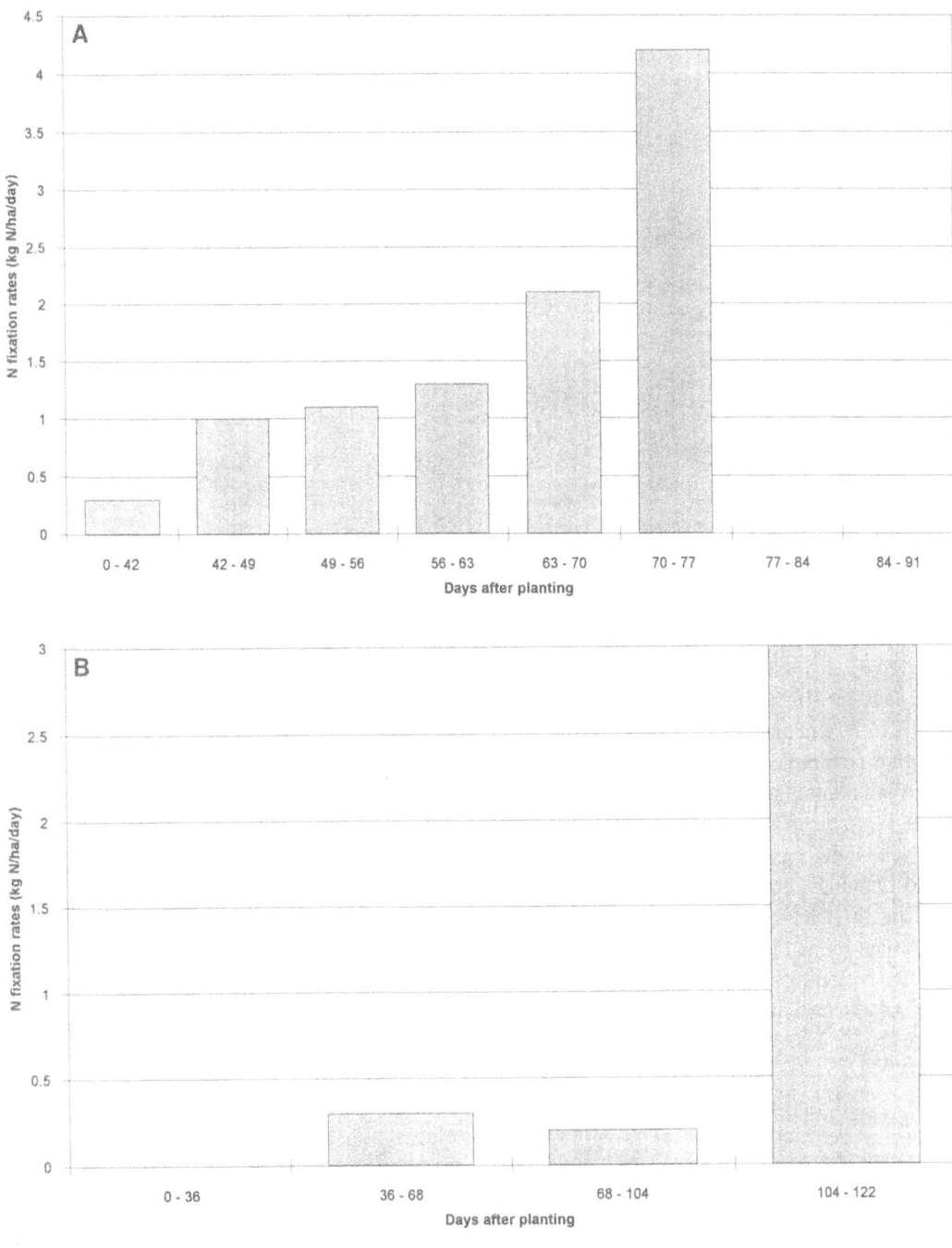

Fig. 2. Time course of nitrogen fixation under field condition by (**A**) common bean (cv. Flor de Mayo RMC) in Mexico (Peña-Cabriales et al., 1993); (**B**) groundnut (cv Ashford) in Sudan (Mukhtar et al. In press) ; (**C**) soybean (cv Chippewa) in Austria (Zapata et al., 1987a) and (**D**) faba bean (cv Wieselburger) in Austria (Zapata et al ., 1987 b).

pasture legumes usually derived higher percentages of their N from the atmosphere than grain legumes (Fig. 7, Bowen and Danso, 1987). %Nfda was therefore not limiting, but total N_2 fixed was constrained by the low

yields of most pastures. Future programmes for the enhancement of forage or pasture production should therefore concentrate on improvement of yield and the

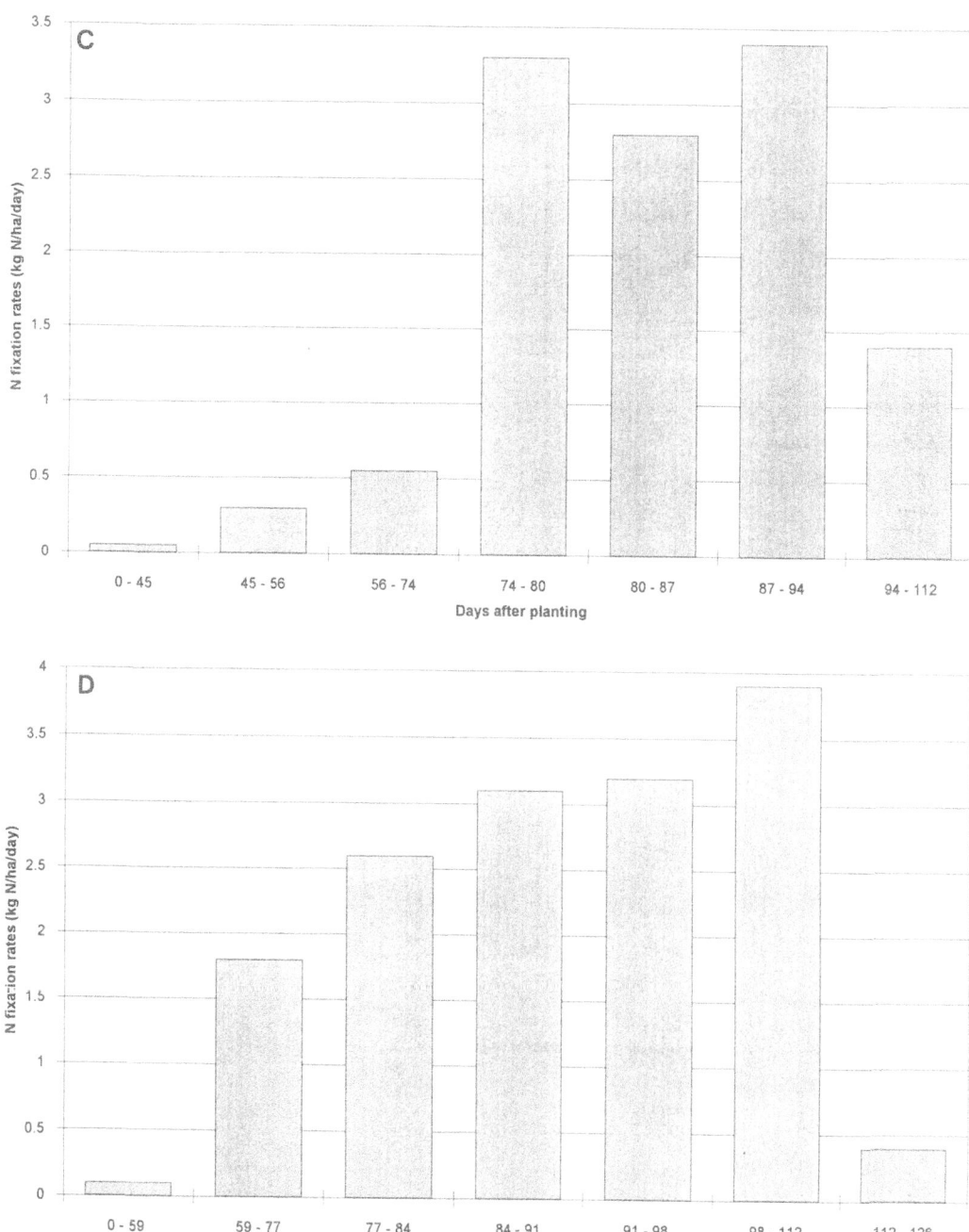

Fig. 2. (continued)

amount of nitrogen fixed rather than enhancing % N_2 derived from atmosphere.

Azolla, 1984–1989

During the five year period of 1984–1989 a CRP on *Azolla* was conducted by the FAO/IAEA (Table 1). The results of this programme have recently been published

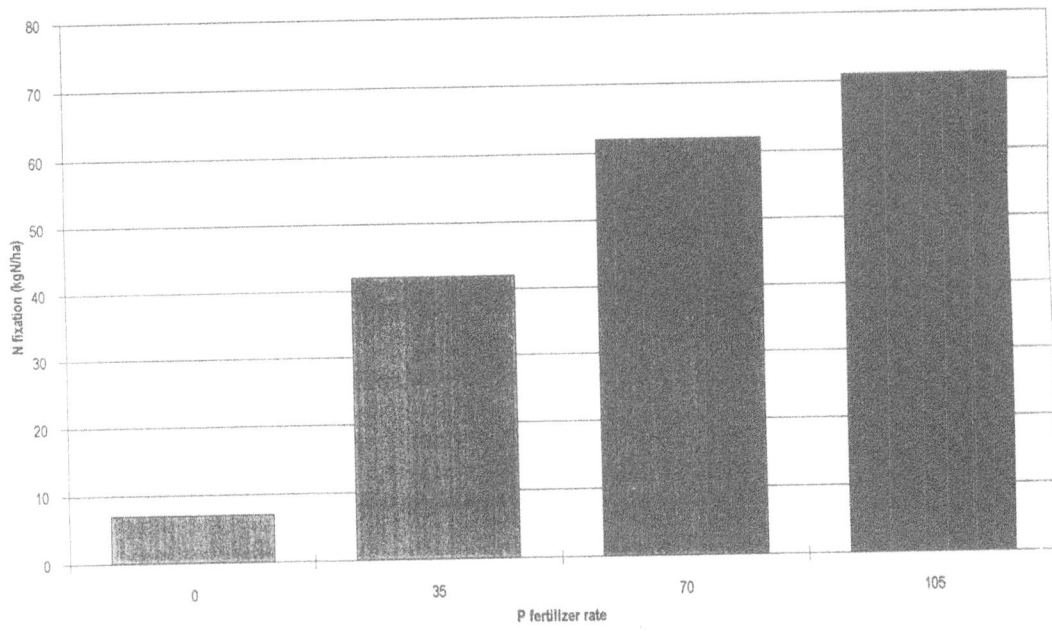

Fig. 3. Field evaluation in Hungary of nitrogen fixation in soybean (Data of J. Dombovari) as affected by P fertilizer rate (Zapata and Baert, 1989).

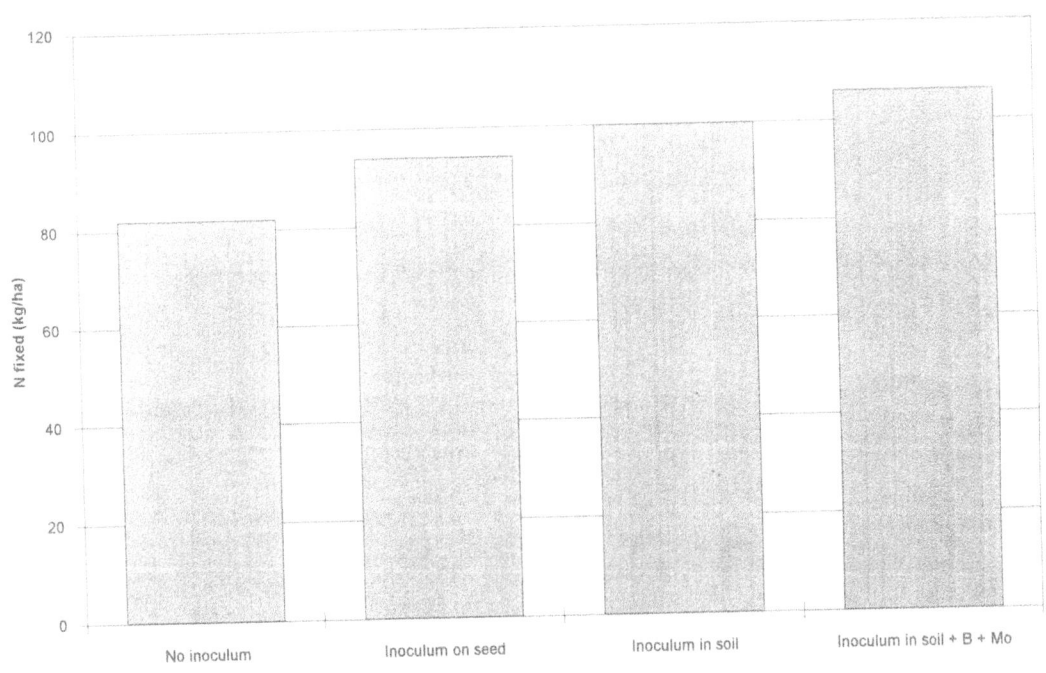

Fig. 4. Field response to rhizobial inoculation of groundnut in Ghana (Data of S. Ofori and P. Kwakye) (Zapata and Baert, 1989).

(Kumarashinge and Eskew, 1993). The main objectives of this programme were to quantify the amount of N_2 fixed by the *Azolla/Anabaena* symbiosis, to evaluate *Azolla* as a N source for rice production, and to develop management techniques to integrate *Azolla* into sustainable rice cropping systems while maintaining high yields.

Using the ^{15}N isotope dilution technique the results from this CRP showed that, on average, 70–80% of N taken up by *Azolla* was derived from the atmosphere.

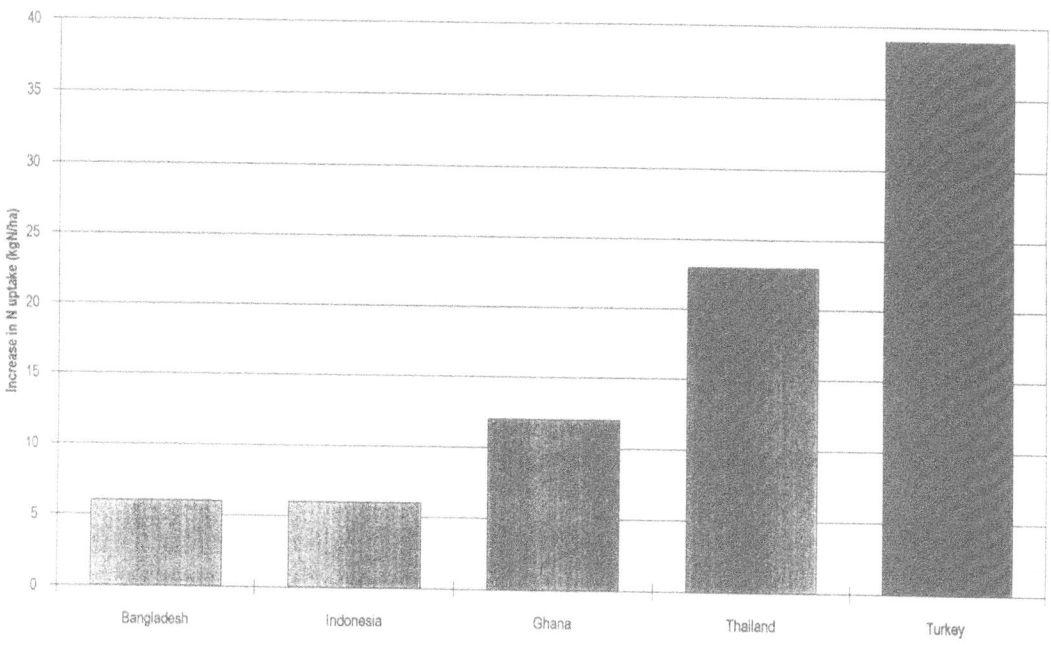

Fig. 5. Increase in N uptake in cereal following chickpea and lentil in Bangladesh, soybean in Indonesia, cowpea in Ghana, groundnut and soybean in Thailand and soybean in Turkey as compared to an identical treatment following cereal (IAEA, 1986).

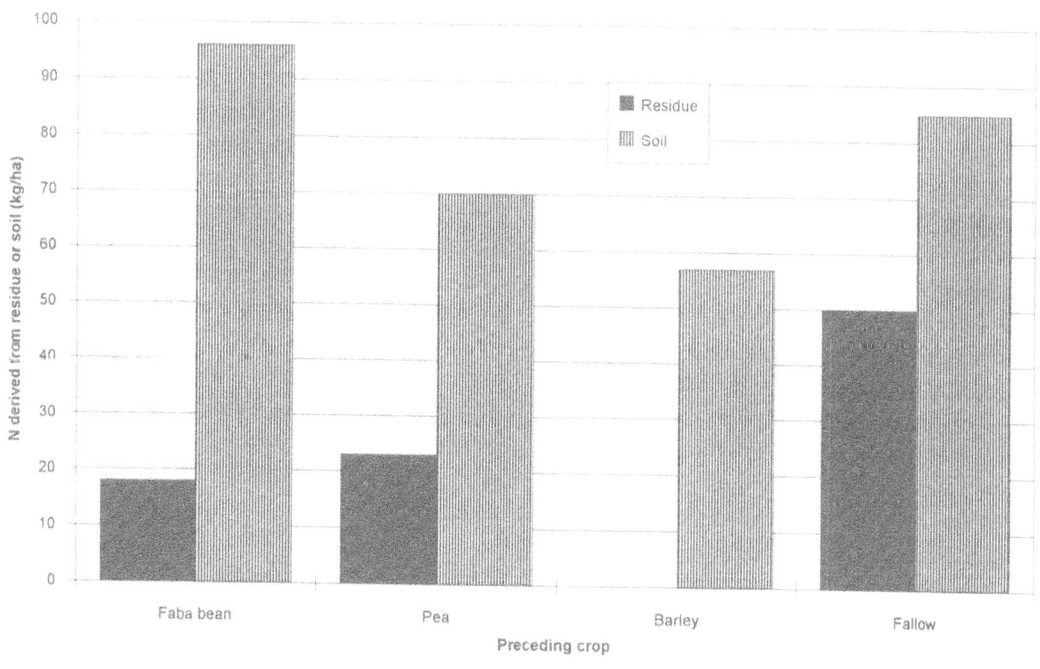

Fig. 6. Nitrogen derived from soil and the residue of faba bean, pea and a fallow treatment as measured by two succeeding cereals grown for two successive seasons. (Senaratne and Hardarson, 1988).

However, the ability to fix nitrogen varied with *Azolla* species, location and environmental conditions. In this programme the nitrogen uptake and yield of rice as affected by the addition of *Azolla* or urea fertilizer were studied using ^{15}N methodology. Results from several different locations and years showed that *Azolla* was equivalent to urea as a N source for rice, both in terms of yield increase and N uptake by rice. It was con-

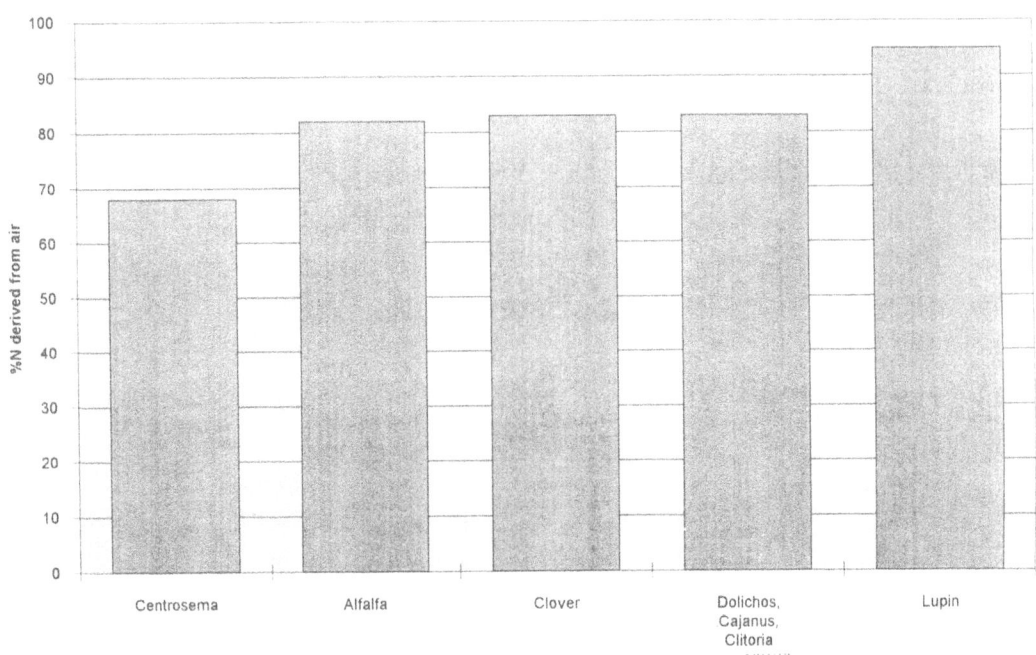

Fig. 7. Biological nitrogen fixation of several forage or pasture legume species under field conditions. The results are averages from several investigators in countries mentioned in Table 1. (Danso and Bowen, 1987).

cluded that with modest inputs of additional research, dramatic advances toward realizing the full benefits of *Azolla* as a green manure are possible (Kumarashinge and Eskew, 1993).

Common bean in Latin America 1986–1991

A consultant's meeting was held in 1983 to plan a programme to enhance nitrogen fixation in common bean. The proceedings of this meeting were published (Hardarson and Lie, 1984). In 1986–1991, an FAO/IAEA CRP was conducted with the objective of enhancing biological nitrogen fixation of common bean in Latin America (Table 1). This programme concentrated on the contribution of plant genotypes to increase N_2 fixation and it was possible to find cultivars and breeding lines that were able to support much higher fixation than usually found in common bean. The detailed summary of the results of this programme has been published by Bliss and Hardarson (1993). One example showing the range of N_2 fixation in common bean as tested in the various countries taking part in this programme is presented in Figure 8. Each participant tested approximately twenty cultivars of common bean under field conditions. Only cultivars of similar growth periods were included in each experiment. The evaluations of nitrogen fixa-

tion were made when the cultivars having the shortest growth cycle reached physiological maturity. At that growth stage all entries were harvested. The % N derived from atmosphere of the poorest and the most effective cultivars are shown in Figure 8. In most cases there was high correlation between %Ndfa and the amount of N_2 fixed (Bliss and Hardarson, 1993). The use of better fixing cultivars of common bean by farmers in Latin America could increase BNF in common bean by at least 10–20%, which is equivalent to about 10–20 kg N ha^{-1}; they would have to apply 30–60 kg N ha^{-1} as fertilizer to provide the same amount of N to the crop. Therefore this increase is economically important for the farmers in that continent.

Grain legumes in Asia, 1987–(1994)

This CRP, which has the objective to enhance nitrogen fixation in various grain legumes in Asia is still being conducted. The following grain legumes are being studied: chickpea, cowpea, lentil, mungbean, peanut and soybean. The programme started with the screening of different lines, cultivars and mutants (gamma rays induced as well as chemical mutagens) for differences in N_2 fixation and yield. Interestingly enough high N_2 fixation was not always associated with high

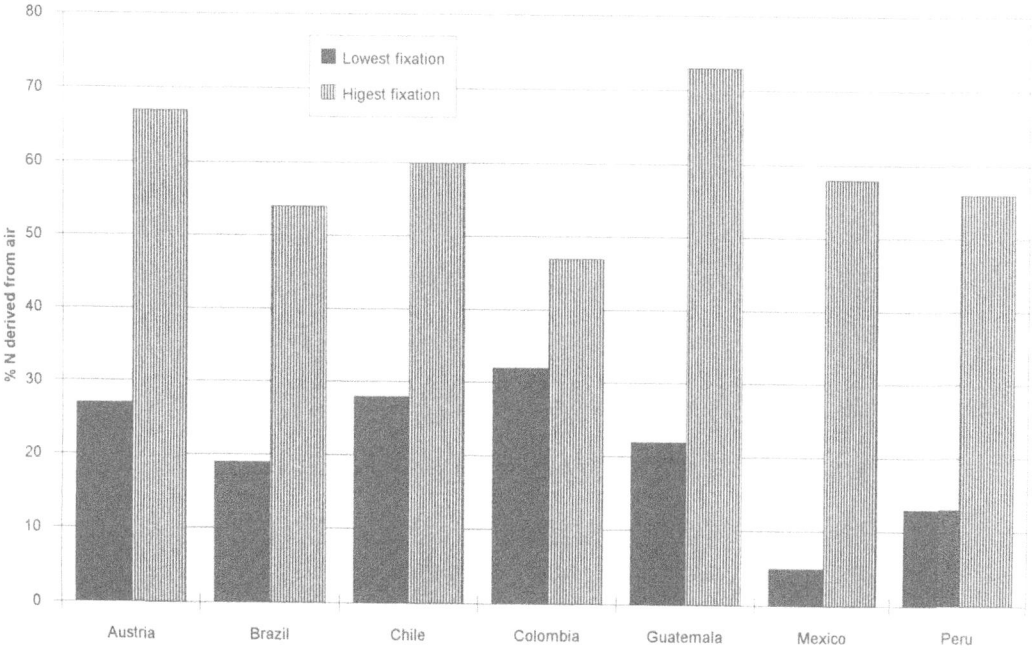

Fig. 8. Range of percentage N derived from atmosphere in common bean when evaluated under field conditions in a number of countries, (Hardarson et al., 1993).

yield, with some cultivars having exceptionally high capacity for N_2 fixation, but yielding poorly and vice versa. Initial results indicate that BNF is a heritable trait and currently lines/cultivars with high fixation are being crossed with those of high yield to obtain cultivars that combine both characteristics. The initial results are encouraging.

Tree legumes, 1989–(1995)

The objectives of the "tree legume" CRP are to examine nitrogen fixation by tree species and study their role in restoring and maintaining soil fertility (Table 1). This is an ongoing programme and initial results have demonstrated large genotypic differences in yield as well as BNF with - in some cases - 10-fold differences in N_2 fixation among provenances of a given species. Several papers have already been published, mostly from the contribution of the FAO/IAEA programme in support of this CRP (Awonaike et al., 1992; Sanginga et al. 1989, 1990a, b, 1991a, b; Sougoufara et al., 1990) Initially this CRP concentrated on adapting the ^{15}N methodology to trees. The main problems being the perennial growth and plant size, which make both labelling with ^{15}N and sampling difficult. These problems and some solutions have been reviewed by Danso et al. (1992).

Microbial ecology, 1992–(1997)

The CRP on "microbial ecology" was recently initiated as a follow up to previous programmes which concentrated on macro-symbionts. Most of the work performed under this programme will be on the micro-symbiont and its ecology. Previous research has reported limited movement of rhizobia in soil and the rhizosphere and emphasized the importance of nodulation of lateral roots to maximize N_2 fixation (Hardarson et al., 1989 ; McDermott and Graham, 1989; Wolyn et al., 1989). It is therefore obvious that only through careful study of the establishment of microbial populations and their migration in the soil will it be possible to optimize root nodulation of nitrogen fixing crops. The methods used to identify rhizobial strains will be both conventional ones as well as those employing molecular biology techniques, such as the use of β-glucuronidase (GUS) marker gene (Sessitch et al., submitted; Wilson et al., In press).

Training

Since 1978 the FAO/IAEA Programme and the IAEA Department of Technical Co-operation has conducted annual interregional training courses at the Seibersdorf Laboratory and several regional courses on the

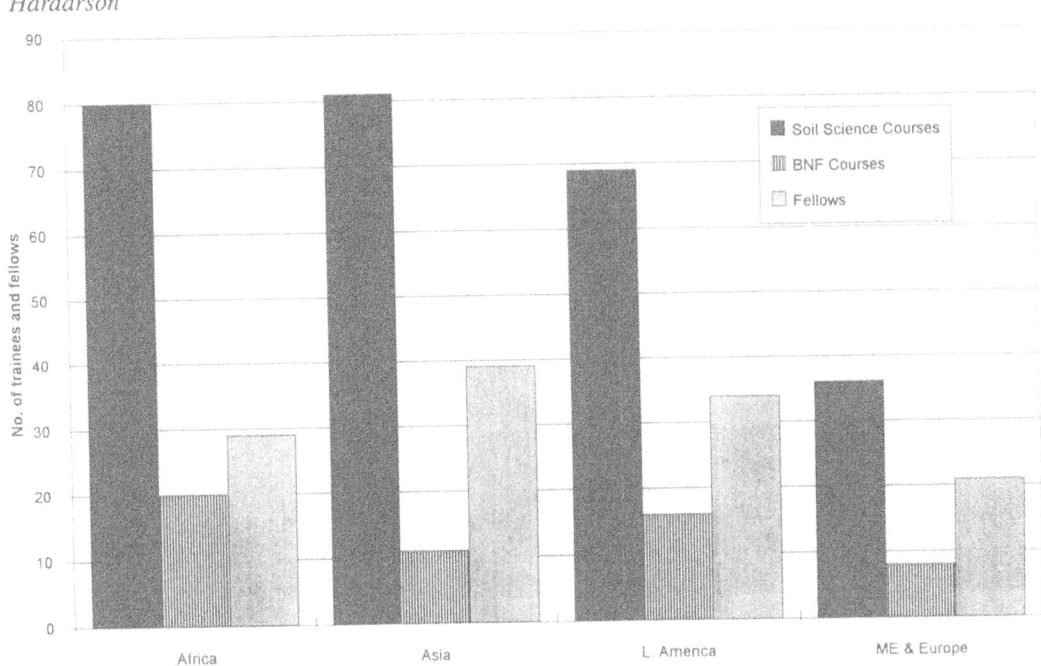

Fig. 9. Numbers of trainees and fellows which have participated in the FAO/IAEA Soils Science training programme during the last two decades.

use of nuclear techniques in studies of soil-plant relationships. Three of these courses were on BNF and all of them had sessions on this subject. The objective has been to give scientists from developing countries a sound working knowledge in the use of ^{15}N to quantify BNF by the legume/*Rhizobium* symbiosis. Other conventional techniques were also covered. A training manual used in these courses was recently published (Hardarson, 1990).

Each training course included approximately twenty participants from all regions (Fig. 9). The average length of training courses is five weeks. Invited lecturers and FAO/IAEA staff gave lectures and practical exercises during these courses.

In addition to training courses the FAO/IAEA conducts fellowship training for scientists from developing countries. The fellows learn to do isotope analyses or perform a research project under the supervision of FAO/IAEA staff. The period of training is from three to twelve months. The Soil Science Unit at the Seibersdorf Laboratory hosts about ten fellows every year. Many of the IAEA fellows are trained in other laboratories around the world.

Support services

In addition to research and training the Soil Science Unit of the Seibersdorf Laboratory performs other support services for the CRP and TC programmes. The largest part of this work is isotope analyses. 15 to 20 thousand ^{15}N analyses are performed every year mostly for projects in developing Member States. It is important for the successful implementation of the CRP and TC programmes that scientists in developing countries, who are initiating work using ^{15}N methodology, can have their samples analyzed. The IAEA can also, through a TC project, support the building of analytical facilities in Member States by supplying emission or mass spectrometers.

Other support services include dispatching of tracers and expert services to projects in developing countries.

Participation in programmes

When a CRP is initiated it is advertised in the Soils Newsletter which is published regularly by the IAEA and sent cost free to subscribers. It is then possible for institutes in developing countries to send Project Proposals to the FAO/IAEA for consideration.

Applications for TCP, training courses or fellowships should be endorsed by and forwarded through the official established channel (the Ministry of Foreign Affairs, the National Atomic Energy Authority, the office of the United Nations Development Programme or the Ministry of Agriculture). Application forms are available from the IAEA.

Acknowledgements

The author greatly appreciates the helpful comments and suggestions made by Drs. S K A Danso, C. Hera, S Kumarasinghe, J Richards, B Sigurbjörnsson and F Zapata during the preparation of this manuscript.

References

Awonaike K O, Hardarson G and Kumarasinghe K S 1991 Biological nitrogen fixation of *Gliricidia sepium /Rhizobium* symbiosis as influenced by plant genotype, bacterial strain and their interactions. Trop. Agric. (Trinidad) 69, 381–385.

Bergersen F J (Ed.) 1980 Methods for Evaluating Biological Nitrogen Fixation. Wiley, Chichester, 701p.

Bliss F A and Hardarson G (Eds.) 1993 Enhancement of Biological Nitrogen Fixation of Common Bean in Latin America. Kluwer Academic Publisher, Dordrecht, 160p.

Bowen G and Danso S K A 1987 Nitrogen nutrition of perennials. IAEA Bulletin 29, 5–8.

Danso S K A, Bowen G D and Sanginga N 1992 Biological nitrogen fixation in trees in agro-ecosystems. Plant and Soil 141, 177–196.

Danso S K A, Hardarson G and Zapata F 1988 Dinitrogen fixation in alfalfa-ryegrass swards using different nitrogen-15 labelling methods. Crop Sci. 28, 106–110.

Fried M and Broeshart H 1975 An independent measurement of the amount of nitrogen fixed by legume crops, Plant and Soil 43, 707–711.

Fried M and Middelboe V 1977 Measurement of amount of nitrogen fixed by a legume crop. Plant and Soil 47, 713–715.

Hardarson G (Ed,) 1990 Use of Nuclear Techniques in Studies of Soil-Plant Relationships. Training Course Series No. 2, IAEA. Vienna, 223p.

Hardarson G and Lie T A (Eds) 1984 Breeding Legumes for Enhanced Symbiotic Nitrogen Fixation. Martinus Nijhoff. Dordrecht. 166p.

Hardarson G and Danso S K A 1993 Methods for measuring biological nitrogen fixation in grain legumes. Plant and Soil 152, 19–23.

Hardarson G, Danso S K A, Zapata F and Reichardt K 1991 Measurement of nitrogen fixation in fababean at different N fertilizer rates using the ^{15}N isotope dilution and "A-value" methods. Plant and Soil 131, 161–168.

Hardarson G, Danso S K A and Zapata F 1988 Dinitrogen fixation measurement in alfalfa-ryegrass sward using nitrogen-15 and influence of the reference crop. Crop Sci. 28, 101–105.

Hardarson G, Golbs M and Danso S K A 1989 Nitrogen fixation in soybean (*Glycine max* L.) as affected by nodulation patterns. Soil Biol. Biochem. 21, 783–787.

Hardarson G, Bliss F A, Cigales-Rivero M R, Henson R A, Kipe-Nolt J A, Longeri L, Manrique A, Pena-Cabriales J J, Pereira P, Sanabria C A and Tsai S M 1993 Genotypic variation in biological nitrogen fixation by common bean. Plant and Soil 152 (1), 59–70.

IAEA 1983 Nuclear Techniques in Improving Pasture Management. IAEA, Vienna, 190p.

IAEA 1986 Nuclear Techniques in the Development of Fertilizer Practices for Multiple Cropping Systems. IAEA-TECDOC-394. IAEA, Vienna, 62p.

Kumarasinghe S and Eskew D 1993 Isotopic Studies of Azolla and Nitrogen Fertilization of Rice. Kluwer Academic Publishers, Dordrecht. 145 p.

McAuliffe C, Chamblee D S, Uribe-Arango H and Woodhouse W W 1958 Influence of inorganic nitrogen on nitrogen fixation by legumes as revealed by N-15. Agron. J 50, 334–337.

McDermott T R and Graham P H 1989 *Bradyrhizobium japonicum* inoculant mobility, nodule occupancy, and acetylene reduction in the soybean root system. Appl. Environ. Microb. 55, 2493–2498.

Mukhtar N O, Hardarson G and Zapata F 1993 Time course of biological nitrogen fixation in field grown groundnut (*Arachis hypogaea* L.). Field Crop Res. (Submitted).

Peña-Cabriales J J, Grageda-Cabrera O A, Kola V and Hardarson G 1993 Time course of N$_2$ fixation in common bean (*Phaseolus vulgaris* L.). Plant and Soil 152, 115–121.

Peoples M B, Faizah A W, Rerkasem B and Herridge D F (Eds.) 1989 Methods for evaluating nitrogen fixation by nodulated legumes in the field. Australian Center for International Agricultural Research, Canberra. 76p.

Sanginga N, Danso S K A and Bowen G D 1989 Nodulation and growth response of *Allocasuarina* and *Casuarina* species to phosphorus fertilization. Plant and Soil 118, 125–132.

Sanginga N, Bowen G D and Danso S K A 1990a Assessment of genetic variability for N$_2$ fixation between and within provenances of and *Leucaena leucocephala* and *Acacia albida* estimated by ^{15}N labelling techniques. Plant and Soil 127, 169–178.

Sanginga N, Bowen G D and Danso S K A 1990b Genetic variability in symbiotic nitrogen fixation within and between provenances of two *Casuarina* species using the ^{15}N labelling methods. Soil Biol. Biochem. 22, 539–547.

Sanginga N, Bowen G D and Danso S K A 1991a Intra-specific variation in growth and P accumulation of *Leucaena leucocephala* and *Gliricidia sepium* as influenced by soil phosphate status. Plant and Soil 133, 201–208.

Sanginga N, Manrique K and Hardarson G 1991b Variation in nodulation and N$_2$ fixation by the *Gliricidia sepium/Rhizobium* spp symbiosis in a calcareous soil. Biol. Fert. Soils. 11, 273–278.

Senaratne R and Hardarson G 1988 Estimation of residual N effect of faba bean and pea on two succeeding cereals using ^{15}N methodology. Plant and Soil 110, 81–89.

Sessitch A, Jjemba P K, Hardarson G, Danso S K A, Akkermans A D L and Wilson K J 1993 Use of the *gusA* marker gene in competition studies of *Rhizobium*. Appl. Environ. Microb. (Submitted)

Sougoufara B, Danso S K A, Diem H G and Dommergues Y R 1990 Estimating N$_2$ fixation and N derived from soil by *Casuarina equisetifolia* using labelled ^{15}N fertilizer: Some problems and solutions. Soil Biol. Biochem. 22, 695–701.

Wilson K J, Sessitsch A and Akkermans A D L 1993 Molecular markers as tools to study the ecology of microorganisms. Proceedings of a Conference on "Beyond the Biomass" held 21–24 March 1993 at Wye College, Ashford. (*In Press*).

Wolyn D J, Attewell J, Ludden P W and Bliss F A 1989 Indirect measure of N$_2$ fixation in common bean (*Phaseolus vulgaris* L.)

under field conditions. Role of lateral root nodules. Plant and Soil 113, 181–187.

Zapata F and Baert L 1989 Air nitrogen as fertilizer. *In* Soils for Development. Publication Series No. 1. 61–84. Van Cleemput O Ed. ITC-Ghent.

Zapata F, Danso S K A, Hardarson G and Fried M 1987a Time course of nitrogen fixation in field-grown soybean using nitrogen-15 methodology. Agron. J. 79, 172–176.

Zapata F, Danso S K A, Hardarson G and Fried M 1987b Nitrogen fixation and translocation in field-grown fababean. Agron. J. 79, 505–509.

Developments in Plant and Soil Sciences

1. J. Monteith and C. Webb (eds.): *Soil Water and Nitrogen in Mediterranean-type Environments*. 1981
    ISBN 90-247-2406-6
2. J. C. Brogan (ed.): *Nitrogen Losses and Surface Run-off from Landspreading of Manures*. 1981
    ISBN 90-247-2471-6
3. J. D. Bewley (ed.): *Nitrogen and Carbon Metabolism*. 1981   ISBN 90-247-2472-4
4. R. Brouwer, I. Gašparíková, J. Kolek and B. C. Loughman (eds.): *Structure and Function of Plant Roots*. 1981
    ISBN 90-247-2510-0
5. Y. R. Dommergues and H. G. Diem (eds.): *Microbiology of Tropical Soils and Plant Productivity*. 1982
    ISBN 90-247-2624-7
6. G. P. Robertson, R. Herrara and T. Rosswall (eds.): *Nitrogen Cycling in Ecosystems of Latin America and the Caribbean*. 1982   ISBN 90-247-2719-7
7. D. Atkinson, K. K. S. Bhat, M. P. Coutts, P. A. Mason and D. J. Read (eds.): *Tree Root Systems and Their Mycorrhizas*. 1983   ISBN 90-247-2821-5
8. M. R. Sarić and B. C. Loughman (eds.): *Genetic Aspects of Plant Nutrition*. 1983   ISBN 90-247-2822-3
9. J. R. Freney and J. R. Simpson (eds.): *Gaseous Loss of Nitrogen from Plant-Soil Systems*. 1983
    ISBN 90-247-2820-7
10. United Nations Economic Commission for Europe (ed.): *Efficient Use of Fertilizers in Agriculture*. 1983
    ISBN 90-247-2866-5
11. J. Tinsley and J. F. Darbyshire (eds.): *Biological Processes and Soil Fertility*. 1984   ISBN 90-247-2902-5
12. A. D. L. Akkermans, D. Baker, K. Huss-Danell and J. D. Tjepkema (eds.): Frankia *Symbioses*. 1984
    ISBN 90-247-2967-X
13. W. S. Silver and E. C. Schröder (eds.): *Practical Application of* Azolla *for Rice Production*. 1984
    ISBN 90-247-3068-6
14. P. G. L. Vlek (ed.): *Micronutrients in Tropical Food Crop Production*. 1985   ISBN 90-247-3085-6
15. T. P. Hignett (ed.): *Fertilizer Manual*. 1985   ISBN 90-247-3122-4
16. D. Vaughan and R. E. Malcolm (eds.): *Soil Organic Matter and Biological Activity*. 1985
    ISBN 90-247-3154-2
17. D. Pasternak and A. San Pietro (eds.): *Biosalinity in Action*. Bioproduction with Saline Water. 1985
    ISBN 90-247-3159-3
18. M. Lalonde, C. Camiré and J. O. Dawson (eds.): Frankia *and Actinorhizal Plants*. 1985
    ISBN 90-247-3214-X
19. H. Lambers, J. J. Neeteson and I. Stulen (eds.): *Fundamental, Ecological and Agricultural Aspects of Nitrogen Metabolism in Higher Plants*. 1986   ISBN 90-247-3258-1
20. M. B. Jackson (ed.): *New Root Formation in Plants and Cuttings*. 1986   ISBN 90-247-3260-3
21. F. A. Skinner and P. Uomala (eds.): *Nitrogen Fixation with Non-Legumes* (Proceedings of the 3rd Symposium, Helsinki, 1984). 1986   ISBN 90-247-3283-2
22. A. Alexander (ed.): *Foliar Fertilization*. 1986   ISBN 90-247-3288-3
23. H. G. v.d. Meer, J. C. Ryden and G. C. Ennik (eds.): *Nitrogen Fluxes in Intensive Grassland Systems*. 1986
    ISBN 90-247-3309-X
24. A. U. Mokwunye and P. L. G. Vlek (eds.): *Management of Nitrogen and Phosphorus Fertilizers in Sub-Saharan Africa*. 1986   ISBN 90-247-3312-X
25. Y. Chen and Y. Avnimelech (eds.): *The Role of Organic Matter in Modern Agriculture*. 1986
    ISBN 90-247-3360-X
26. S. K. De Datta and W. H. Patrick Jr. (eds.): *Nitrogen Economy of Flooded Rice Soils*. 1986
    ISBN 90-247-3361-8
27. W. H. Gabelman and B. C. Loughman (eds.): *Genetic Aspects of Plant Mineral Nutrition*. 1987
    ISBN 90-247-3494-0
28. A. van Diest (ed.): *Plant and Soil: Interfaces and Interactions*. 1987   ISBN 90-247-3535-1

Developments in Plant and Soil Sciences

29. United Nations Economic Commission for Europe and FAO (eds.): *The Utilization of Secondary and Trace Elements in Agriculture.* 1987   ISBN 90-247-3546-7
30. H. G. v.d. Meer, R. J. Unwin, T. A. van Dijk and G. C. Ennik (eds.): *Animal Manure on Grassland and Fodder Crops.* Fertilizer or Waste? 1987   ISBN 90-247-3568-8
31. N. J. Barrow: *Reactions with Variable-Charge Soils.* 1987   ISBN 90-247-3589-0
32. D. P. Beck and L. A. Materon (eds.): *Nitrogen Fixation by Legumes in Mediterranean Agriculture.* 1988   ISBN 90-247-3624-2
33. R. D. Graham, R. J. Hannam and N. C. Uren (eds.): *Manganese in Soils and Plants.* 1988   ISBN 90-247-3758-3
34. J. G. Torrey and J. L. Winship (eds.): *Applications of Continuous and Steady-State Methods to Root Biology.* 1989   ISBN 0-7923-0024-6
35. F. A. Skinner, R. M. Boddey and I. Fendrik (eds.): *Nitrogen Fixation with Non-Legumes* (Proceedings of the 4th Symposium, Rio de Janeiro, 1987). 1989   ISBN 0-7923-0059-9
36. B. C. Loughman, O. Gašparíková and J. Kolek (eds.): *Structural and Functional Aspects of Transport in Roots.* 1989   ISBN 0-7923-0060-2; Pb 0-7923-0061-0
37. P. Plancquaert and R. Haggar (eds.): *Legumes in Farming Systems.* 1990   ISBN 0-7923-0134-X
38. A. E. Osman, M. M. Ibrahim and M. A. Jones (eds.): *The Role of Legumes in the Farming Systems of the Mediterranean Areas.* 1990   ISBN 0-7923-0419-5
39. M. Clarholm and L. Bergström (eds.): *Ecology of Arable Land – Perspectives and Challenges.* 1989   ISBN 0-7923-0424-1
40. J. Vos, C. D. van Loon and G. J. Bollen (eds.): *Effects of Crop Rotation on Potato Production in the Temperate Zones.* 1989   ISBN 0-7923-0495-0
41. M. L. van Beusichem (ed.): *Plant Nutrition – Physiology and Applications.* 1990   ISBN 0-7923-0740-2
42. N. El Bassam, M. Dambroth and B.C. Loughman (eds.): *Genetic Aspects of Plant Mineral Nutrition.* 1990   ISBN 0-7923-0785-2
43. Y. Chen and Y. Hadar (eds.): *Iron Nutrition and Interactions in Plants.* 1991   ISBN 0-7923-1095-0
44. J. J. R. Groot, P. de Willigen and E. L. J. Verberne (eds.): *Nitrogen Turnover in the Soil-Crop System.* 1991   ISBN 0-7923-1107-8
45. R. J. Wright, V.C. Baligar and R. P. Murrmann (eds.): *Plant-Soil Interactions at Low pH.* 1991   ISBN 0-7923-1105-1
46. J. Kolek and V. Kozinka (eds.): *Physiology of the Plant Root System.* 1992   ISBN 0-7923-1205-8
47. A. U. Mokwunye (ed.): *Alleviating Soil Fertility Constraints to Increased Crop Production in West Africa.* 1991   ISBN 0-7923-1221-X; Pb 0-7923-1222-8
48. M. Polsinelli, R. Materassi and M. Vincenzini (eds.): *Nitrogen Fixation* (Proceedings of the 5th Symposium, Florence, 1990). 1991   ISBN 0-7923-1410-7
49. J.K. Ladha, T. George and B.B. Bohlool (eds.): *Biological Nitrogen Fixation for Sustainable Agriculture.* 1992   ISBN 0-7923-1774-2
50. P.J. Randall, E. Delhaze, R.A. Richards and R. Munns (eds.): *Genetic Aspects of Plant Mineral Nutrition.* 1993   ISBN 0-7923-2118-9
51. K.S. Kumarasinghe and D.L. Eskew (eds.): *Isotopic Studies of Azolla and Nitrogen Fertilization of Rice.* 1993   ISBN 0-7923-2274-6
52. F.A. Bliss and G. Hardarson (eds.): *Enhancement of Biological Nitrogen Fixation of Common Baen in Latin America.* 1993   ISBN 0-7923-2451-X
53. M.A.C. Fragoso and M.L. van Beusichem (eds.): *Optimization of Plant Nutrition.* 1993   ISBN 0-7923-2519-2
54. N.J. Barrow (ed.) *Plant Nutrition - From Genetic Engineering to Field Practice.* 1993   ISBN 0-7923-2540-0
55. A.D. Robson (ed.): *Zinc in Soils and Plants.* 1993   ISBN 0-7923-2631-8
56. A.D. Robson, L.K. Abbott and N. Malajczuk (eds.): *Management of Mycorrhizas in Agriculture, Horticulture and Forestry.* 1994   ISBN 0-7923-2700-4

Developments in Plant and Soil Sciences

57. P.H. Graham, M.J. Sadowsky and C.P. Vance (eds.): *Symbiotic Nitrogen Fixation*. 1994
ISBN 0-7923-2781-0

58. F. Baluška, M. Čiamporová, O. Gašparíkova and P.W. Barlow (eds.): *Structure and Function of Roots*. 1994
ISBN 0-7923-2832-9

59. J. Abadía (ed.): *Iron Nutrition in Soils and Plants*. 1994
ISBN 0-7923-2900-7

60. P.S. Curtis, E.G. O'Neill, J.A. Teeri, D.R. Zak and K.S. Pregitzer (eds.): *Belowground Responses to Rising Atmospheric CO_2. Implications for Plants, Soil Biota, and Ecosystem Processes*. 1994   ISBN 0-7923-2901-5

61. P.C. Struik, W.J. Vredenburg, J.A. Renkema and J.E. Parlevliet (eds.): *Plant Production on the Threshold of a New Century*. 1994
ISBN 0-7923-2903-1

Kluwer Academic Publishers – Dordrecht / Boston / London

The manufacturer's authorised representative in the EU is Springer
Nature Customer Service Centre GmbH, Europaplatz 3, 69115 Heidelberg,
Germany. If you have any concerns regarding our products, please
contact ProductSafety@springernature.com

Printed and bound by CPI Group (UK) Ltd, Croydon, CR0 4YY

23/04/2026

02095657-0001